MASSIVE NEUTRINOS

Flavor Mixing of Leptons and Neutrino Oscillations

ADVANCED SERIES ON DIRECTIONS IN HIGH ENERGY PHYSICS

ISSN: 1793-1339

Published

Vol. 1 – High Energy Electron–Positron Physics (*eds. A. Ali and P. Söding*)
Vol. 2 – Hadronic Multiparticle Production (*ed. P. Carruthers*)
Vol. 3 – CP Violation (*ed. C. Jarlskog*)
Vol. 4 – Proton–Antiproton Collider Physics (*eds. G. Altarelli and L. Di Lella*)
Vol. 5 – Perturbative QCD (*ed. A. Mueller*)
Vol. 6 – Quark–Gluon Plasma (*ed. R. C. Hwa*)
Vol. 7 – Quantum Electrodynamics (*ed. T. Kinoshita*)
Vol. 9 – Instrumentation in High Energy Physics (*ed. F. Sauli*)
Vol. 10 – Heavy Flavours (*eds. A. J. Buras and M. Lindner*)
Vol. 11 – Quantum Fields on the Computer (*ed. M. Creutz*)
Vol. 12 – Advances of Accelerator Physics and Technologies (*ed. H. Schopper*)
Vol. 13 – Perspectives on Higgs Physics (*ed. G. L. Kane*)
Vol. 14 – Precision Tests of the Standard Electroweak Model (*ed. P. Langacker*)
Vol. 15 – Heavy Flavours II (*eds. A. J. Buras and M. Lindner*)
Vol. 16 – Electroweak Symmetry Breaking and New Physics at the TeV Scale (*eds. T. L. Barklow, S. Dawson, H. E. Haber and J. L. Siegrist*)
Vol. 17 – Perspectives on Higgs Physics II (*ed. G. L. Kane*)
Vol. 18 – Perspectives on Supersymmetry (*ed. G. L. Kane*)
Vol. 19 – Linear Collider Physics in the New Millennium (*eds. K. Fujii, D. J. Miller and A. Soni*)
Vol. 20 – Lepton Dipole Moments (*eds. B. Lee Roberts and William J. Marciano*)
Vol. 21 – Perspectives on Supersymmetry II (*ed. G. L. Kane*)
Vol. 22 – Perspectives on String Phenomenology (*eds. B. Acharya, G. L. Kane and P. Kumar*)
Vol. 23 – 60 Years of CERN Experiments and Discoveries (*eds. H. Schopper and L. Di Lella*)
Vol. 24 – The High Luminosity Large Hadron Collider: The New Machine for Illuminating the Mysteries of Universe (*eds. O. Brüning and L. Rossi*)
Vol. 25 – Massive Neutrinos: Flavor Mixing of Leptons and Neutrino Oscillations (*ed. H. Fritzsch*)

Forthcoming

Vol. 8 – Standard Model, Hadron Phenomenology and Weak Decays on the Lattice (*ed. G. Martinelli*)

Advanced Series on
Directions in High Energy Physics — Vol. 25

MASSIVE NEUTRINOS

Flavor Mixing of Leptons and Neutrino Oscillations

Editor

Harald Fritzsch

Ludwig Maximilian University of Munich, Germany

NEW JERSEY • LONDON • SINGAPORE • BEIJING • SHANGHAI • HONG KONG • TAIPEI • CHENNAI • TOKYO

Published by

World Scientific Publishing Co. Pte. Ltd.
5 Toh Tuck Link, Singapore 596224
USA office: 27 Warren Street, Suite 401-402, Hackensack, NJ 07601
UK office: 57 Shelton Street, Covent Garden, London WC2H 9HE

Library of Congress Cataloging-in-Publication Data
Massive neutrinos : flavor mixing of leptons and neutrino oscillations / edited by Harald Fritzsch
(Ludwig Maximilian University of Munich, Germany).
pages cm -- (Advanced series on directions in high energy physics ; v. 25)
Includes bibliographical references and index.
ISBN 978-9814704762 (hardcover : alk. paper) -- ISBN 9814704768 (hardcover : alk. paper)
1. Neutrinos. 2. Lepton interactions. 3. Quantum flavor dynamics. 4. Leptons (Nuclear physics).
I. Fritzsch, Harald, 1943– editor. II. Series: Advanced series on directions in high energy physics ; v. 25.
QC793.5.N42M37 2015
539.7'215--dc23

2015026505

British Library Cataloguing-in-Publication Data
A catalogue record for this book is available from the British Library.

Preface

In 1911, Otto Hahn and Lise Meitner observed that in the radioactive beta decays of atomic nuclei the energy was not conserved. In the beta decay of a nucleus into another nucleus, an electron was emitted. One expected that the energy of the electron is given by the energy of the outgoing nucleus. However, the energy of the electron was not fixed, but continuous and always less than the expected energy. Apparently, energy was not conserved in the weak decays.

In 1930, Wolfgang Pauli suggested that in the beta decay a second neutral particle, which could not be observed, was emitted and that this particle carried the missing energy. This elusive particle was later called "neutrino". Pauli did not publish his idea, since he was convinced that the neutrino could not be observed in the experiments. However, in 1956, Clyde Cowan and Frederick Reines observed the neutrinos in an experiment conducted close to a big nuclear reactor. A neutrino emitted from the reactor collided with a proton and turned into a positron, while the proton changed into a neutron. Both the neutron and the positron were observed.

Later, it was discovered that for each charged lepton there is a neutrino — thus there are electron-neutrinos, muon-neutrinos and tau-neutrinos. The leptons come in three families, an electron-family, a muon-family and a tau-family, analogous to the three families of the quarks, the (u,d)-family, the (c,s)-family and the (t,b)-family.

It was assumed that the neutrinos are left-handed fermions and do not have a mass. But, already in 1957, Bruno Pontecorvo suggested that the neutrinos have a small mass and that they can oscillate. Since only the electron-neutrino was known at that time, he studied the possibility of oscillations between the neutrino and the antineutrino. Forty-one years later, the neutrino oscillations were discovered.

In the late 1960s, there began the Homestake experiment in Lead, South Dakota (United States) by astrophysicists Raymond Davis and John Bahcall. Its purpose was to collect and count the neutrinos emitted by the nuclear fusion in the Sun. Bahcall did the theoretical calculations and Davis designed the experiment. The solar neutrinos collided with the nuclei of Chlorine-37 atoms and produced nuclei of Argon-37 atoms. These atoms were collected later on. By this way Davis determined the number of solar neutrinos arriving on Earth. Bahcall had calculated the rate at which the detector should capture neutrinos, but Davis found only one third of the expected rate. This created the "solar neutrino problem".

The Homestake experiment was followed by other experiments such as Kamiokande in Kamioka, Japan and the Sudbury Neutrino Observatory (SNO) in Ontario, Canada. When cosmic rays collide with the nuclei in the upper atmosphere, many pions are created, which decay into neutrinos, in particular. These atmospheric neutrinos were investigated in Kamioka. In particular, the flux of neutrinos created in the atmosphere above Kamioka was compared to the flux of neutrinos coming from the other side of the Earth. Through this way the neutrino oscillations were discovered in 1998 and therefore neutrinos must have small masses. The atmospheric mixing angle, describing these oscillations, turned out to be rather large, close to 40°.

Oscillations of solar neutrinos were discovered in 2001 with the SNO detector, which could investigate both charged and neutral current interactions of neutrinos. The solar mixing angle was also large, about 34°. In the neutrino experiments using reactor neutrinos, the Daya Bay experiment in China, in particular, the third mixing angle was determined in 2012 to about 9°. Thus two neutrino mixing angles are large, while the flavor mixing angles for the quarks are quite small. Here the largest angle is the Cabibbo angle, about 13°. So far this phenomenon is not understood.

We still do not know if the CP symmetry is violated for leptons, analogous to the CP violation for the quarks. CP violation could be discovered by observing a difference between the oscillations of neutrinos and of antineutrinos. At the Massive Neutrinos conference, four speakers discussed the various models for CP violation.

In the neutrino oscillations, only the mass differences of the neutrinos can be measured, not the absolute masses. They are less than 0.1 eV. The masses of the neutrinos might be in the range 0.01 eV–0.1 eV. Presumably the flavor mixing angles are functions of the quark or lepton masses. The experimental data are in good agreement with the predictions of the "texture zero models", which were discussed at the conference by several speakers. In particular, there might be a connection between the large mixing angles and the very small masses of the neutrinos. These masses might be Majorana masses, implying that lepton number is not conserved. Thus the neutrino mass could be measured by the neutrinoless double beta decay experiments.

The International Conference on Massive Neutrinos was held from 9–13 February 2015 at the Nanyang Executive Centre of the Nanyang Technological University in Singapore (group photo on page vii). It was organized by Prof. Phua, Prof. Xing and by me. There were 56 speakers all together. Twelve lectures were on the recent or future results of the neutrino experiments. The various theories for the neutrinos were discussed by 26 speakers. Other talks were on neutrinoless double beta decay, flavor physics of quarks, cosmic neutrino background, cosmology, dark matter and astrophysics. We thank all speakers and participants for their contributions to this volume, which provide an overview of the present state of research, and for their participation in the discussions.

H. Fritzsch

Institute of Advanced Studies

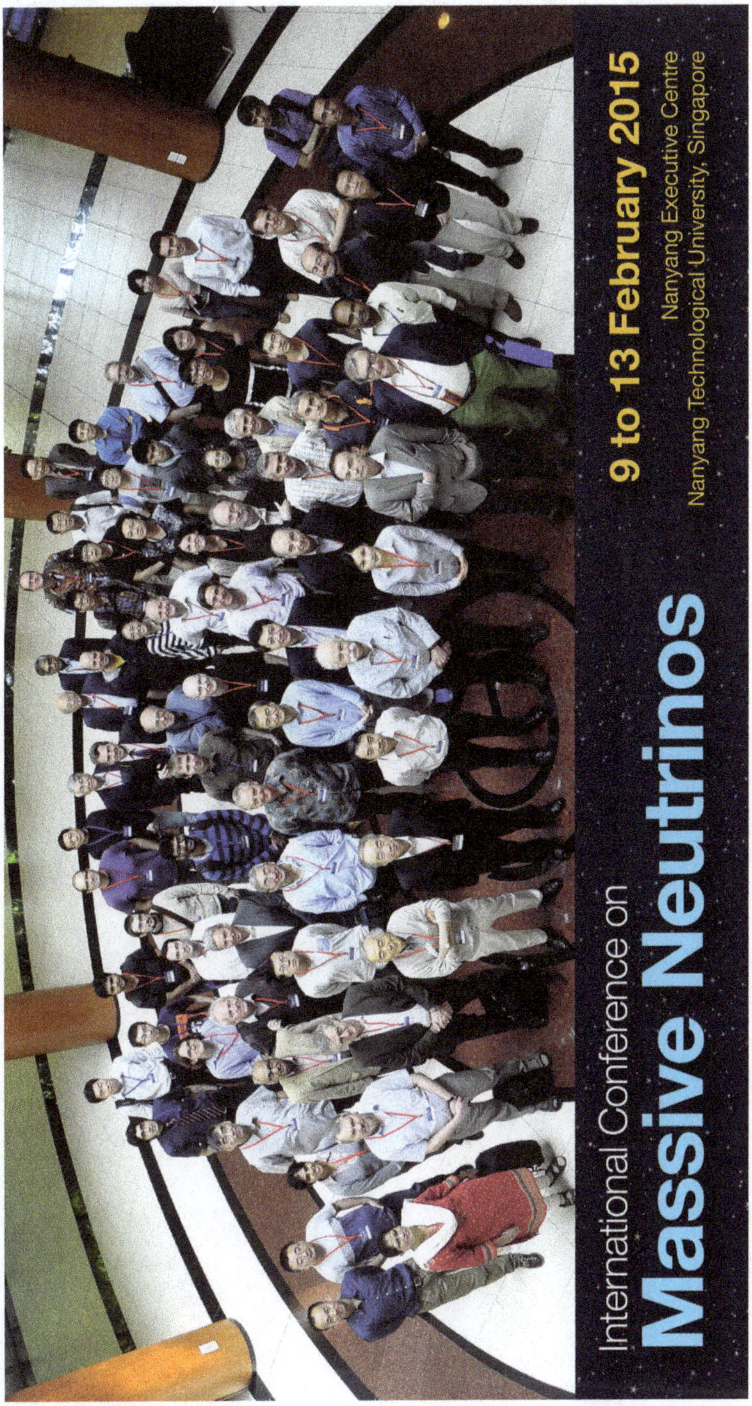

International Conference on
Massive Neutrinos

9 to 13 February 2015
Nanyang Executive Centre
Nanyang Technological University, Singapore

Contents

Birth of Lepton Flavor Mixing

Makoto Kobayashi
KEK,
1-1 Oho, Tsukuba, Ibaraki 305-0801, Japan

The history of the lepton flavor mixing could be traced back to the early 60s, when Maki, Nakagawa and Sakata (MNS) discussed the neutrino mixing. Their work emerged in the course of the developments of the composite model of elementary particles which was initiated by Sakata. In Sakata's model, the weak interaction of the hadrons can be described by two types of transitions among the fundamental triplet baryons. This pattern of the weak interaction of the hadrons is similar to that of leptons provided that the neutrino consists of a single species. From this similarity, Maki, Nakagawa, Ohnuki and Sakata proposed the so-called Nagoya model, in which the fundamental triplet baryons are regarded as composite states of the leptons and a hypothetical object called B-matter. Although the Nagoya model did not make a remarkable success, when existence of two kinds of neutrinos was discovered in 1962, Maki, Nakagawa and Sakata precisely formulated lepton flavor mixing to associate leptons with the fundamental baryons in the framework of the Nagoya model. To recognize their contributions, the flavor mixing matrix of the lepton sector is called the MNS matrix. See also: M. Kobayashi, "Neutrino mass and mixing — The beginning and future", *Nucl. Phys. B (Proc. Suppl.)* Vol. 235–236, (2013), pp. 4–7.

Neutrino Masses and Flavor Mixing

Harald Fritzsch

Department für Physik, Universität München,
Theresienstraße 37, 80333 München, Germany
fritzsch@mppmu.mpg.de

We discuss the neutrino oscillations, using texture zero mass matrices for the leptons. The reactor mixing angle θ_l is calculated. The ratio of the masses of two neutrinos is determined by the solar mixing angle. We can calculate the masses of the three neutrinos: $m_1 \approx 0.003$ eV, $m_2 \approx 0.012$ eV, $m_3 \approx 0.048$ eV.

The flavor mixing of the quarks is parametrized by the CKM-matrix. There are several ways to describe the CKM-matrix in terms of three angles and one phase parameter. I prefer the parametrization, given below, which Z. Xing and I introduced years ago,[1,2] given by the angles θ_u, θ_d and θ:

$$U = \begin{pmatrix} c_u & s_u & 0 \\ -s_u & c_u & 0 \\ 0 & 0 & 1 \end{pmatrix} \times \begin{pmatrix} e^{-i\delta} & 0 & 0 \\ 0 & c & s \\ 0 & -s & c \end{pmatrix} \times \begin{pmatrix} c_d & -s_d & 0 \\ s_d & c_d & 0 \\ 0 & 0 & 1 \end{pmatrix}, \tag{1}$$

where $c_{u,d} \sim \cos\theta_{u,d}$, $s_{u,d} \sim \sin\theta_{u,d}$, $c \sim \cos\theta$ and $s \sim \sin\theta$.

The angle θ_u describes the mixing between the quarks "u–c," the angle θ_d the mixing between the quarks "d–s" and the angle θ the mixing among the heavy quarks "t, c–b, s." The three angles have been determined by the experiments:

$$\theta_u \simeq 5.4^\circ\,, \quad \theta_d \simeq 11.7^\circ\,, \quad \theta \simeq 2.4^\circ\,.$$

Presumably the flavor mixing angles are not fixed values, but functions of the quark masses. If the masses change, the mixing angles will also change. For example, the Cabibbo angle $\theta_C \simeq 13^\circ$ could be given by the ratio of the quark masses:

$$\tan\theta_C \simeq \sqrt{m_d/m_s}\,. \tag{2}$$

This relation works very well:

$$\tan\theta_C \simeq 0.23\,, \tag{3}$$

$$\sqrt{m_d/m_s} \simeq 0.23\,. \tag{4}$$

Such a relation can be derived, if the quark mass matrices have "texture zeros," as shown by S. Weinberg and me in 1977.[3–5]

Let me discuss a simple example, using only four quarks: u, d–c, s. Their mass matrices have a zero in the $(1,1)$-position:

$$M = \begin{pmatrix} 0 & A \\ A^* & B \end{pmatrix} . \tag{5}$$

These mass matrices can be diagonalized by a rotation. The rotation angles are:

$$\theta_u \simeq \sqrt{m_u/m_c} \simeq 0.09 \,, \quad \theta_d \simeq \sqrt{m_d/m_s} \simeq 0.23 \,. \tag{6}$$

The Cabibbo angle is given by the difference:

$$\theta_C \simeq \left| \sqrt{m_d/m_s} - e^{i\phi} \sqrt{m_u/m_c} \right| . \tag{7}$$

In the complex plane this relation describes a triangle. The phase parameter is unknown, however it must be close to 90°, since the Cabibbo angle is given by the ratio m_d/m_s:

$$\theta_C \simeq \sqrt{m_d/m_s} \,. \tag{8}$$

Thus the triangle is close to a rectangular triangle.

For six quarks the "texture zero" mass matrices for the quarks of charge (2/3) and of charge (−1/3) are:

$$M = \begin{pmatrix} 0 & A & 0 \\ A^* & B & C \\ 0 & C^* & D \end{pmatrix} . \tag{9}$$

We can calculate the angles θ_u and θ_d as functions of the mass eigenvalues:

$$\theta_u \simeq \sqrt{m_u/m_c} \,, \quad \theta_d \simeq \sqrt{m_d/m_s} \,. \tag{10}$$

Using the observed mass values for the quarks, we find:

$$\theta_d \simeq (13.0 \pm 0.4)^\circ \,, \quad \theta_u \simeq (5.0 \pm 0.7)^\circ \,.$$

The experimental values agree very well with the theoretical results:

$$\theta_d \simeq (11.7 \pm 2.6)^\circ \,, \quad \theta_u \simeq (5.4 \pm 1.1)^\circ \,.$$

Now we consider the flavor mixing of the leptons. The neutrinos, emitted in weak decays, are mixtures of different mass eigenstates. This leads to neutrino oscillations — at least two neutrinos must have finite masses.

The lepton flavor mixing is described by a 3×3 unitary matrix U, analogous to the CKM mixing matrix for the quarks. It can be parametrized in terms of three angles and three phases. I use a parametrization, introduced by Z. Xing and me:[6]

$$U = \begin{pmatrix} s_l s_\nu c + c_l c_\nu e^{-i\varphi} & s_l c_\nu c - c_l s_\nu e^{-i\varphi} & s_l s \\ c_l s_\nu c - s_l c_\nu e^{-i\varphi} & c_l c_\nu c + s_l s_\nu e^{-i\varphi} & c_l s \\ -s_\nu s & -c_\nu s & c \end{pmatrix} P_\nu \,, \tag{11}$$

where $c_{l,\nu} \sim \cos\theta_{l,\nu}$, $s_{l,\nu} \sim \sin\theta_{l,\nu}$, $c \sim \cos\theta$ and $s \sim \sin\theta$. The angle θ_ν is the solar angle $\theta_{\rm sun}$, the angle θ is the atmospheric angle $\theta_{\rm at}$, and the angle θ_l is the "reactor angle." The phase matrix $P_\nu = {\rm Diag}\{e^{i\rho}, e^{i\sigma}, 1\}$ is relevant only, if the neutrino masses are Majorana masses.

The neutrino oscillations are described by two large mixing angles:

$$\theta_{\rm sun} = \theta_\nu \simeq 34^\circ\,, \quad \theta_{\rm at} = \theta \simeq 45^\circ\,.$$

The reactor angle θ_l is much smaller: $\theta_l \simeq 13°$.

We assume that the mass matrices of the leptons have "texture zeros":

$$M = \begin{pmatrix} 0 & A & 0 \\ A^* & B & C \\ 0 & C^* & D \end{pmatrix}. \tag{12}$$

In this case we can calculate two leptonic mixing angles as functions of mass ratios:

$$\tan\theta_l \simeq \sqrt{m_e/m_\mu}\,, \quad \tan\theta_\nu \simeq \sqrt{m_1/m_2}\,. \tag{13}$$

From the solar mixing angle we obtain for the neutrino mass ratio:

$$m_1/m_2 \simeq 0.42\,. \tag{14}$$

This relation and the experimental results for the mass differences of the neutrinos, measured by the neutrino oscillations, allow us to determine the three neutrino masses:

$$\begin{aligned} m_1 &\simeq 0.003 \text{ eV}\,, \\ m_2 &\simeq 0.012 \text{ eV}\,, \\ m_3 &\simeq 0.048 \text{ eV}\,. \end{aligned} \tag{15}$$

We expect that the mass matrices of the quarks and leptons are not exactly given by texture zero matrices. Radiative corrections of the order of the fine-structure constant α will contribute — the zeros will be replaced by small numbers.

The ratios of the masses of the quarks with the same electric charge seem to be universal:

$$\begin{aligned} \frac{m_u}{m_c} &\simeq \frac{m_c}{m_t} \simeq 0.005\,, \\ \frac{m_d}{m_s} &\simeq \frac{m_s}{m_b} \simeq 0.044\,. \end{aligned} \tag{16}$$

The dynamical reason for this universality is unclear. It might follow from specific properties of the texture zero mass matrices. But in the case of the charged leptons there is no universality:

$$\begin{aligned} \frac{m_e}{m_\mu} &\simeq 0.005\,, \\ \frac{m_\mu}{m_\tau} &\simeq 0.06\,. \end{aligned} \tag{17}$$

If the ratios of the charged lepton masses were universal, the mass of the electron would have to be about 6 MeV.

The universality is not expected to be exact, due to radiative corrections. A radiative correction of the order of $\pm(\alpha/\pi)m_\tau \simeq \pm 4$ MeV would have to be added to the charged lepton masses. Such a contribution is relatively small for the muon and the tauon, but large in comparison to the observed electron mass. One expects that the physical electron mass is the sum of a bare electron mass M_e, due to the texture zero mass matrix, and a radiative correction R_e:

$$m_e = M_e - R_e \,.$$

If we assume, that the bare electron mass is given by the universality, we obtain:

$$M_e = 5.51 \text{ MeV}\,, \quad R_e = 5.00 \text{ MeV}\,.$$

Radiative corrections also contribute to the muon and the tauon mass, but here the corrections are small in comparison to the bare masses and will be neglected.

We calculate the angle θ_l (see Eq. (13)):

$$\tan\theta_l \simeq \sqrt{M_e/m_\mu} \simeq 0.23\,, \quad \theta_l \simeq 13^\circ\,. \tag{18}$$

This angle agrees with the experimental result.

If we would not have taken into account the radiative corrections for the electron mass, we would obtain a value for the reactor angle, which is much smaller than the experimental value:

$$\tan\theta_l \simeq \sqrt{m_e/m_\mu} \simeq 0.07\,, \quad \theta_l \simeq 4^\circ\,. \tag{19}$$

The neutrino masses, given in Eq. (15), are very small, much smaller than the masses of the charged leptons. The ratio of the mass of the tau neutrino and of the mass of the tauon is only about 2.7×10^{-11}.

Are the neutrino masses normal Dirac masses, as the masses of the charged leptons and quarks, or are they Majorana masses? In this case the smallness of the neutrino masses can be understood by the "seesaw"-mechanism.[8–12] Here the neutrino masses are mixtures of Majorana and Dirac masses.

The mass matrix of the neutrinos is a matrix with one "texture zero" in the $(1,1)$-position. The two off-diagonal terms are given by the Dirac mass term D — a large Majorana mass term appears in the $(2,2)$-position:

$$M_\nu = \begin{pmatrix} 0 & D \\ D & M \end{pmatrix}. \tag{20}$$

After diagonalization one obtains a large Majorana mass M and a small neutrino mass:

$$m_\nu \simeq D^2/M\,. \tag{21}$$

One expects that the Dirac term D is similar to the corresponding charged lepton mass. As an example we consider a Dirac mass matrix D with the eigenvalues:

$$D = \begin{pmatrix} M_1 & 0 & 0 \\ 0 & M_2 & 0 \\ 0 & 0 & M_3 \end{pmatrix} . \tag{22}$$

We assume that these Dirac masses are proportional to the neutrino masses, given in Eq. (15). As an example we consider the mass values:

$$M_1 \simeq 30 \text{ MeV}\,, \quad M_2 \simeq 120 \text{ MeV}\,, \quad M_3 \simeq 480 \text{ MeV}\,.$$

For the heavy Majorana mass M we find in this case:

$$M = 4.8 \times 10^9 \text{ GeV}\,.$$

The only way to test the nature of the neutrino masses is to study the neutrinoless double beta decay. Some heavy nuclei decay by double beta decay — two neutrons emit electrons and antielectron–neutrinos, e.g. the decay of selenium into krypton, which is a normal double beta decay.

If neutrinos are Majorana particles, the two antielectron–neutrinos can annihilate. Only two electrons are emitted — this is the neutrinoless double beta decay, which violates lepton number conservation.

Thus far the neutrinoless double beta decay has not been observed. For example, in the Cuoricino experiment in the Grand Sasso Laboratory one searches for the neutrinoless double beta decay of tellurium.

If neutrinos mix, all three neutrino masses will contribute to the decay rate for neutrinoless double beta decay. Their contributions are given by the mass of the neutrino, multiplied by the square of the transition element, including the CP-violation phases.

Using the neutrino masses, given in Eq. (15) and the transition elements, given by the mixing angles, I find for the effective neutrino mass, relevant for the neutrinoless double beta decay:

$$\tilde{m} \simeq 0.016 \text{ eV}\,. \tag{23}$$

The present limit for this effective mass is provided by the Cuoricino experiment: $\tilde{m} < 0.23$ eV.

The texture zero idea provides a coherent framework to understand the flavor mixing of quarks and leptons. Two of the three mixing angles for the quarks are given by quark mass ratios. For the leptons the solar mixing angle is determined by the ratio of the masses of two neutrinos. Thus the absolute masses of the neutrinos are fixed and very small. The reactor angle is given by the ratio of the bare electron mass and the muon mass.

References

1. H. Fritzsch and Z. Z. Xing, *Phys. Lett. B* **413**, 396 (1997).
2. H. Fritzsch and Z. Z. Xing, *Phys. Rev. D* **57**, 594 (1998).
3. S. Weinberg, *Trans. New York Acad. Sci.* **38**, 185 (1977).
4. H. Fritzsch, *Phys. Lett. B* **70**, 436 (1977).
5. H. Fritzsch, *Phys. Lett. B* **73**, 317 (1978).
6. H. Fritzsch and Z. Z. Xing, *Phys. Lett. B* **682**, 220 (2009).
7. H. Fritzsch, *Mod. Phys. Lett. A* **27**, 1250079 (2012).
8. H. Fritzsch, M. Gell-Mann and P. Minkowski, *Phys. Lett. B* **59**, 256 (1975).
9. P. Minkowski, *Phys. Lett. B* **67**, 421 (1977).
10. M. Gell-Mann, P. Ramond and R. Slansky, *Supergravity* (North-Holland, 1979), p. 315.
11. T. Yanagida, *Prog. Theor. Phys.* **64**, 1103 (1980).
12. R. N. Mohapatra and G. Senjanovic, *Phys. Rev. Lett.* **44**, 912 (1980).

Fermion Mass Matrices, Textures and Beyond

Manmohan Gupta*, Priyanka Fakay, Samandeep Sharma[@] and Gulsheen Ahuja
Department of Physics, Panjab University, Chandigarh, India
[@]*Department of Physics, GGDSD College, Chandigarh, India*
**mmgupta@pu.ac.in*

The issue of texture specific fermion mass matrices have been examined briefly from the 'bottom-up' perspective. In case no conditions are imposed, the texture *ansätze* leads to a large number of viable possibilities. However, besides textures, if in case one incorporates the ideas of 'natural mass matrices' and uses the facility of Weak Basis Transformations, then one is able to arrive at a minimal finite set of viable mass matrices in the case of quarks.

Understanding fermion masses and mixings is one of the biggest challenges in the present day High Energy Physics. One of the key difficulties in this area is the fact that the fermion masses and mixings span several orders of magnitude. In the case of charged fermions, the range of masses is from 10^5 eV to 10^{12} eV, corresponding respectively to the electron mass and the mass of the top quark. Further, the absolute masses of the neutrinos are not known, however, two of the lightest neutrino masses can be of the order of a fraction of an eV, with no lower limit for the third neutrino mass. In case the theory requires the existence of right-handed neutrinos, responsible for see-saw mechanism[1–6] with the mass range of $10^{12} - 10^{15}$ GeV, the fermion masses would then cover almost 25 orders of magnitude.

The problem gets further complicated when one notices that the pattern of mixings are also quite different in case of quarks and leptons. In fact, in the case of quarks we have clearly hierarchical structure of the mixing angles, for example, $s_{12} \sim 0.22, s_{23} \sim 0.04, s_{13} \sim 0.004$. In contrast, the two of the mixing angles in case of neutrinos are quite large, whereas the third angle although small as compared to the other two angles yet it is of the order of the Cabibbo angle. Similarly, the pattern of masses in the case of charged leptons has a very well defined hierarchy, whereas in the case of neutrino we may have normal/inverted hierarchy or degenerate scenario of neutrino masses. Since the mixing matrices are related to the corresponding mass matrices therefore formulating viable fermion mass matrices becomes all the more complicated.

In the absence of fundamental theory of flavor physics wherein fermion masses and mixings can be understood, the present day phenomenological approaches can be broadly categorized as "top-down" and "bottom-up". The top-down approach essentially starts with the formulation of mass matrices at the GUT scale, whereas, the bottom-up approach starts with the phenomenological mass matrices at the weak scale. Despite large number of attempts from the top-down perspective[7] yet we are not in a position to incorporate the vast amount of data related to fermion mixing within a consistent framework. In this context, therefore, it is desirable to look at bottom-up approach[8–11] consisting of finding the phenomenological fermion mass matrices which are in tune with the low energy data, i.e., observables like quark and lepton masses, mixing angles in both the sectors, angles of the unitarity triangle in the quark sector, etc.. Also, successful phenomenological formulation of mass matrices may provide clues for appropriate dynamical models, in particular, important clues for their formulation at the GUT scale.

The purpose of the present work is to explore the essentials, from a "bottom-up" approach perspective, needed to arrive at a minimal set of fermion matrices which are compatible with the latest mixing data. To this end, we have not gone into a detailed and comprehensive analysis rather would like to present a brief overview related to the issue mentioned above. Further, we would like to discuss the possibility of arriving at a minimal set of viable mass matrices using textures and other ideas.

To begin with, we discuss the earliest *ansätz* made in the context of quark mass matrices. The first step in this direction was taken by Fritzsch,[12,13] essentially laying down the path for future investigations in this direction. According to his hypothesis, the 3×3 mass matrices for the up and down sectors, M_U and M_D, are hermitian and are given by

$$M_U = \begin{pmatrix} 0 & A_U & 0 \\ A_U^* & 0 & B_U \\ 0 & B_U^* & C_U \end{pmatrix}, \qquad M_D = \begin{pmatrix} 0 & A_D & 0 \\ A_D^* & 0 & B_D \\ 0 & B_D^* & C_D \end{pmatrix}. \tag{1}$$

Another *ansätz* proposed by Stech[14] has the following form for the mass matrices in the up and down sectors,

$$M_U = S\,, \qquad M_D = \beta S + A\,, \tag{2}$$

where S and A are symmetric and antisymmetric 3×3 matrices respectively. Yet another *ansätz*, proposed by Gronau,[15] had the features of both Fritzsch's and Stech's *ansätze*, e.g,

$$M_U = \begin{pmatrix} 0 & A & 0 \\ A & 0 & B \\ 0 & B & C \end{pmatrix}, \qquad M_D = \beta \begin{pmatrix} 0 & A & 0 \\ A & 0 & B \\ 0 & B & C \end{pmatrix} + \begin{pmatrix} 0 & ia & 0 \\ -ia & 0 & ib \\ 0 & -ib & 0 \end{pmatrix}. \tag{3}$$

Interestingly, these *ansätze* were ruled out by the "high" value of the t quark mass and these continue to be ruled out even with subsequent refinements in the data. To this end, we discuss, in somewhat detail, the case of Fritzsch *ansätz*. The essentials of the methodology usually used to carry out the analysis include diagonalizing the mass matrices M_U and M_D by unitary transformations and obtaining a Cabibbo–Kobayashi–Maskawa (CKM) matrix from these transformations. To ensure the viability of the considered mass matrices, this CKM matrix should be compatible with the quark mixing data, for details regarding this we refer the readers to Ref. 11. Following this methodology for the above mentioned *ansätz* considered by Fritzsch, the CKM matrix so obtained by considering latest inputs from PDG 2014[17] is given by

$$V_{\rm CKM} = \begin{pmatrix} 0.9837 - 0.9872 & 0.2248 - 0.2268 & 0.0053 - 0.0075 \\ 0.2203 - 0.2264 & 0.9160 - 0.9721 & 0.0601 - 0.2037 \\ 0.0302 - 0.0308 & 0.0043 - 0.0194 & 0.9991 - 0.9999 \end{pmatrix}. \quad (4)$$

A look at this matrix immediately reveals that the ranges of most of the CKM elements show no overlap with those obtained by recent global analyses.[17] This, therefore, leads to the conclusion that the Fritzsch *ansätz* is not compatible with the recent quark mixing data.

The above conclusion can be explicitly understood by studying the analytical expressions of the elements $|V_{ub}|$ and $|V_{cb}|$, e.g.,

$$V_{ub} = -\sqrt{\frac{m_d}{m_s}}\left(\frac{m_s}{m_d}\right)^{\frac{3}{2}} e^{i\phi_1} - \sqrt{\frac{m_u}{m_c}}\sqrt{\frac{m_s}{m_b}} + \sqrt{\frac{m_u}{m_c}}\sqrt{\frac{m_c}{m_t}} e^{i\phi_2}, \quad (5)$$

$$V_{cb} = \sqrt{\frac{m_u}{m_c}}\sqrt{\frac{m_d}{m_s}}\left(\frac{m_s}{m_b}\right)^{\frac{3}{2}} e^{i\phi_1} - \sqrt{\frac{m_s}{m_b}} + \sqrt{\frac{m_c}{m_t}} e^{i\phi_2}, \quad (6)$$

where phases ϕ_1 and ϕ_2 are related to the phases associated with the elements of the mass matrices.[10] In Fig. 1, we have plotted the dependence of these elements with respect to the strange quark mass m_s. While plotting the allowed ranges of the matrix elements $|V_{ub}|$ and $|V_{cb}|$, all other parameters have been given full variation within the allowed ranges. A general look at the figure immediately shows that the plotted values both $|V_{ub}|$ and $|V_{cb}|$ have no overlap with the allowed experimental ranges of these. Thus, one can again conclude that Fritzsch *ansätz* is not viable.

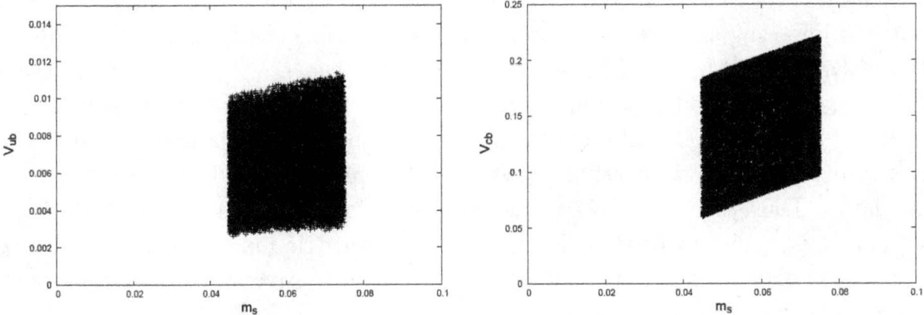

Fig. 1. Plots showing the allowed range of $|V_{ub}|$ and $|V_{cb}|$ w.r.t the light quark mass m_s for the Fritzsch mass matrix.

The generalization of the Fritzsch *ansätze* led to the idea of textures. *A particular texture structure is said to be texture n zero, if it has n number of non-trivial zeros, for example, if the sum of the number of diagonal zeros and half the number of the symmetrically placed off diagonal zeros is n.* Therefore, if both M_U and M_D have n texture zeros each, together these are called texture $2n$ zero mass matrices. For example, the Fritzsch *ansätz*, mentioned in Eq. (1), corresponds to texture 6 zero quark mass matrices.

Apart from texture 6 zero mass matrices considered by Fritzsch, some other versions of these were also analyzed and consequently ruled out by Ramond *et al.*,[16] these continue to be ruled out even by the present quark mixing data. In this context, Ramond *et al.*[16] also arrived at an important conclusion that the texture structure of a matrix as well as its hermiticity property are not "affected" when one scales down from GUT scale to weak scale, justifying the formulation of texture specific mass matrices. This important conclusion also leads to the fact that the texture zeros of fermion mass matrices can be considered as phenomenological zeros, thereby implying that at all energy scales the corresponding matrix elements are sufficiently suppressed in comparison with their neighboring counterparts. This, therefore, opens the possibility of considering less than six texture zeros[10] for the quark mass matrices.

Extending their analysis of texture 6 zero mass matrices, Ramond *et al.*[16] have also examined the viability of a few texture 5 zero quark mass matrices. Recently, the compatibility of texture 5 zero mass matrices with the latest mixing data has also been examined in detail.[10] Interestingly, even in this case one finds that there is only marginal compatibility, in particular, out of the large number of possibilities for texture 5 zero mass matrices, only Fritzsch-like mass matrices have limited compatibility with the experimental data. As an extension of texture 5 zero mass matrices, several authors have carried out the study of the implications of the Fritzsch-like texture 4 zero mass matrices[18–21]. These analyses reveal that the texture 4 zero mass matrices, undoubtedly, are able to accommodate the quark mixing data quite well.

Very recently, Ludl and Grimus[22] have performed a detailed and comprehensive analysis for general as well as symmetric texture specific quark mass matrices. Without imposing any restrictions on textures and using the facility of Weak Basis Transformations, Ludl and Grimus arrive at 243 classes of texture specific mass matrices, related through permutations. To reduce the number of possibilities they use the concept of maximally restrictive classes (one cannot place another texture zero into one of the two mass matrices while keeping the model compatible with the data). Thus, they found 27 viable classes for general mass matrices, however, without any predictive powers. In the case of symmetric mass matrices they have found 15 maximally restrictive textures which are predictive with respect to one or more light quark masses.

The above analysis indicates that in the absence of any additional conditions on textures, even texture 5 zero mass matrices could also be viable and the number of viable possibilities increases rapidly as one goes to lower textures. This therefore, brings us to the conclusion that in case we have to arrive at finite set of mass matrices which may serve as clues for their formulation at fundamental level, one needs to go beyond texture *ansätze.* In this context, two important ideas for the quark matrices have been considered in the literature, e.g., the concept of "natural mass matrices", advocated by Peccei and Wang[23] and that of Weak Basis (WB) transformations, considered by Fritzsch and Xing[24] as well as Branco *et al.*[25]

The essential idea of "natural mass matrices" consists of formulating quark mass matrices which are able to reproduce hierarchical mixing angles without resorting to fine tuning. This results in considerably constraining the parameter space available to the elements of the mass matrices. Using this concept Peccei and Wang[23] have attempted to reconstruct mass matrices at M_z as well as GUT scale, however without invoking any other condition they are not able to find any finite or viable set of mass matrices. In the context of texture specific mass matrices, the idea of "natural mass matrices" has been found to be useful in reproducing the data when the following hierarchy is imposed on the elements of the quark mass matrices

$$(1, i) \leq (2, j) \leq (3, 3) \quad i = 1, 2, 3; j = 2, 3. \tag{7}$$

As mentioned earlier, Weak Basis transformations is an another idea to go beyond texture *ansätze*, considered by by Fritzsch and Xing[24] as well as Branco *et al.*[25] Within the framework of the SM, the hermitian quark mass matrices, which encode all the information about the quark masses and mixings, have a total of 18 real free parameters, which is a large number compared to only ten physical parameters corresponding to six quark masses and four physical parameters of the CKM matrix. In this context, it is interesting to note that one has the freedom to make a unitary transformation, e.g., $q_L \rightarrow Wq_L$, $q_R \rightarrow Wq_R$, $q'_L \rightarrow Wq'_L$, $q'_R \rightarrow Wq'_R$ under which the gauge currents

$$-\mathcal{L}_W^{cc} = \frac{g}{\sqrt{2}}\overline{(u,c,t)}_L \gamma^\mu \begin{pmatrix} d \\ s \\ b \end{pmatrix}_L W_\mu + \text{h.c.} \tag{8}$$

remain real and diagonal but the mass matrices transform as

$$M_U \longrightarrow M'_U = W^\dagger M_U W, \;\; M_D \longrightarrow M'_D = W^\dagger M_D W \tag{9}$$

where W is an arbitrary unitary matrix. Such transformations are referred to as 'Weak Basis (WB) Transformations'.

The WB transformations broadly lead to two possibilities for the texture zero fermion mass matrices. In the first possibility, as observed by Fritzsch and Xing,[24] one ends up with texture 2 zero fermion mass matrices, wherein both the fermion mass matrices assume a texture 1 zero hermitian structure of the following form

$$M_q^{'} = \begin{pmatrix} * & * & 0 \\ * & * & * \\ 0 & * & * \end{pmatrix}, \qquad q^{'} = U, D. \tag{10}$$

In the second possibility, as observed by Branco *et al.*[25] one ends up with texture 3 zero fermion mass matrices M_U and M_D wherein one of the matrix among these pairs is a texture 2 zero Fritzsch-like hermitian mass matrix given by

$$M_q = \begin{pmatrix} 0 & * & 0 \\ * & * & * \\ 0 & * & * \end{pmatrix}, \qquad q = U, D, \tag{11}$$

while the other mass matrix is a texture 1 zero hermitian mass matrix of the following form

$$M_q^{'} = \begin{pmatrix} 0 & * & * \\ * & * & * \\ * & * & * \end{pmatrix}, \qquad q^{'} = U, D. \tag{12}$$

Further, we would like to emphasize here that although the two approaches for WB transformations are equivalent, but the approach by Branco *et al.* leads to non parallel texture three zero structure while the approach by Fritzsch and Xing leads to parallel texture two zero structure.

Recently an analysis by Costa and Simoes[26] shows that starting from arbitrary matrices M_U and M_D, it is always possible to perform a WB transformation that renders them Hermitian with a particular texture, therefore, resulting in reducing the number of free parameters of general mass matrices. The obtained quark matrices are confronted with the experimental data, reconstructing them at the electroweak scale and at a high scale where the Froggatt–Nielsen mechanism can be implemented. However, in the absence of any constraints on the elements of the mass matrices, it leads to a large number of viable texture zero matrices.

It is therefore evident from the above discussion that neither texture *ansätze* nor Weak Basis transformations or "naturalness" criteria, on their own, are able to lead to a finite set of viable texture specific mass matrices. In order to obtain the same, perhaps one needs to combine the three as discussed recently by Sharma *et al.*[27] This analysis shows that one can start with the most general mass matrices and consequently explore the possibility of obtaining a finite set of viable texture specific mass matrices formulated by using weak basis transformations as well as the constraints imposed due to "naturalness". Interestingly, the analysis reveals that a particular set of texture 4 zero quark mass matrices can be considered to be a unique viable option for the description of quark mixing data.

A corresponding analysis in the lepton sector, wherein one explores the possibility of arriving at a minimal set of lepton texture specific mass matrices, reveals that this is not possible because of a large number of viable possibilities. The analysis pertaining to texture 4 zero Fritzsch-like mass matrices in the Dirac as well as Majorana neutrino case indicates that these matrices are compatible with the normal hierarchy and degenerate scenario of neutrino masses whereas for inverted hierarchy such matrices are ruled out in case the naturalness conditions are imposed. In conclusion, we can perhaps say that the texture 4 zero Fritzsch-like mass matrices provide an almost unique class of viable fermion mass matrices giving vital clues towards unified textures for model builders.

Acknowledgments

M.G. and P.F. would like to acknowledge CSIR, Government of India (Grant No:03:(1313)14/EMR-II) for financial support. S.S. acknowledges the Principal, GGDSD College, Sector 32, Chandigarh. G.A. would like to acknowledge DST, Government of India (Grant No: SR/FTP/PS-017/2012) for financial support. P.F. and S.S. acknowledge the Chairperson, Department of Physics, P.U., for providing facilities to work.

References

1. H. Fritzsch, M. Gell-Mann and P. Minkowski, *Phys Lett. B* **59**, 256 (1975).
2. P. Minkowski, *Phys. Lett. B* **67**, 421 (1977).
3. T. Yanagida, in *Proceedings of the Workshop on Unified Theory and the Baryon Number of the Universe*, eds. O. Sawada and A. Sugamoto (KEK, Tsukuba, 1979), p. 95.
4. M. Gell-Mann, P. Ramond and R. Slansky, in *Supergravity*, eds. F. van Nieuwenhuizen and D. Freedman (North Holland, Amsterdam, 1979), p. 315.
5. S. L. Glashow, in *Quarks and Leptons*, ed. M. Lévy *et al.* (Plenum, New York, 1980), p. 707.
6. R. N. Mohapatra and G. Senjanovic, *Phys. Rev. Lett.* **44**, 912 (1980).
7. M.-C. Chen and K. T. Mahanthappa, *Int. J. Mod. Phys. A* **18**, 5819 (2003).
8. H. Fritzsch and Z. Z. Xing, *Nucl. Phys. B* **556**, 49 (1999) and references therein.
9. Z. Z. Xing and H. Zhang, *J. Phys. G* **30**, 129 (2004) and refrences therein.
10. M. Gupta and G. Ahuja, *Int. J. Mod. Phys. A* **26**, 2973 (2011) and refrences therein.
11. M. Gupta and G. Ahuja, *Int. J. Mod. Phys. A* **27**, 1230033 (2012) and references therein.
12. H. Fritzsch, *Phys. Lett. B* **70**, 436 (1977).
13. H. Fritzsch, *Phys. Lett. B* **73**, 317 (1978).
14. B. Stech, *Phys. Lett. B* **130**, 189 (1983).
15. M. Gronau, R. Johnson and J. Schechter, *Phys. Rev. Lett.* **54**, 2176 (1985).
16. P. Ramond, R. G. Roberts and G. G. Ross, *Nucl. Phys. B* **406**, 19 (1993).
17. K. A. Olive *et al.* (Particle Data Group), *Chin. Phys. C* **38**, 090001 (2014).
18. D. Du and Z. Z. Xing, *Phys. Rev. D* **48**, 2349 (1993).
19. P. S. Gill and M. Gupta, *Pramana* **45**, 333 (1995).
20. P. S. Gill and M. Gupta, *Phys. Rev. D* **57**, 3971 (1998).
21. M. Randhawa and M. Gupta, *Phys. Rev. D* **63**, 097301 (2001).
22. P. Ludl and W. Grimus, arXiv:hep-ph/1501.04942.
23. R. D. Peccei and K. Wang, *Phys. Rev. D* **53**, 5 (1996).
24. H. Fritzsch and Z. Z. Xing, *Phys. Lett. B* **413**, 396 (1997) and references therein.

25. G. C. Branco *et al.*, *Phys. Rev. Lett.* **82**, 683 (1999).
26. D. Emmanuel-Costa and C. Simoes, *Phys. Rev. D* **79**, 073006 (2009).
27. S. Sharma, P. Fakay, G. Ahuja and M. Gupta, *Phys. Rev. D* **91**, 053004 (2015).

General Lepton Textures and Their Implications

Gulsheen Ahuja*, Samandeep Sharma@, Priyanka Fakay and Manmohan Gupta

Department of Physics, Panjab University, Chandigarh, India

@*Department of Physics, GGDSD College, Chandigarh, India*

gulsheen@pu.ac.in

The present work attempts to provide an overview of texture specific lepton mass matrices. In particular, we summarize the findings of some recent analyses carried out within non-flavor basis, wherein a parallel texture structure for the lepton and neutrino mass matrices is considered.

1. Introduction

In the last decade, we have almost reached 'precision' level for the measurement of neutrino oscillation parameters, including the recently measured mixing angle θ_{13}. This has led to the need of a more intense activity towards understanding the pattern of neutrino masses and mixings which is quite different from the corresponding quark mixing case. In the absence of a theory providing a viable understanding of these issues, most of the phenomenological work is carried out within the general premises of 'bottom-up' approach. As an example of this approach, texture specific lepton mass matrices have been tried with a good deal of success. In particular, several attempts[1–8] have been made to understand the neutrino mixing data by formulating the phenomenological mass matrices with charged lepton matrix being diagonal, usually referred to as the flavor basis case. In addition, for both Majorana as well as Dirac neutrinos, some attempts[9–10] have also been made to explain the neutrino mixing data by considering texture specific structures for both the charged lepton and the neutrino mass matrices, referred to as the non-flavor basis case. It may be noted that the non-flavor basis enables quarks and leptons to be treated at the same footing and also to explore the possibility to arrive at a minimal set of fermion mass matrices which are compatible with the latest mixing data.

It is now well known, that, in the leptonic sector, the search for viable mass matrices is complicated by the 'smallness' of neutrino masses. The most popular explanation for this smallness is the see-saw mechanism[11–16] which requires the neutrinos to be Majorana fermions. However, at present, neither the Majorana nature is established nor can we rule out the Dirac nature of neutrinos. On theoretical grounds, the existence of small Dirac masses requires the corresponding Yukawa couplings to be exceptionally small compared to their charged counterparts. The Dirac neutrino mass, although seemingly 'unnatural', can be explained by additional $U(1)_{B-L}$ symmetry of the Lagrangian which forbids Majorana mass term for the neutrinos. Apart from SM, this possibility can be realized in many of the models[10] such as supersymmetry, superstring, supergravity and large extra dimensions. Keeping in mind that Dirac neutrinos are still not ruled out, in the present work we discuss texture specific mass matrices for both Dirac as well as Majorana neutrinos.

In particular, we have presented an overview of some recent analyses[9,17,18] wherein texture specific lepton mass matrices have been considered in the non-flavor basis for Dirac as well as Majorana neutrinos. In the following section, we first present the relation between lepton mass matrices and mixing matrix. The present experimental status of the neutrino mixing parameters have been given in Sec. 3. A brief summary of texture 6, 5 and 4 zero lepton mass matrices has been presented in Sec. 4. Finally, Sec. 5 summarizes our conclusions.

2. Lepton Mass Matrices and PMNS Matrix

For the case of neutrinos, it is important to note that these may have either the Dirac masses or the more general Dirac–Majorana masses. A Dirac mass term can be generated by the Higgs mechanism with the standard Higgs doublet. In this case, the neutrino mass term can be written as

$$\overline{\nu}_{a_L} M_{\nu D} \nu_{a_R} + \text{h.c.}, \tag{1}$$

where $a = e, \mu, \tau$. ν_e, ν_μ, ν_τ are the flavor eigenstates and $M_{\nu D}$ is a complex 3×3 Dirac mass matrix. As mentioned earlier, in the non-flavor basis, both the charged lepton and the neutrino mass matrices are considered having the same texture structure,[9] e.g.,

$$M_l = \begin{pmatrix} 0 & A_l & 0 \\ A_l^* & D_l & B_l \\ 0 & B_l^* & C_l \end{pmatrix}, \qquad M_{\nu D} = \begin{pmatrix} 0 & A_\nu & 0 \\ A_\nu^* & D_\nu & B_\nu \\ 0 & B_\nu^* & C_\nu \end{pmatrix}, \tag{2}$$

M_l and $M_{\nu D}$ respectively corresponding to hermitian Dirac-like charged lepton and neutrino mass matrices. It may be noted that each of the above matrix is texture 2 zero type with $A_{l(\nu)} = |A_{l(\nu)}|e^{i\alpha_{l(\nu)}}$ and $B_{l(\nu)} = |B_{l(\nu)}|e^{i\beta_{l(\nu)}}$, in case these are symmetric then $A^*_{l(\nu)}$ and $B^*_{l(\nu)}$ should be replaced by $A_{l(\nu)}$ and $B_{l(\nu)}$, as well as $C_{l(\nu)}$ and $D_{l(\nu)}$ should respectively be defined as $C_{l(\nu)} = |C_{l(\nu)}|e^{i\gamma_{l(\nu)}}$ and $D_{l(\nu)} = |D_{l(\nu)}|e^{i\omega_{l(\nu)}}$.

The texture 6 zero mass matrices can be obtained from the above mentioned matrices by taking both D_l and D_ν to be zero, which reduces the matrices M_l and $M_{\nu D}$ each to texture 3 zero type. Texture 5 zero matrices can be obtained by taking either $D_l = 0$ and $D_\nu \neq 0$ or $D_\nu = 0$ and $D_l \neq 0$, thereby, giving rise to two possible cases of texture 5 zero matrices, referred to as texture 5 zero $D_l = 0$ case pertaining to M_l texture 3 zero type and $M_{\nu D}$ texture 2 zero type and texture 5 zero $D_\nu = 0$ case pertaining to M_l texture 2 zero type and $M_{\nu D}$ texture 3 zero type.

It should be noted that in the case of texture 6 zero and texture 4 zero mass matrices we can have parallel structures for both the neutrino mass matrix and the charged lepton mass matrix, however, for the case of texture 5 zero mass matrices, one cannot have parallel structures. To consider all possible textures, we have considered only those possibilities which are compatible with the 'Weak Basis' transformations.[19,20] To this end, in Table 1, we have presented all possible texture 2 zero mass matrices, from which we can derive texture 6 zero, 5 zero and 4 zero mass matrices for the discussion.

Table 1. Table showing various 'Weak Basis' transformation compatible texture 2 zero possibilities categorized into four distinct classes and their permutations given by a, b, c, d, e and f.

	Class I	Class II	Class III	Class IV
a	$\begin{pmatrix} 0 & Ae^{i\alpha} & 0 \\ Ae^{-i\alpha} & D & Be^{i\beta} \\ 0 & Be^{-i\beta} & C \end{pmatrix}$	$\begin{pmatrix} D & Ae^{i\alpha} & 0 \\ Ae^{-i\alpha} & 0 & Be^{i\beta} \\ 0 & Be^{-i\beta} & C \end{pmatrix}$	$\begin{pmatrix} 0 & Ae^{i\alpha} & De^{i\gamma} \\ Ae^{-i\alpha} & 0 & Be^{i\beta} \\ De^{-i\gamma} & Be^{-i\beta} & C \end{pmatrix}$	$\begin{pmatrix} A & 0 & 0 \\ 0 & D & Be^{i\beta} \\ 0 & Be^{-i\beta} & C \end{pmatrix}$
b	$\begin{pmatrix} 0 & 0 & Ae^{i\alpha} \\ 0 & C & Be^{i\beta} \\ Ae^{-i\alpha} & Be^{-i\beta} & D \end{pmatrix}$	$\begin{pmatrix} D & 0 & Ae^{i\alpha} \\ 0 & C & Be^{i\beta} \\ Ae^{-i\alpha} & Be^{-i\beta} & 0 \end{pmatrix}$	$\begin{pmatrix} 0 & De^{i\gamma} & Ae^{i\alpha} \\ De^{-i\gamma} & C & Be^{i\beta} \\ Ae^{-i\alpha} & Be^{-i\beta} & 0 \end{pmatrix}$	$\begin{pmatrix} C & 0 & Be^{i\alpha} \\ 0 & A & 0 \\ Be^{-i\alpha} & 0 & D \end{pmatrix}$
c	$\begin{pmatrix} D & Ae^{i\alpha} & Be^{i\beta} \\ Ae^{-i\alpha} & 0 & 0 \\ Be^{-i\beta} & 0 & C \end{pmatrix}$	$\begin{pmatrix} 0 & Ae^{i\alpha} & Be^{i\beta} \\ Ae^{-i\alpha} & D & 0 \\ Be^{-i\beta} & 0 & C \end{pmatrix}$	$\begin{pmatrix} 0 & Ae^{i\alpha} & Be^{i\beta} \\ Ae^{-i\alpha} & 0 & De^{i\gamma} \\ Be^{-i\beta} & De^{-i\gamma} & C \end{pmatrix}$	$\begin{pmatrix} C & Be^{i\alpha} & 0 \\ Be^{-i\alpha} & D & 0 \\ 0 & 0 & A \end{pmatrix}$
d	$\begin{pmatrix} C & Be^{i\beta} & 0 \\ Be^{-i\beta} & D & Ae^{i\alpha} \\ 0 & Ae^{-i\alpha} & 0 \end{pmatrix}$	$\begin{pmatrix} C & Be^{i\alpha} & 0 \\ Be^{-i\alpha} & 0 & Ae^{i\beta} \\ 0 & Ae^{-i\beta} & D \end{pmatrix}$	$\begin{pmatrix} 0 & Be^{i\alpha} & Ce^{i\gamma} \\ Be^{-i\alpha} & 0 & Ae^{i\beta} \\ Ce^{-i\gamma} & Ae^{-i\beta} & D \end{pmatrix}$	$\begin{pmatrix} A & 0 & 0 \\ 0 & C & Be^{i\beta} \\ 0 & Be^{-i\beta} & D \end{pmatrix}$
e	$\begin{pmatrix} D & Be^{i\beta} & Ae^{i\alpha} \\ Be^{-i\beta} & C & 0 \\ Ae^{-i\alpha} & 0 & 0 \end{pmatrix}$	$\begin{pmatrix} C & 0 & Be^{i\alpha} \\ 0 & D & Ae^{i\beta} \\ Be^{-i\alpha} & Ae^{-i\beta} & 0 \end{pmatrix}$	$\begin{pmatrix} 0 & Ce^{i\gamma} & Be^{i\alpha} \\ Ce^{-i\gamma} & D & Ae^{i\beta} \\ Be^{-i\alpha} & Ae^{-i\beta} & 0 \end{pmatrix}$	$\begin{pmatrix} D & 0 & Be^{i\alpha} \\ 0 & A & 0 \\ Be^{-i\alpha} & 0 & C \end{pmatrix}$
f	$\begin{pmatrix} C & 0 & Be^{i\beta} \\ 0 & 0 & Ae^{i\alpha} \\ Be^{-i\beta} & Ae^{-i\alpha} & D \end{pmatrix}$	$\begin{pmatrix} 0 & Be^{i\alpha} & Ae^{i\beta} \\ Be^{-i\alpha} & C & 0 \\ Ae^{-i\beta} & 0 & D \end{pmatrix}$	$\begin{pmatrix} 0 & Be^{i\alpha} & Ae^{i\beta} \\ Be^{-i\alpha} & 0 & Ce^{i\gamma} \\ Ae^{-i\beta} & Ce^{-i\gamma} & D \end{pmatrix}$	$\begin{pmatrix} C & Be^{i\alpha} & 0 \\ Be^{-i\alpha} & D & 0 \\ 0 & 0 & A \end{pmatrix}$

Coming to the diagonalization of lepton mass matrices, similar to the quark sector, these can also be diagonalized by bi-unitary transformations, e.g.,

$$M_{\nu D}^{\rm diag} = U_{\nu L}^{\dagger} M_{\nu D} U_{\nu R} = {\rm Diag}(m_1,\ m_2,\ m_3), \tag{3}$$

where $U_{\nu L}$ and $U_{\nu R}$ are unitary matrices and $M_{\nu D}^{diag}$ is a diagonal matrix. The corresponding mixing matrix obtained, known as Pontecorvo–Maki–Nakagawa–Sakata (PMNS) or lepton mixing matrix $V_{\rm PMNS}$, is given as

$$V_{\rm PMNS} = V_{l_L}^{\dagger} V_{\nu L}, \tag{4}$$

where $V_{l_L}^{\dagger}$ and V_{ν_L} correspond to the diagonalization transformations of lepton and neutrino mass matrices respectively. The $V_{\rm PMNS}$ expresses the relationship between the neutrino mass eigenstates and the flavor eigenstates, e.g.,

$$\begin{pmatrix} \nu_e \\ \nu_\mu \\ \nu_\tau \end{pmatrix} = \begin{pmatrix} V_{e1} & V_{e2} & V_{e3} \\ V_{\mu 1} & V_{\mu 2} & V_{\mu 3} \\ V_{\tau 1} & V_{\tau 2} & V_{\tau 3} \end{pmatrix} \begin{pmatrix} \nu_1 \\ \nu_2 \\ \nu_3 \end{pmatrix}, \tag{5}$$

where ν_e, ν_μ, ν_τ are the flavor eigenstates ; ν_1, ν_2, ν_3 are the mass eigenstates and the 3×3 mixing matrix is leptonic mixing matrix. For the case of three Dirac neutrinos, in the Particle Data Group (PDG) parameterization, involving three angles θ_{12}, θ_{23}, θ_{13} and the Dirac-like CP violating phase δ_l the mixing matrix has the form

$$V_{\rm PMNS} = \begin{pmatrix} c_{12}c_{13} & s_{12}c_{13} & s_{13}e^{-i\delta_l} \\ -s_{12}c_{23} - c_{12}s_{23}s_{13}e^{i\delta_l} & c_{12}c_{23} - s_{12}s_{23}s_{13}e^{i\delta_l} & s_{23}c_{13} \\ s_{12}s_{23} - c_{12}c_{23}s_{13}e^{i\delta_l} & -c_{12}s_{23} - s_{12}c_{23}s_{13}e^{i\delta_l} & c_{23}c_{13} \end{pmatrix}, \tag{6}$$

with $s_{ij} = \sin\theta_{ij}$, $c_{ij} = \cos\theta_{ij}$.

The neutrino might be a Majorana particle which is defined as is its own anti particle and is characterized by only two independent particle states of the same mass ($\nu_{\rm L}$ and $\bar{\nu}_{\rm R}$ or $\nu_{\rm R}$ and $\bar{\nu}_{\rm L}$). A Majorana mass term which violates both the law of total lepton number conservation and that of individual lepton flavor conservation can be written either as

$$\frac{1}{2}\overline{\nu}_{a_L} M_L \nu_{a_R}^c + \text{h.c.}, \tag{7}$$

or as

$$\frac{1}{2}\overline{\nu}_{a_L}^c M_R \nu_{a_R} + \text{h.c.}, \tag{8}$$

where M_l and M_R are complex symmetric matrices leading to the famous see-saw mechanism,[11–16] given by

$$M_\nu = -M_{\nu D}^T \left(M_R\right)^{-1} M_{\nu D}, \tag{9}$$

where $M_{\nu D}$ and M_R are respectively the Dirac neutrino mass matrix and the right-handed Majorana neutrino mass matrix. This mechanism requires the inclusion of right-handed neutrinos with very large Majorana masses, therefore inducing a very small mass for the left-handed neutrinos. Thus, the generation of masses in neutrinos is not straight-forward as they may have either the Dirac masses or the more general Dirac–Majorana masses. Further, when discussing texture possibilities textures are imposed on $M_{\nu D}$, unlike many other attempts[1–8] in the literature where texture is imposed on M_ν.

In the case of the Majorana neutrinos, there are extra phases which cannot be removed, therefore, the above matrix V_{PMNS} takes the following form

$$\begin{pmatrix} c_{12}c_{13} & s_{12}c_{13} & s_{13}e^{-i\delta_l} \\ -s_{12}c_{23} - c_{12}s_{23}s_{13}e^{i\delta_l} & c_{12}c_{23} - s_{12}s_{23}s_{13}e^{i\delta_l} & s_{23}c_{13} \\ s_{12}s_{23} - c_{12}c_{23}s_{13}e^{i\delta_l} & -c_{12}s_{23} - s_{12}c_{23}s_{13}e^{i\delta_l} & c_{23}c_{13} \end{pmatrix} \begin{pmatrix} e^{i\alpha_1/2} & 0 & 0 \\ 0 & e^{i\alpha_2/2} & 0 \\ 0 & 0 & 1 \end{pmatrix}, \tag{10}$$

where δ_l is the Dirac-like CP violating phase in the leptonic sector and α_1 and α_2 are the Majorana phases which do not play any role in neutrino oscillations.

3. Experimental Status of Neutrino Masses and Mixing Parameters

While carrying out an analysis regarding exploring the compatibility of neutrino mass matrices with the recent data, one needs to keep in mind the experimental constraints imposed by the relationship between mass matrices and their corresponding mixing matrices. To facilitate our discussion in this regard, we present the status of relevant data in the lepton sector. The 3σ confidence level ranges of the neutrino oscillation parameters obtained in a latest global three neutrino oscillation analysis carried out by Fogli *et al.*[21] have been presented in Table 2.

Table 2. Current data for neutrino mixing parameters from global fits.[21]

Parameter	3σ range
$\Delta m^2_{\rm sol}$ $[10^{-5}{\rm eV}^2]$	(6.99–8.18)
$\Delta m^2_{\rm atm}$ $[10^{-3}{\rm eV}^2]$	(2.19–2.62)(NH); (2.17–2.61)(IH)
$\sin^2\theta_{13}$ $[10^{-2}]$	(1.69–3.13)(NH); (1.71–3.15) (IH)
$\sin^2\theta_{12}$ $[10^{-1}]$	(2.59–3.59)
$\sin^2\theta_{23}$ $[10^{-1}]$	(3.31–6.37)(NH);(3.35–6.63)(IH)

While carrying out the analysis, the magnitudes of atmospheric and solar neutrino mass square differences, defined as $m_2^2 - m_1^2$ and $m_3^2 - \frac{(m_1^2+m_2^2)}{2}$ respectively, are allowed full variation within their 3σ ranges. The lightest neutrino mass, m_1 for the case of normal hierarchy (NH) and m_3 for the case of inverted hierarchy (IH), is considered as the free parameter while the other two masses are obtained using the following relations,

$$\text{NH}: \quad m_2^2 = \Delta m^2_{\rm sol} + m_1^2, \quad m_3^2 = \Delta m^2_{\rm atm} + \frac{(m_1^2 + m_2^2)}{2}, \tag{11}$$

$$\text{IH}: \quad m_2^2 = \frac{2(m_3^2 + \Delta m^2_{\rm atm}) + \Delta m^2_{\rm sol}}{2}, \quad m_1^2 = \frac{2(m_3^2 + \Delta m^2_{\rm atm}) - \Delta m^2_{\rm sol}}{2}. \tag{12}$$

It should be noted that while carrying out analyses of different texture specific mass matrices, we have also imposed the condition of 'naturalness'[9] so as to keep the quark-lepton similarity in this regards. Further, the phases $\phi_1 = \alpha_{\nu D} - \alpha_l$, $\phi_2 = \beta_{\nu D} - \beta_l$ and the elements $D_{l,\nu}$, $C_{l,\nu}$ are considered to be free parameters. In the absence of any constraint on the phases, ϕ_1 and ϕ_2 have been given full variation from 0 to 2π. Although $D_{l,\nu}$ and $C_{l,\nu}$ are free parameters, however, they have been constrained such that diagonalizing transformations O_l and O_ν always remain real.

Before presenting the results, we would like to mention that unlike the quark case, wherein it has been shown that texture 4 zero Fritzsch like matrices are perhaps the only compatible matrices with data,[9,10,22–25] in the case of leptons, we cannot arrive at this kind of conclusion. In the sequel, we present an overview of the viability of different textures for Dirac as well as Majorana nature of neutrinos.

4. Viable Texture Specific Lepton Mass Matrices

In the context of quarks it is well known that texture 6 zero mass matrices are completely ruled out by the existing data. Interestingly, in case we consider Dirac like neutrinos, texture 6 zero or minimal texture is also ruled out for normal/inverted hierarchy and degenerate scenario of neutrino masses. However, for Majorana neutrinos inverted hierarchy and degenerate scenario are ruled out whereas in the case of normal hierarchy, there are several compatible combinations with the current neutrino oscillation data. For a detailed discussion, we refer the readers to Ref. 10.

Coming to the cases of non-minimal textures, i.e., the texture 5 zero and texture 4 zero mass matrices, we present our conclusions from our recent analyses.[17,18] To begin with, we first discuss the texture 5 zero and texture 4 zero lepton mass matrices for the case of Dirac neutrinos. Corresponding to this, a detailed and comprehensive analysis has been carried out for normal/inverted hierarchy and degenerate scenario of neutrino masses. In this context, for texture 5 zero mass matrices, the analysis has been carried out for $D_l = 0$, $D_\nu \neq 0$ as well as $D_l \neq 0$, $D_\nu = 0$ cases, corresponding to all the viable classes. For Class I, mentioned in Table 1, inverted hierarchy is ruled out for both the cases, whereas normal hierarchy is viable for the $D_l = 0$, $D_\nu \neq 0$ case. For Class II, normal hierarchy is viable for both the cases while the inverted hierarchy is ruled out for the case $D_l = 0$, $D_\nu \neq 0$. Finally, for Class III we find that inverted hierarchy is viable for the case $D_l = 0$, $D_\nu \neq 0$, while the normal hierarchy is compatible with the $D_l \neq 0$, $D_\nu = 0$ case. It may be mentioned that Class IV is not phenomenologically viable due to de-coupling of one of the generations.

Coming to the texture 4 zero case, due to the availability of an additional parameter large number of viable possibilities emerge. Without getting into the details of these possibilities, we would like to mention only broad conclusions in this regard. Interestingly, unlike the case of texture 6 zero mass matrices, both inverted hierarchy and degenerate scenario are not ruled out for all the classes of texture specific mass matrices mentioned in Table 1. For the case of normal hierarchy, it seems mass matrices corresponding to all the classes are compatible with the data. However, inverted hierarchy is ruled out for Class I but compatible with Class II and Class III. Similarly, degenerate scenario of neutrino masses is compatible only with Class III.

Coming to the case of texture 5 zero and texture 4 zero mass matrices for neutrinos being Majorana particles. To begin with, we consider texture 5 zero lepton mass matrices, for both the cases, viz. $D_l = 0$, $D_\nu \neq 0$ as well as $D_l \neq 0$, $D_\nu = 0$ for matrices mentioned in Class II and Class III of Table 1. For Class II, normal hierarchy is viable for both the cases, while the inverted hierarchy seems to be ruled out for the case, $D_l = 0$, $D_\nu \neq 0$. Finally, for texture 5 zero mass matrices pertaining to Class III, we find that inverted hierarchy is viable for the case $D_l \neq 0$, $D_\nu = 0$, while the normal hierarchy is compatible with the $D_l = 0$, $D_\nu \neq 0$ case.

It may be mentioned that the number of viable possibilities is understandably quite large. The analysis reveals that the Fritzsch like texture two zero lepton mass matrices are compatible with the recent lepton mixing data pertaining to normal as well as inverted neutrino mass hierarchies. Interestingly, one finds that both the normal as well as inverted neutrino mass hierarchies are compatible with texture four zero mass matrices pertaining to Class II and Class III of Table 1 contrary to the case for texture four zero mass matrices pertaining to Class I wherein inverted hierarchy seems to be ruled out. Interestingly for Classes I and II, the degenerate neutrino mass scenario is incompatible, whereas it is compatible for mass matrices in Class III.

It is interesting to add that in the context of quarks, it has been recently shown[25] that texture 4 zero Fritzsch like mass matrices and their permutations, compatible with the Weak Basis transformations, provides a unique texture in agreement with the data. In case of leptons also, we have seen that texture 4 zero matrices are compatible with data for Dirac as well as Majorana neutrinos. Therefore, we can conclude that Fritzsch like texture 4 zero matrices may provide vital clues for the fundamental theories of flavor physics.

5. Summary and Conclusions

A broad based survey of the texture specific lepton mass matrices has been presented. It seems that in the case of Dirac neutrinos, texture 6 zero mass matrices are ruled out. However, this is not true in the case of Majorana neutrinos. Lesser than texture 6 zeros, we find compatibility of the mass matrices with data for both the kind of neutrinos and for all kind of neutrino mass hierarchies.

Acknowledgments

G.A. would like to acknowledge DST, Government of India (Grant No: SR/FTP/PS-017/2012) for financial support. M.G. and P.F. would like to acknowledge CSIR, Govt. of India,(Grant No:03:(1313)14/EMR-II) for financial support. S.S. acknowledges the Principal, GGDSD College, Sector 32, Chandigarh. P.F. and S.S. acknowledge the Chairperson, Department of Physics, P.U., for providing facilities to work.

References

1. P. H. Frampton, S. L. Glashow and D. Marfatia, *Phys. Lett. B* **536**, 79 (2002).
2. Z. Z. Xing, *Phys. Lett. B* **530**, 159 (2002).
3. Z. Z. Xing, *Int. J. Mod. Phys. A* **19**, 1 (2004) and references therein.
4. A. Merle and W. Rodejohann, *Phys. Rev. D* **73**, 073012 (2006).
5. S. Dev, S. Kumar, S. Verma and S. Gupta, *Nucl. Phys. B* **784**, 103 (2007).
6. S. Dev, S. Verma and S. Gupta, *Phys. Lett. B* **687**, 53–56 (2010).
7. S. Dev, R. R. Gautam and L. Singh, *Phys. Rev. D* **87**, 073011 (2013).
8. G. Blankenburg and D. Meloni, *Nucl. Phys. B* **867**, 749 (2013).
9. M. Gupta and G. Ahuja, *Int. J. Mod. Phys. A* **26**, 2973 (2011) and refrences therein.
10. M. Gupta and G. Ahuja, *Int. J. Mod. Phys. A* **27**, 1230033 (2012) and references therein.
11. H. Fritzsch, M. Gell-Mann and P. Minkowski, *Phys Lett. B* **59**, 256 (1975).
12. P. Minkowski, *Phys. Lett. B* **67**, 421 (1977).
13. T. Yanagida, in *Proceedings of the Workshop on Unified Theory and the Baryon Number of the Universe*, edited by O. Sawada and A. Sugamoto (KEK, Tsukuba, 1979), p. **95**.
14. M. Gell-Mann, P. Ramond and R. Slansky, in *Supergravity*, edited by F. van Nieuwenhuizen and D. Freedman (North Holland, Amsterdam, 1979), p. **315**.
15. S. L. Glashow, in *Quarks and Leptons*, edited by M. Lévy *et al.* (Plenum, New York, 1980), p. **707**.
16. R. N. Mohapatra and G. Senjanovic, *Phys. Rev. Lett.* **44**, 912 (1980).
17. S. Sharma, P. Fakay, G. Ahuja and M. Gupta, arXiv:hep-ph/1402.0628.
18. S. Sharma, P. Fakay, G. Ahuja and M. Gupta, arXiv:hep-ph/1402.1598.
19. G. C. Branco, D. Emmanuel-Costa and R. Felipe, *Phys. Rev. Lett.* **82**, 683 (1999).
20. G. C. Branco, D. Emmanuel-Costa, R. Felipe and H. Serodio, *Phys. Rev. Lett.* **670**, 340 (2009).
21. G. L. Fogli, E. Lisi, A. Marrone, D. Montanino, A. Palazzo and A. M. Rotunno, *Phys. Rev. D* **86**, 013012 (2012).
22. P. S. Gill and M. Gupta, *Pramana* **45**, 333 (1995).
23. P. S. Gill and M. Gupta, *Phys. Rev. D* **57**, 3971 (1998).
24. M. Randhawa and M. Gupta, *Phys. Rev. D* **63**, 097301 (2001).
25. S. Sharma, P. Fakay, G. Ahuja and M. Gupta, *Phys. Rev. D* **91**, 053004 (2015).

Status and Implications of Neutrino Masses: A Brief Panorama

José W. F. Valle

Instituto de Física Corpuscular (CSIC-UV),
Parc Científic de la Universitat de València,
C/Catedrático José Beltrán, 2, E-46980 Paterna, Valencia, Spain
https://www.astroparticles.es/

With the historic discovery of the Higgs boson our picture of particle physics would have been complete were it not for the neutrino sector and cosmology. I briefly discuss the role of neutrino masses and mixing upon gauge coupling unification, electroweak breaking and the flavor sector. Time is ripe for new discoveries such as leptonic CP violation, charged lepton flavor violation and neutrinoless double beta decay. Neutrinos could also play a role in elucidating the nature of dark matter and cosmic inflation.

Keywords: Neutrino mixing and oscillations; seesaw mechanism; quark-lepton unification; flavor symmetry; electroweak symmetry breaking; neutrinoless double beta decay; dark matter; inflation.

1. Introduction

Neutrinos are the most ubiquitous particles in the universe, over 300/cm^3 coming from the Big Bang cross us every second. If cosmological neutrinos were the only ones available probably there would be no neutrino physics, given their incredibly tiny interaction cross-sections. Fortunately nature is more generous and stars, such as our Sun, are copious sources of higher energy neutrinos that can be detected say, in gigantic underground detectors like Super-Kamiokande. Likewise, neutrinos arising from cosmic ray interactions in the upper atmospheric arrive the Earth from all directions of the sky. Here too, the agreement between theory and experiments requires the oscillation hypothesis, characterized by a nearly maximal angle θ_{23}, surprisingly at odds from expectations based upon the quark sector.

The resolution of the long-standing discrepancies between theoretical expectation and experimental measurements of solar and atmospheric neutrinos has opened this century with a revolution in particle physics, by providing the first solid evidence for new physics and the need to revise Standard Model of particle physics (see extensive discussion in Ref. 1). The latter assembles the fundamental constituents in three generations of quarks and leptons whose interactions are dictated by the principle of $SU(3)_c \otimes SU(2)_L \otimes U(1)_Y$ gauge invariance. It provides a precise theory of particle interactions, well tested up the highest energies so far explored in particle accelerators. While the photon and the gluon, carriers of electromagnetic and the strong force, are massless, the weak interaction messengers, the W and the Z are massive, along with all of the quarks and leptons. The basic theory relies on the principle of gauge invariance and this forbids mass. The simplest way out is the spontaneous electroweak symmetry breaking mechanism, which implies the existence of a physical Higgs boson. Its historic discovery three years ago led many to say that the standard model is now complete. However, the long-standing discrepancies between theoretical expectations and experimental measurements of solar and atmospheric neutrinos requires the existence of neutrino flavor oscillations,[2] and hence the existence of nonzero neutrino masses.[3] This discovery has triggered a revolution in particle physics, as it provides the first solid evidence for new physics and the need to revise Standard Model. Indeed, particle physics would have been "completed" with the Higgs boson discovery, were it not for the need to account for neutrino oscillations as well as the cosmological puzzles such as dark matter, baryon asymmetry and inflation. In this talk I will give a brief summary of the current landscape of particle physics in view of these issues.

2. Neutrino Mixing and Oscillations

The long-standing discrepancy between theoretical expectation and experimental measurements of solar neutrinos has finally been resolved in favor of the oscillation mechanism, characterized by an angle θ_{12}, substantially larger than its CKM analogue, the Cabbibo angle θ_C. Similarly, the agreement between theory predictions and measurements of atmospheric neutrinos at underground experiments, both event yields and angular distributions, indicates the need for neutrino oscillations, characterized by a nearly maximal mixing angle θ_{23}, quite different from its quark sector analogue.

Both solar and atmospheric neutrino discrepancies were crucially confirmed by the results of Earth-bound experiments based at reactors and accelerators. For example, the reactor experiment KamLAND pinned down that oscillations is the mechanism underlying the conversion of solar neutrinos and identified the relevant region of oscillation parameters, characterized by a "small" angle θ_{12}, as opposed to oscillations in vacuum. Recent reactor and accelerator experiments have also provided a good measurement of the third lepton flavor mixing parameter θ_{13}, with a first hint of leptonic CP violation just emerging, characterized by a CP phase δ which promises to open a new era in neutrino physics.

The basic ingredient needed to describe neutrino oscillations is the lepton mixing matrix K, which comes from the mismatch between the charged and neutral mass matrices arising after the spontaneous electroweak and lepton number breaking. If neutrinos get mass *a la seesaw* (see below) then one expects that the heavy neutrino

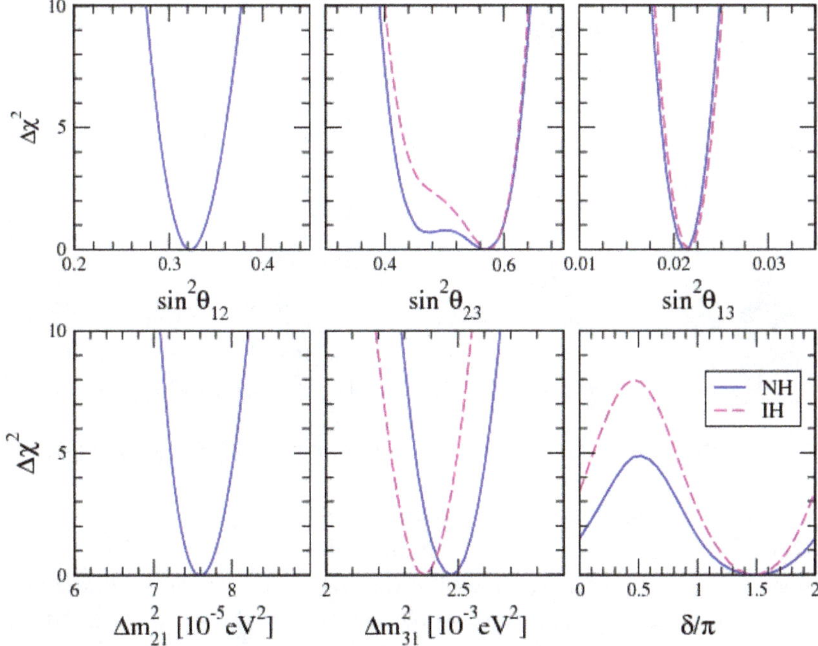

Fig. 1. Global picture of neutrino oscillation parameters after Neutrino 2014, from Ref. 2.

messengers will couple, subdominantly, in the charged current weak interaction leading to a rectangular form[3] for the matrix K.

To analyze the current solar, atmospheric, reactor and accelerator neutrino oscillation data one normally assumes the simplest unitary form for K. The two extra physical CP phases present in K are called Majorana phases and are most transparently expressed in terms of the original symmetric parametrization.[3] However they appear only in neutrinoless double beta decay and other lepton number violation processes. Hence they are omitted in neutrino oscillation analyses, for which the symmetric and the PDG presentations coincide.

The summary of the oscillation parameters after the Neutrino 2014 conference are presented in Fig. 1 (more discussion in Lisi's talk). Clearly one has good determinations of all the oscillation parameters except for the leptonic CP phase, which is just making its first appearance in the scene. The squared mass splitting parameters are tiny, without any counterpart in the charged fermion sector. Likewise, the values obtained for the solar and atmospheric angles θ_{12} and θ_{23}, are much larger than their CKM counterparts. The nonzero value of the reactor angle θ_{13} opens the door to future leptonic CP violations studies at the upcoming reactor and accelerator neutrino experiments, such as LBNF-DUNE. The measurement of the leptonic CP-phase using atmospheric neutrinos has been discussed in Smirnov's talk.

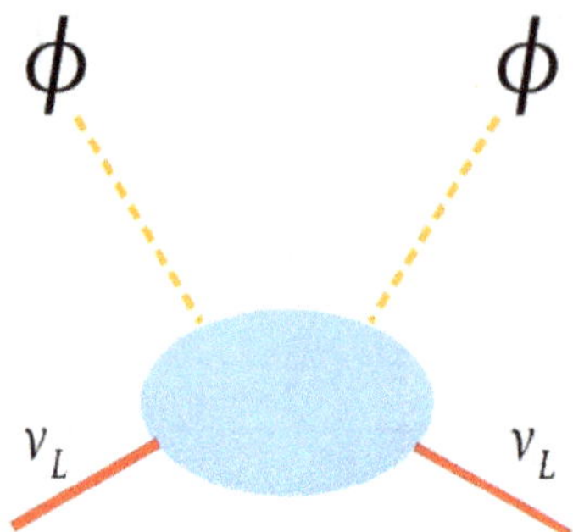

Fig. 2. Weinberg dimension five lepton number violation operator leading to neutrino mass.

3. Effective Neutrino Mass, Seesaw Mechanism and Unification

In the standard model neutrinos are massless so we need new physics in order to account for neutrino mixing and oscillations. As noted by Weinberg, one can add nonrenormalizable operators, such as the dimension five operator shown in Fig. 2, that break lepton number and which would account for the small observed neutrino masses. We have no clues as to the characteristic scale, the underlying mechanism or the flavor structure of the relevant operator. If anything, the neutrino oscillation observations indicate a very special pattern of mixing parameters, unlikely to be accidental. Its explanation from first principles, along with the other fermion masses and mixing parameters, constitutes the so-called *flavor problem*, one of the most stubborn problems in particle physics, and one for which the simplest gauge paradigm falls short at addressing. Here we stress the challenge of reconciling small CKM mixing parameters with large lepton mixing angles within a predictive framework.

The most popular way to induce the operator in Fig. 2 is through the exchange of heavy messengers, as present in SO(10) Grand unified theories (GUTS). In this case the smallness of neutrino mass is dynamically explained by minimizing the Higgs potential through a simple "1-2-3" VEV (vacuum expectation value) seesaw relation of the type

$$v_3 v_1 \sim {v_2}^2 \quad \text{with the hierarchy} \quad v_1 \gg v_2 \gg v_3 \,. \tag{1}$$

The isosinglet VEV v_1 drives the spontaneous breaking of lepton number symmetry and induces also a small but nonzero isotriplet VEV v_3 which generates the $\nu\nu$ entry in the neutrino mass matrix. Since the isodoublet v_2 fixes the masses of the weak gauge bosons, W and Z, one sees that $v_3 \to 0$ as $v_1 \to \infty$. The most popular messengers are heavy "right-handed" neutrinos (type-I seesaw) and heavy triplet scalar with a small induced VEV (type-II seesaw). Although these arise naturally in the framework of SO(10) GUTS, they may be introduced simply in terms of the $\mathrm{SU(3)_c \otimes SU(2)_L \otimes U(1)_Y}$ structure.

4. New Physics: To Unify or Not to Unify?

Despite the solid evidence for physics beyond-the-Standard Model in the neutrino sector, most theoretical extensions, such as grand unification, have so far been mainly driven by aesthetical principles. GUTS realize one of the most elegant ideas in particle physics. The three observed gauge interactions of the Standard Model which describe electromagnetic, weak, and strong forces merge into a single one at high energies. GUTS bring a *rationale* to charge quantization and the quantum numbers of the Standard Model. They are thought of as an intermediate step towards the ultimate *theory of everything*, which would also include gravity. As a generic feature, GUTS break the baryon number symmetry, allowing protons to decay in many ways. To date, all attempts to observe proton decay have failed. Here we stress three attractive features of GUTS:

- Simplest GUTS embed $SU(3)_c \otimes SU(2)_L \otimes U(1)_Y$ in an enlarged simple Lie group, characterized by a single unified gauge coupling constant.
- GUTS open the door to the possibility of relating quark and lepton masses.
- GUTS like SO(10) require the existence of right-handed neutrinos and the required breaking of B-L implies massive neutrinos.

Here we show how nonunified extended electroweak models with massive neutrinos may unify the gauge couplings as well as relate quark and lepton masses.

4.1. *Neutrino masses without GUTS*

Given that the number and properties of the messengers leading to Fig. 2 are to a large extent arbitrary, one can devise a variety of low-scale realizations of the seesaw paradigm, putting it literally "upside-down." In particular, the seesaw may be naturally realized at low scale, for example, the inverse and the linear seesaw mechanism. While these can be formulated in a GUT context,[4] this is not necessary at all.[5]

An alternative low-scale approach to neutrino masses is to assume that they arise only radiatively, typically as a result of extended symmetry breaking sectors. However, interesting examples have recently been suggested where neutrino masses arise from new gauge interactions, as illustrated in Fig. 3. The crosses denote VEV insertions of the relevant scalar multiplets responsible for symmetry breaking in the relevant extended $SU(3)_C \otimes SU(3)_L \otimes U(1)_X$ electroweak setup.[6]

4.2. *Gauge coupling unification without GUTS*

Within the standard $SU(3)_c \otimes SU(2)_L \otimes U(1)_Y$ gauge theory gauge coupling unification constitutes a "near miss." What kind of new physics could make the gauge coupling constants unify "exactly"? The first possibility is having a full-fledged Grand Unified Theory, as described above. This, however, entails as phenomenological implication the existence of proton decay, so far unobserved.

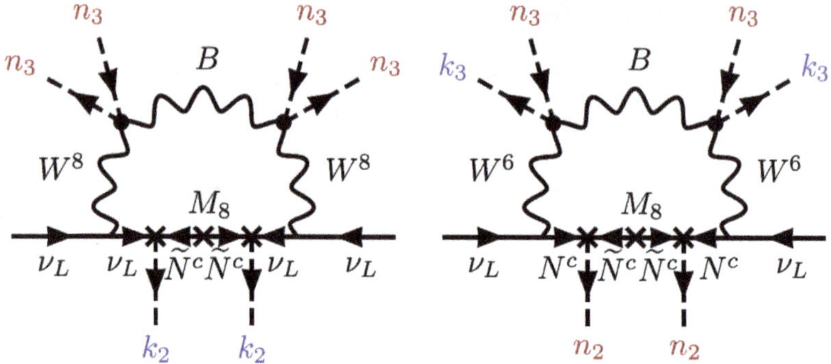

Fig. 3. Two diagrams which contribute to the light neutrinos mass matrix.[6]

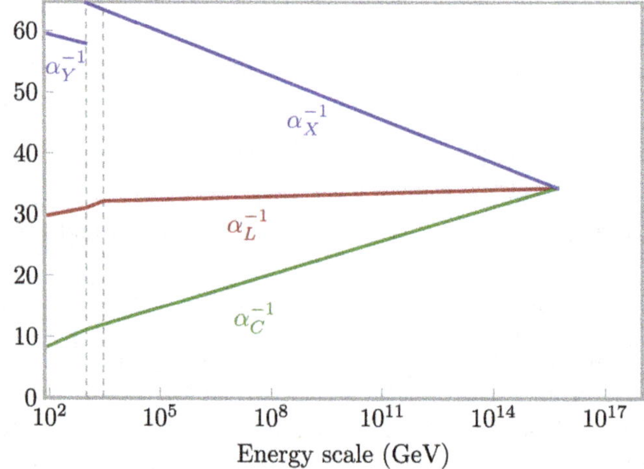

Fig. 4. Gauge coupling unification in $SU(3)_C \otimes SU(3)_L \otimes U(1)_X$ model at 3 TeV, from Ref. 6.

Alternatively, low energy supersymmetry would provide a simple way to account for gauge coupling unification. Such "completion" would however require "sparticles" accessible at the LHC, so far not seen. While we look forward to possible signs of supersymmetry in the next run of the LHC, we note that the physics responsible for gauge coupling unification may be *the same inducing small neutrino masses.*

A realization of such "GUT-less" unification scenario employs the $SU(3)_C \otimes SU(3)_L \otimes U(1)_X$ electroweak gauge structure, "explaining" why there are three generations from anomaly cancellation.[7] Neutrino masses arise radiatively in the presence of three fermion octets as illustrated in Fig. 4. Altogether, one finds that such "neutrino completion" scheme unifies the gauge couplings thanks to the existence of new states providing neutrino mass. These may lie in the TeV range and hence be accessible to the LHC.

4.3. *Generalized $b - \tau$ unification without GUTS*

Flavor symmetries have been suggested as a way to put order in the "flavor chaos." Here we stress the striking fact that such symmetries have the potential of relating quark and charged lepton masses, in the absence of unification. Indeed, in a class of such $SU(3)_c \otimes SU(2)_L \otimes U(1)_Y$ models one can obtain a canonical mass relation[8–11]

$$\frac{m_b}{\sqrt{m_d m_s}} \approx \frac{m_\tau}{\sqrt{m_e m_\mu}} \,. \tag{2}$$

between down-type quark and charged lepton masses. This formula can be understood from the group structure, when there are three vacuum expectation values but only two invariant contractions determining the Yukawa couplings. Note that Eq. (2) provides a successful multi-generation generalized b-tau unification scenario which, moreover, does not require the existence of grand-unification. Note also that it relates mass ratios instead of absolute masses.

5. Predicting Neutrino Oscillation Parameters

A remarkable feature, which came as a surprise, is that the smallest of the lepton mixing angles is similar to the largest of the CKM mixing parameters, the Cabibbo angle, while the solar and atmospheric mixing parameters are rather large.[2] One phenomenological approach is to take the reactor angle, similar to the Cabibbo angle, as the universal seed for quark and lepton mixing. Such *bi-large* schemes point towards Abelian flavor symmetry groups and Frogatt–Nielsen-type schemes.[12,13] It has been noted however that the observed neutrino mixing angles take very special values, atmospheric mixing being nearly *bi-maximal* with solar mixing nearly *tri-maximal*. Hence a *tri-bimaximal* pattern seems reasonable as a starting point.[14] Although the full pattern might occur accidentally, it seems that nature is telling its message here: (i) observations seem to suggest some symmetry, and (ii) we must also redefine our strategy in flavor model-building. The challenge is to reconcile the large lepton mixing with the small CKM parameters in a predictive way.

As a first step one can assign the three lepton families to a three-dimensional irreducible representation of a non-Abelian flavor symmetry group, the smallest one being A_4. This opens the way to the possibility of predicting the pattern on neutrino oscillation parameters. As simplest zeroth-order predictions one obtains[15] a maximum atmospheric mixing parameter $\theta_{23} = \pi/4$ and zero reactor angle $\theta_{13} = 0$, with a possible prediction as also for the solar angle, *a la tri-bimaximal*.

However recent neutrino oscillation data from reactors and accelerators measure a nonzero θ_{13} value, requiring the early models to be revamped so as to induce a nonzero θ_{13}, without spoiling the other prediction(s). This has been done in a minimal way in Ref. 16, leading to a striking predicted correlation between the magnitude of CP violation in neutrino oscillations and the octant of the atmospheric mixing parameter θ_{23} illustrated in Fig. 5. One sees that, at face value, the left octant necessarily violates CP. Time will tell whether this predicted *correlations* is

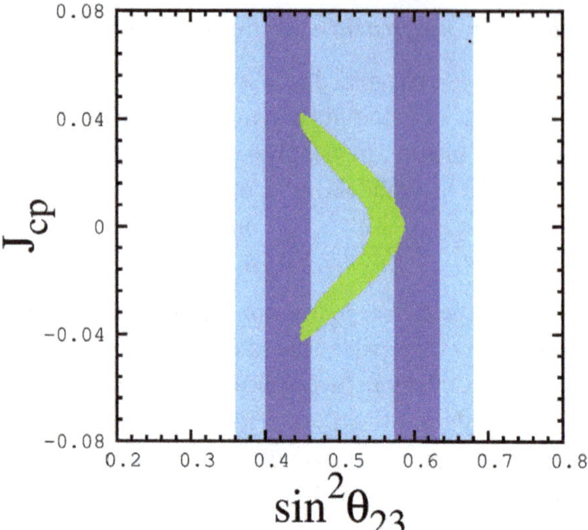

Fig. 5. Correlating CP violation in neutrino oscillations with the octant of the atmospheric mixing parameter θ_{23}, adapted from Ref. 16.

right. Finally we note that flavor-symmetry-based models often predict the pattern charged lepton flavor violation processes.[17]

6. Neutrinos and Electroweak Symmetry Breaking

After the Higgs boson discovery at CERN it is natural to imagine that all symmetries in nature are broken spontaneously. It is also reasonable to imagine that the smallness of neutrino mass is due to the feeble breaking of lepton number, which can be realized in many ways, see below. This requires an extension of the standard model Higgs sector and, if the minimal $\mathrm{SU(3)_c \otimes SU(2)_L \otimes U(1)_Y}$ structure is kept, there must be a physical Nambu–Goldstone boson, generically called majoron.[1]

Although the detailed properties of the majoron in general depend on the model, the existence of new invisible Higgs decays is generically expected, if lepton number violation takes place at or below the weak scale. This is easy to arrange, leading to missing momentum signals at colliders.[19–21] Given the good agreement of the results from ATLAS and CMS with the standard model Higgs scenario[22] one can place limits on the presence of invisible channels. Within the simplest $\mathrm{SU(3)_c \otimes SU(2)_L \otimes U(1)_Y}$ spontaneous low-scale lepton number violation scenario one finds that the current LHC restrictions on the Higgs boson decay branchings can be summarized as in Fig. 6, where the parameters μ_{ZZ} and $\mu_{\gamma\gamma}$ are "signal-strength" parameters. This restriction still leaves an important chunk of Higgs boson mass and mixing parameters to be explored at the next run of the LHC. Many alternative richer electroweak breaking sectors leading to the double breaking of electroweak and lepton number symmetries can be envisaged.

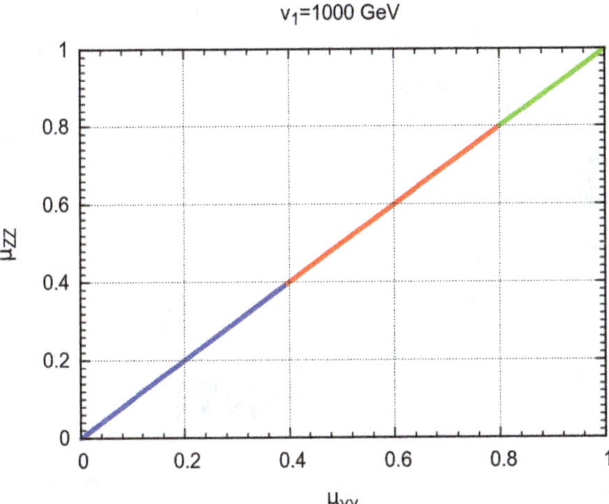

Fig. 6. Correlation between μ_{ZZ} and $\mu_{\gamma\gamma}$. The points in green pass all constraints, from Ref. 18.

7. Neutrinoless Double Beta Decay

As we saw neutrino oscillations are insensitive to the absolute neutrino mass scale. This can be probed using cosmological data as well as tritium beta decay endpoint studies.[1] A specially interesting complementary approach is the search for neutrinoless double beta decay. While the two-neutrino double beta decay has been experimentally observed in many isotopes, so far we have only experimental lower bounds on the half-lives for the neutrinoless mode.[23] However the latter is expected, on general grounds, to take place at some level, due to the existence of neutrino mass. Using the previous oscillation parameters and leaving the values of the Majorana phases free, one obtains the two broad branches corresponding to the cases of normal and inverted hierarchies indicated in Fig. 7. The horizontal and vertical lines indicate future expected sensitivities. Models based upon flavor symmetries often lead, as phenomenological predictions, to correlations between the neutrino oscillation parameters. In a large class of such models these translate as lower bounds for the effective mass parameter $|m_{ee}|$ characterizing $0\nu\beta\beta$ decay even for the normal mass ordering. This is seen as the two dark subregions in Fig. 7. Many other models leading to a lower bound on the $0\nu\beta\beta$ decay rate can be constructed.[24]

In gauge theories $0\nu\beta\beta$ can be induced in many ways other than the neutrino exchange or "mass mechanism." For example, there can be short-range mechanisms involving the exchange of heavy particles such as present in left–right or supersymmetric extensions of the standard model.[26,27] The significance of neutrinoless double beta decay comes from the fact that, whatever the mechanism responsible for $0\nu\beta\beta$ in a gauge theory one can always "dress" the corresponding amplitude with W bosons, showing that a Majorana neutrino mass is necessarily induced,[25] as illustrated in Fig. 8. This theorem holds under very general assumptions, as shown by Lindner and collaborators.[28]

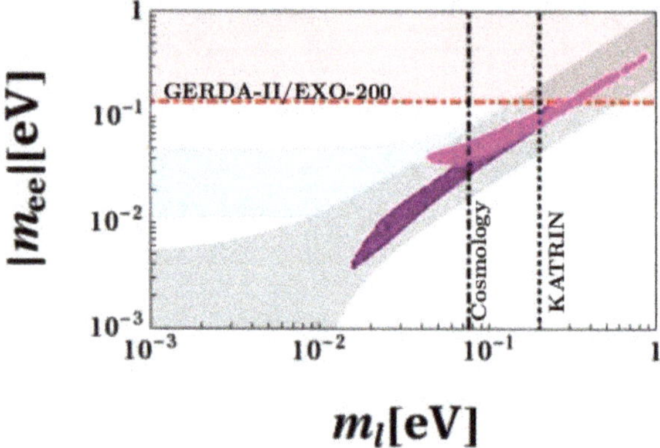

Fig. 7. Neutrinoless double beta decay effective amplitude parameter versus the lightest neutrino mass, in a generic model versus a flavor-symmetry-based model, from Refs. 8–11.

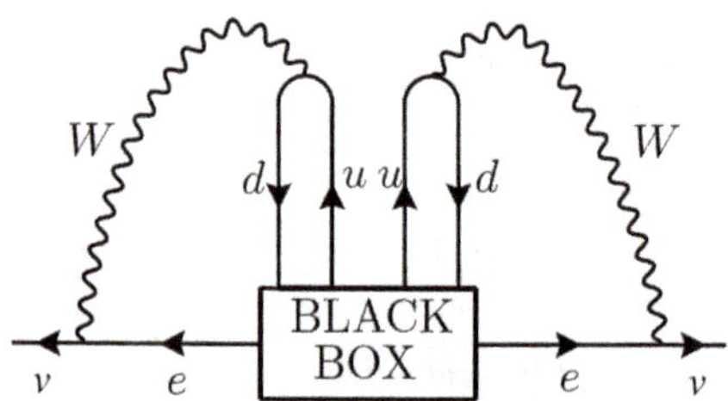

Fig. 8. Neutrinoless double beta decay implies Majorana neutrinos, from Ref. 25.

8. Neutrinos and Cosmology

Neutrinos affect the cosmic microwave background (CMB) and large scale structure in the universe, playing a key role in the synthesis of light elements, which takes place when the universe is about a few minutes old. The feeble interaction of neutrinos allows us to use them as cosmic probes, down to epochs far earlier than we can probe with optical telescopes. The current cosmological puzzles associated with the baryon number of the universe, inflation and dark matter are probably associated with epochs earlier than the electroweak phase transition at $\sim 10^{-12}$ sec. It is not inconceivable that (some of) these puzzles may have a common origin with the physics driving neutrino masses,[1] as schematized in Fig. 9.

Here we focus on the possibility that neutrino masses arise from spontaneous breaking of ungauged lepton number. The associated Nambu–Goldstone boson may acquire mass from lepton number violation by quantum gravity effects at the Planck scale.[29,30] If its mass lies in the keV range, the weakly interacting majoron can play the role of dark matter particle, providing both the required relic density as well as the scale for galaxy formation.[31] Since the majoron couples to neutrinos proportionally to their tiny mass, it is expected to be very long-lived, stable on cosmological

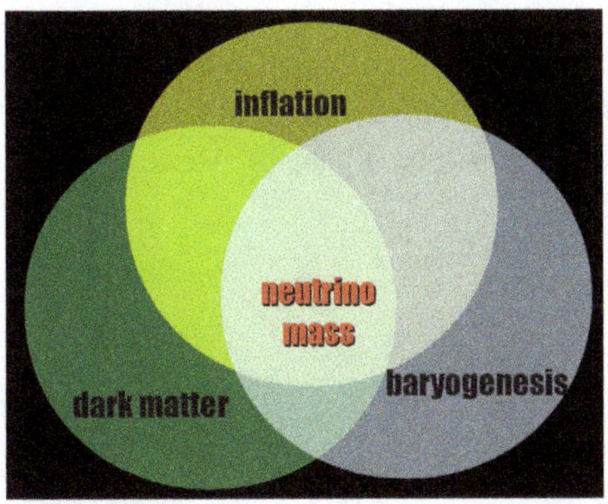

Fig. 9. Cosmological puzzles possibly associated to the same physics that drives neutrino mass.

scales.[32] Though model-dependent, its coupling to the charged leptons is expected to be very weak so the majorons produced during the phase transition may never be in thermal equilibrium during the history of the universe. Alternatively they could be in thermal equilibrium only for some period. The lifetime and mass of the late-decaying dark matter majoron consistent with the cosmic microwave background observations can be determined.[34,35] Moreover, as a pseudoscalar, like the π^0, the dark matter majoron will have a (sub-dominant) decay to two photons, leading to a characteristic mono-energetic X-ray emission line.[35,36] These features fit nicely in models where neutrino masses arise from a type-II seesaw mechanism.

A recent twist along these lines was the proposal that inflation and dark matter have a common origin (similar idea was suggested by Smoot in arXiv:1405.2776 [astro-ph]), with the inflaton identified to the real part of the complex singlet containing the majoron and breaking lepton number through its VEV.[33] The resulting inflationary scenario is consistent with the recent CMB observations, including the B-mode observation by the BICEP2 experiment re-analyzed jointly with the

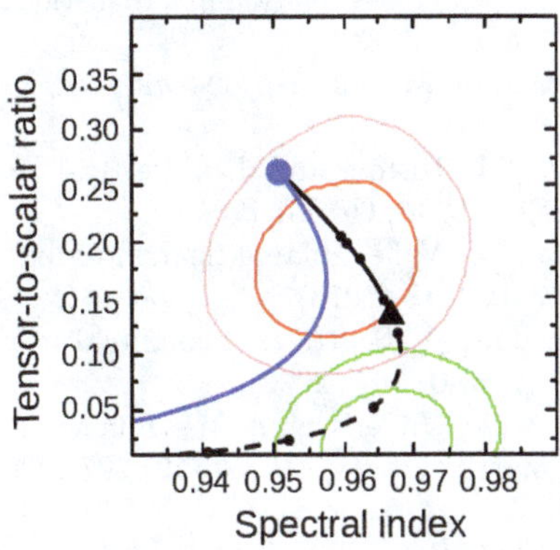

Fig. 10. (Color online) Cosmological predictions of seesaw inflation and majoron dark matter model of Ref. 33.

Planck data, as illustrated in Fig. 10. The upper (red) contours correspond to the BICEP2 results, while the lower ones (green) follow from the new analysis released jointly with PLANCK.[37] The lines correspond to 68 and 95% C.L. contours. Further restrictions on the majoron dark matter scenario should follow from structure formation considerations.

Acknowledgment

Work supported by MINECO grants FPA2014-58183-P, Multidark CSD2009-00064, and the PROMETEOII/2014/084 grant from Generalitat Valenciana.

References

1. J. W. F. Valle and J. C. Romao, *Neutrinos in High Energy and Astroparticle Physics*, 1st edn. (Wiley-VCH, Berlin, 2015).
2. D. Forero, M. Tortola and J. W. F. Valle, Neutrino oscillations refitted, *Phys. Rev. D* **90**, 093006 (2014) arXiv:1405.7540 [hep-ph].
3. J. Schechter and J. W. F. Valle, Neutrino masses in $SU(2) \times U(1)$ theories, *Phys. Rev. D* **22**, 2227 (1980).
4. M. Malinsky, J. Romao and J. W. F. Valle, Novel supersymmetric SO(10) seesaw mechanism, *Phys. Rev. Lett.* **95**, 161801 (2005).
5. S. M. Boucenna, S. Morisi and J. W. F. Valle, The low-scale approach to neutrino masses, *Adv. High Energy Phys.* **2014**, 831598 (2014).
6. S. M. Boucenna *et al.*, Small neutrino masses and gauge coupling unification, *Phys. Rev. D* **91**, 031702 (2015).
7. M. Singer, J. Valle and J. Schechter, Canonical neutral current predictions from the weak electromagnetic gauge group $SU(3) \times u(1)$, *Phys. Rev. D* **22**, 738 (1980).
8. S. Morisi *et al.*, Relating quarks and leptons without grand-unification, *Phys. Rev. D* **84**, 036003 (2011).
9. S. Morisi *et al.*, Quark-lepton mass relation and CKM mixing in an A_4 extension of the minimal supersymmetric standard model, *Phys. Rev. D* **88**, 036001 (2013).
10. S. F. King *et al.*, Quark-lepton mass relation in a realistic A_4 extension of the standard model, *Phys. Lett. B* **724**, 68 (2013).
11. C. Bonilla *et al.*, Relating quarks and leptons with the T_7 flavour group, *Phys. Lett. B* **742**, 99 (2015).
12. S. Boucenna, S. Morisi, M. Tortola and J. Valle, Bi-large neutrino mixing and the Cabibbo angle, *Phys. Rev. D* **86**, 051301 (2012).
13. G.-J. Ding, S. Morisi and J. Valle, Bilarge neutrino mixing and Abelian flavor symmetry, *Phys. Rev. D* **87**, 053013 (2013).
14. P. Harrison, D. Perkins and W. Scott, Tri-bimaximal mixing and the neutrino oscillation data, *Phys. Lett. B* **530**, 167 (2002).
15. K. Babu, E. Ma and J. Valle, Underlying A(4) symmetry for the neutrino mass matrix and the quark mixing matrix, *Phys. Lett. B* **552**, 207 (2003), arXiv:hep-ph/0206292.
16. S. Morisi, D. Forero, J. C. Romao and J. W. F. Valle, Neutrino mixing with revamped A4 flavour symmetry, *Phys. Rev. D* **88**, 016003 (2013), arXiv:1305.6774 [hep-ph].
17. S. Morisi and J. W. F. Valle, Neutrino masses and mixing: A flavour symmetry roadmap, *Fortschr. Phys.* **61**, 466 (2013).

18. C. Bonilla, J. W. F. Valle and J. C. Romão, Neutrino mass and invisible Higgs decays at the LHC, arXiv:1502.01649 [hep-ph].
19. A. S. Joshipura and J. W. F. Valle, Invisible Higgs decays and neutrino physics, *Nucl. Phys. B* **397**, 105 (1993).
20. F. de Campos *et al.*, Searching for invisibly decaying Higgs bosons at LEP II, *Phys. Rev. D* **55**, 1316 (1997).
21. DELPHI Collab. (J. Abdallah *et al.*), Searches for invisibly decaying Higgs bosons with the DELPHI detector at LEP, *Eur. Phys. J. C* **32**, 475 (2004), arXiv:hep-ex/0401022.
22. ATLAS, CMS Collabs. (G. Aad *et al.*), Combined measurement of the Higgs boson mass in pp collisions at $\sqrt{s} = 7$ and 8 TeV with the ATLAS and CMS experiments, arXiv:1503.07589 [hep-ex].
23. A. Barabash, 75 years of double beta decay: yesterday, today and tomorrow, arXiv:1101.4502 [nucl-ex].
24. L. Dorame, D. Meloni, S. Morisi, E. Peinado and J. Valle, Constraining neutrinoless double beta decay, *Nucl. Phys. B* **861**, 259 (2012).
25. J. Schechter and J. Valle, Neutrinoless double beta decay in $\mathrm{SU}(2) \times \mathrm{U}(1)$ theories, *Phys. Rev. D* **25**, 2951 (1982).
26. F. Bonnet, M. Hirsch, T. Ota and W. Winter, Systematic decomposition of the neutrinoless double beta decay operator, *J. High Energy Phys.* **1303**, 055 (2013).
27. S. Das, F. Deppisch, O. Kittel and J. Valle, Heavy neutrinos and lepton flavour violation in left-right symmetric models at the LHC, *Phys. Rev. D* **86**, 055006 (2012), arXiv:1206.0256 [hep-ph].
28. M. Duerr, M. Lindner and A. Merle, On the quantitative impact of the Schechter-Valle theorem, *J. High Energy Phys.* **1106**, 091 (2011).
29. S. B. Giddings and A. Strominger, Loss of incoherence and determination of coupling constants in quantum gravity, *Nucl. Phys. B* **307**, 854 (1988).
30. T. Banks and N. Seiberg, Symmetries and strings in field theory and gravity, *Phys. Rev. D* **83**, 084019 (2011), arXiv:1011.5120 [hep-th].
31. V. Berezinsky and J. Valle, The KeV majoron as a dark matter particle, *Phys. Lett. B* **318**, 360 (1993), arXiv:hep-ph/9309214.
32. J. Schechter and J. Valle, Neutrino decay and spontaneous violation of lepton number, *Phys. Rev. D* **25**, 774 (1982).
33. S. Boucenna, S. Morisi, Q. Shafi and J. Valle, Inflation and majoron dark matter in the seesaw mechanism, *Phys. Rev. D* **90**, 055023 (2014).
34. M. Lattanzi and J. Valle, Decaying warm dark matter and neutrino masses, *Phys. Rev. Lett.* **99**, 121301 (2007), arXiv:0705.2406 [astro-ph].
35. M. Lattanzi, S. Riemer-Sorensen, M. Tortola and J. Valle, Constraints on majoron dark matter from cosmic microwave background and astrophysical observations, *Nucl. Instrum. Methods A* **742**, 154 (2014).
36. M. Lattanzi, S. Riemer-Sorensen, M. Tortola and J. W. F. Valle, Updated CMB and x- and γ-ray constraints on Majoron dark matter, *Phys. Rev. D* **88**, 063528 (2013), arXiv:1303.4685 [astro-ph.HE].
37. BICEP2, Planck Collabs. (P. Ade *et al.*), Joint analysis of BICEP2/Keckarray and Planck data, *Phys. Rev. Lett.* **114**, 101301 (2015).

Neutrino Masses and SO10 Unification

P. Minkowski

Albert Einstein Center for Fundamental Physics – ITP,
University of Bern, Bern, 3006 Bern, Switzerland
PH-TH division, CERN, Meyrin, 1211 Meyrin 23, Switzerland
mink@itp.unibe.ch, Peter.Minkowski@cern.ch

We present the embedding of the SM gauge group in SO10, a simple, compact unifying gauge group, with each of the three basic spin 1/2 families forming a unitary, irreducible 16-dimensional representation of spin10, which is complex, i.e. chiral. Subtle differences to the mixed representations of SU5, contained in the SO10 scheme, are pointed out. These have consequences for neutrino flavors, which become paired in a light $SU2_L$-active doublet mode and a heavy SM singlet mode, one ν, $\mathcal{N}$-pair per family.

Keywords: Neutrino masses; SO10.

1. Introduction

We propose to pass "en revue" the concepts of gauge field theories based on the standard model gauge group: "minimally the neutrino mass extended standard model with gauge group $\mathcal{G}_{\rm SM} = {\rm SU3}_c \times {\rm SU2}_L \times {\rm U1}_{\mathcal{Y}_w}$ and one scalar doublet with respect to $SU2_L$ — acting in a chargelike and perturbatively renormalizable manner through (pseudo)-vectorial connections in rigid 3+1 space–time dimensions as lowest broken level of a simple unifying compact one: $\mathcal{G}_U \supset \mathcal{G}_{\rm SM}$, viz.

$$\mathcal{G}_U = {\rm SO10} \equiv {\rm spin10}\,. \tag{1}$$

This puts in evidence the regularity and boundary conditions to be imposed on these connections, discussed in Sec. 2 and sets apart the limit of vanishing gravitational constant, while the extension to include quantized gravity remains a task whose implementation is for the future.

1.1. *Problems with SU5 anomalies and charge quantization*

1.1.1. *The left-chiral basis for spin 1/2 fields*

The following deductions date back to the year 2005 in Ref. 8 and continued in Ref. 9. In this process we shall make use of the left- (or right-)chiral basis for spin

1/2 fields. The interested shall be rendered attentive that the dates of the works cited here, in particular Refs. 7 and 8, do not correspond to aspects discussed here, whence they were first derived by the author.

The left-chiral notation shall be

$$\begin{aligned} (f_k)^{\dot\gamma}_F; \quad {}^{\dot\gamma} = 1,2: &\quad \text{spin projection}, \\ {}_F = \text{I, II, III}: &\quad \text{family label}, \\ {}_k = 1,\ldots,16: &\quad \text{SO10 label}. \end{aligned} \tag{2}$$

The basis containing one family of the basic 16 irreducible representations of SO10 is displayed in Eq. (3) below:

$$\left(\begin{array}{cccc|cccc} u^1 & u^2 & u^3 & \nu & \mathcal{N} & \hat{u}^3 & \hat{u}^2 & \hat{u}^1 \\ d^1 & d^2 & d^3 & e^- & e^+ & \hat{d}^3 & \hat{d}^2 & \hat{d}^1 \end{array}\right)^{\dot\gamma \to L} = (f)^{\dot\gamma}. \tag{3}$$

The left-chiral basis as defined in Eq. (3) together with the associated right-chiral one were used in Ref. 4 and in relation with the present $B-L$ anomaly structure in a common paper in 1991 with W. Buchmüller and C. Greub.[10]

We consider the $B-L$ current and its associated intrinsic SO10 current orthogonal to all 24 currents pertaining to SU5, which must not be identical to $B-L$ only equivalent as far as cancellation of anomalies is concerned. We call this current within the Cartan subalgebra of SO10

$$j^{\varrho}(x); \text{ with charge operator: } \hat{J} = \int \Big|_t d^3x \, j^0(t, \vec{x}). \tag{4}$$

We consider the eigenvalues of $\hat{J}$ in the limit of unbroken SO10 gauge group within the irreducible 16-representation in the left-chiral basis in Eq. (3).

1.1.2. *The Majorana logic and mass from mixing — setting within the "tilt to the left" or "seesaw" of type I* $(\cdots)$

Within the subgroup decompositions of SO10 the "tilt to the left" does not appear obvious

$$\begin{array}{ccc} \swarrow & \text{spin10} & \searrow \\ \text{spin6} \equiv \text{SU4} & \times & \text{spin4} \equiv \text{SU2}_L \times \text{SU2}_R \\ \text{lepton number as 4th color}^5 & & \\ \downarrow & & \downarrow \\ \text{SU3}_c \times \text{U1}_{B-L} & \times & \text{SU2}_L \times \text{U1}_{I_{3R}} \\ \searrow & \text{SU3}_c \times \text{U1}_{Q_{\text{e.m.}}} & \swarrow \\ & Q_{\text{e.m.}}/e = I_{3L} + I_{3R} + \frac{1}{2}(B-L) & \end{array} \tag{5}$$

The large scale breaking of *gauged* $B - L$ or "tilt to the left" was not assumed essential in Refs. 4–12 and brings about a definite "mass from mixing" scenario,[13,14] to which we turn below.

1.1.3. *The electromagnetic, operator-valued, conserved Noether current*

We consider the space–time independent phase transformations leaving $\mathcal{L}$ in Eq. (23) invariant

$$\begin{aligned}\{\psi_j^{n_j}\}_A &\to \exp\left(\frac{1}{i} n_j e_{1/3}\chi\right)\{\psi_j^{n_j}\}_A \to [\{\psi_j^{n_j}\}_A]^* \\ &\to \exp\left(-\frac{1}{i} n_j e_{1/3}\chi\right)[\{\psi_j^{n_j}\}_A]^*, \\ &\chi: \quad \text{constant, real}\end{aligned} \tag{6}$$

defining the χ-variation

$$\delta = \partial_\chi|_{\chi=0}\,, \quad \delta\partial_\varrho = \partial_\varrho\delta\,, \tag{7}$$

we obtain for the variation of $\mathcal{L}_\psi$ in Eq. (23)

$$\begin{aligned}0 = \delta\mathcal{L}_\psi = \frac{1}{2}\Big(&-(\partial_\varrho(\delta[\{\bar{\psi}_j^{n_j}\}_B]))(\gamma^\varrho)_{BA}\{\psi_j^{n_j}\}_A \\ &+ [\{\bar{\psi}_j^{n_j}\}_B](\gamma^\varrho)_{BA}(\partial_\varrho(\delta(\{\psi_j^{n_j}\}_A)))\Big)\,.\end{aligned} \tag{8}$$

In Eq. (8) appropriate summation over the indices j, n_j, avoiding counting the states of the same fields multiply, is understood.

Using Eq. (6) it follows that

$$0 = \delta\mathcal{L}_\psi = \partial_\varrho\left[\sum_{j,n_j} n_j e_{1/3}\{\bar{\psi}_j^{n_j}\}_B(\gamma^\varrho)_{BA}\{\psi_j^{n_j}\}_A\right]. \tag{9}$$

The implementation of the regularity of the expression contained in Eq. (9) implies the existence of the conserved current relative to the electromagnetic gauge invariance and conserved operator-valued charge:

$$\begin{aligned}&\text{e.m. current:} \\ &\quad j^\varrho_{\text{e.m.}}(x) = \sum_{j,n_j} n_j e_{1/3}\{\bar{\psi}_j^{n_j}\}_B(\gamma^\varrho)_{BA}\{\psi_j^{n_j}\}_A\,, \\ &\text{e.m. current conservation:} \quad \partial_{x\varrho} j^\varrho_{\text{e.m.}}(x) = 0\,, \\ &\text{e.m. charge in units of the proton charge } e\text{:} \\ &\quad \hat{Q}_{\text{e.m.}} = \frac{1}{e}\int\Big|_t d^3x\, j^0_{\text{e.m.}}(t,x)\,, \\ &\text{e.m. charge conservation:} \quad \frac{d}{dt}\hat{Q}_{\text{e.m.}}(t) = 0\,.\end{aligned} \tag{10}$$

The form of $\hat{Q}_{\text{e.m.}}$ becomes by the Ansatz for the e.m. current, in Eq. (10),

$$\begin{aligned} \hat{Q}_{\text{e.m.}} &= \frac{1}{e} \sum_{j,n_j} \hat{Q}_{\text{e.m.},\, j,n_j} \\ &= \frac{1}{e} \int \bigg|_t d^3x \sum_{j,n_j} n_j e_{1/3} j^0_{\text{e.m.},\, j,n_j}(t,x)\,. \end{aligned} \tag{11}$$

In the second relation in Eq. (11), we have introduced the e.m. current pertaining to the charge flavor j, n_j:

$$j^\varrho_{\text{e.m.},\, j,n_j}(x) = \{\bar{\psi}^{n_j}_j\}_B(x)(\gamma^\varrho)_{BA}\{\psi^{n_j}_j\}_A(x)\,. \tag{12}$$

There are a couple of subtleties in the relations in Eqs. (10)–(12), towards which we turn one by one.

(1) Local current conservation $\leftrightarrow$ Local conservation of currents.

From the local e.m. current conservation

$$\begin{aligned} &\partial_{x\varrho} j^\varrho_{\text{e.m.}}(x) = 0\,, \\ &j^\varrho_{\text{e.m.}}(x) = \frac{1}{e} \sum_{j,n_j} n_j e_{1/3} \{\bar{\psi}^{n_j}_j\}_B(\gamma^\varrho)_{BA}\{\psi^{n_j}_j\}_A \end{aligned} \tag{13}$$

and the decomposition

$$j^\varrho_{\text{e.m.}}(x) = \frac{1}{e} \sum_{j,n_j} n_j e_{1/3} j^\varrho_{\text{e.m.},\, j,n_j}(x)\,, \tag{14}$$

where the individual e.m. currents are independent of each other, it follows

$$\partial_{x\varrho} j^\varrho_{\text{e.m.},\, j,n_j}(x) = 0 \quad \forall j,\, n_j \tag{15}$$

as long as *exclusively* the exactly conserved gauge interactions $\mathcal{G}_{\text{cons}} = \text{SU3}_c \times \text{U1}_{\text{e.m.}}$ are retained.

(2) Rigid phase transformations imply regularity conditions.

The rigid phase transformations defined in Eq. (6) repeated below

$$\begin{aligned} \{\psi^{n_j}_j\}_A &\to \exp\left(\frac{1}{i} n_j e_{1/3} \chi\right) \{\psi^{n_j}_j\}_A \to \left[\{\psi^{n_j}_j\}_A\right]^* \\ &\to \exp\left(-\frac{1}{i} n_j e_{1/3} \chi\right) \left[\{\psi^{n_j}_j\}_A\right]^* \quad \chi\text{: constant, real} \end{aligned} \tag{16}$$

imply regularity and boundary conditions on all fields enforcing charge conservation, also in the form in Eq. (10), repeated below

$$\frac{d}{dt} \hat{Q}_{\text{e.m.}}(t) = 0 \tag{17}$$

ensuring exact charge conservation upon extension to all fields including those taking part in broken gauge transformations.

(3) The eigenvalues $Q_{\text{e.m.}}$ of $\hat{Q}_{\text{e.m.}}$

From Eqs. (13) and (14) it follows

$$
\begin{gathered}
Q_{\text{e.m.}} = \sum_{j,n_j} n_j \frac{1}{3}\,; \\
e_{1/3} = \frac{1}{3} e\,; \quad n_j = (0),\, \pm 1, \pm 2, \ldots\,; \\
e = \sqrt{4\pi\alpha} = 0.302822120882(49)\,; \quad (\alpha)^{-1} = 137.035999074(44)\,.
\end{gathered} \tag{18}
$$

(4) Equation (18) expresses the quantization of e.m. charge, but at this stage this has the status of an *assumption*, while we look for an *explanation* of this quantization.

On the experimental side one check consists in measuring the charges of neutrino and antineutrino flavors, for which *a priori* any, relatively negative, real value is possible.

A look at the PDG particle property listings[3] under the heading neutrino charge reveals, that hitherto no value nor upper limit for the respective absolute charges could be assigned to any neutrino or antineutrino flavor.

(5) The minimal *simple* unifying gauge group is SO10.

With the dynamical nature of a *simple*, compact Lie group, to dynamically account for the quantization of exactly conserved e.m. charge, the determination of the *minimal* such group, containing $\mathcal{G}_{\text{SM}} = \text{SU3}_c \times \text{SU2}_L \times \text{U1}_{\mathcal{Y}_w}$ is unique, as discussed in the Introduction (Eq. (1), repeated below)

$$
\mathcal{G}_U = \text{SO10} \equiv \text{spin10}\,. \tag{19}
$$

These were the initial deliberations, whence with Harald Fritzsch we undertook *our* study "in quest of unification," end of 1973 at Caltech, which eventually led to Ref. 4, first announced as a preprint at the end of 1974.

(5a) The first papers "in the quest of unification."

A short account follows here:

The first paper was by J. Pati and A. Salam,[5] where the authors point out that in trying to unify $\mathcal{G}_{\text{SM}}$ with QCD anomaly cancellation must proceed through leptons and quarks as far as spin 1/2 fermions are concerned, and in any case baryon- and lepton-numbers are not going to be conserved. In the second paper by H. Georgi and S. L. Glashow in Ref. 6 a simple gauge group, viz. SU5, is achieved, yet with problems to be discussed in the next subsections with respect to charge quantization. Finally restricting ourselves to the first three contributions, all in 1974, by H. Poggio, H. Quinn and S. Weinberg in Ref. 7, the renormalization group is used to search for a unification scale of the strong- electromagnetic- and weak-coupling-constants, with the then valid knowledge compatible with all known phenomena, including the upper limit of the proton lifetime, yielding a unification scale

$$M_U \sim 10^{16\pm 1}\ \text{GeV}\,. \tag{20}$$

2. The Electroweak Gauge Breaking Tree

The electroweak gauge group

$$\mathcal{G}_{\text{SM}} = \text{SU3}_c \times \text{SU2}_L \times \text{U1}_{\mathcal{Y}_w} \tag{21}$$

defined in the Introduction, develops its own gauge breaking tree along the path:

$$\begin{array}{c} \mathcal{G}_{\text{SM}} \\ \downarrow \\ \text{SU3}_c \times \text{U1}_{\text{e.m.}} \end{array} \quad . \tag{22}$$

In Eq. (22) U1_{em} refers to the compact, Abelian, local gauge group associated with the *conserved* electromagnetic current and its scalar commensurable charges.

To assess the consequences of gauge invariance with respect to a space–time independent phase transformation of a set of charged spin 1/2 fields, denoted $\{\psi_j^{n_j}\}_A(x)$ with electric charges $Q_j = e_{1/3} n_j$; $n_j = 0, \pm 1, \pm 2, \ldots$ of the form

$$\begin{gathered} \mathcal{L}_\psi = \sum_j \{\bar{\psi}_j^{n_j}\}_B \left[\left(\frac{i}{2} \overleftrightarrow{\partial}_\mu - n_j e_{1/3} W_\varrho \right) (\gamma^\varrho)_{BA} - m_j \delta_{BA} \right] \{\psi_j^{n_j}\}_A\,, \\ A, B = 1, \ldots, 4: \ \text{spinor indices}; \quad \{\bar{\psi}_j^{n_j}\}_D = \{\psi_j^{n_j}\}_A^* \gamma^0_{AD}\,. \end{gathered} \tag{23}$$

Here we consider the alternative subgroup decomposition

$$\text{spin10} \to \text{SU5} \times \text{U1}_J\,. \tag{24}$$

Among the three generators of spin10 commuting with SU3_c, I_{3L}, I_{3R}, $B - L$ and forming part of the Cartan subalgebra of spin10, there is one combination, denoted $\hat{J} \to J$ in Eqs. (4) and (24), commuting with its largest unitary subgroup SU5.

The 16 representations in the left-chiral basis display the charges pertinent to $\hat{J}$ normalized to integer values *modulo an overall sign* — as in the discussion of genuinely chiral U1-charges in Eqs. (5)–(24) — but here referring to $N = 16$.

While the Majorana logic indeed opens a "path" to trace the origin of the "tilt to the left," the origin of three families remains unexplained at this stage.

The associative Clifford algebras $\{\Gamma_{p,q}; \mathbb{C}\} \supset \{\Gamma_{\tilde{p},\tilde{q}}; \mathbb{R}\}$ are not constructed explicitly here. They are reported from material supplementary to the present outline. p, q denote timelike (p) and spacelike (q) dimensions of space–time.

Figure 1 shows the repartition of real (Maj-r) and complex (Maj-c) character of irreducible, *associative*, real (Majorana) Clifford algebras with their characteristic mod-8 property relative to q, p.[15]

These representations form the roots of the "Majorana logic." The details pertaining to left- and right-chiral bases of spin 1/2 fields can be derived from the short collection of identities in the subsection "the Majorana representation of $\{\Gamma_{p,q}; \mathbb{R}\}$ for $p = 1$, $q = 3$" below.

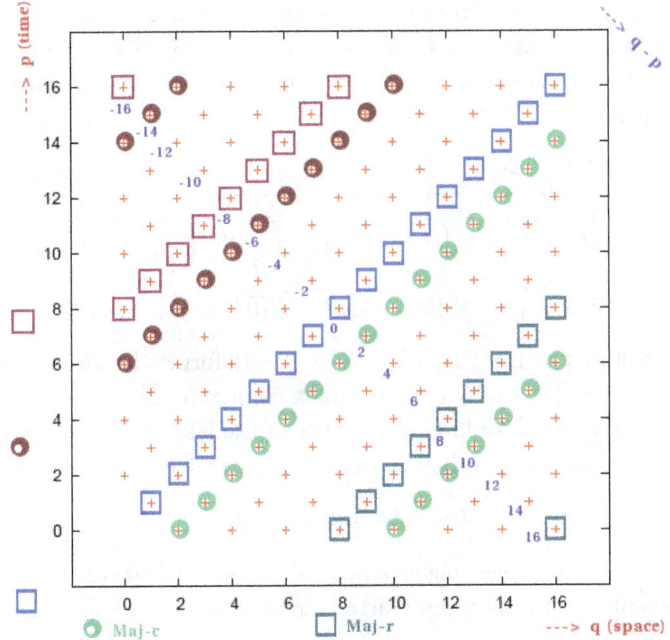

Fig. 1. The complex and real Majorana representations MajCR (p, q).

The Majorana logic characterized by $\mathcal{N}_F$

$$
(f)^{\dot{\gamma}} = \left(\begin{array}{cccc|cccc} u^1 & u^2 & u^3 & \nu & \mathcal{N} & \hat{u}^3 & \hat{u}^2 & \hat{u}^1 \\ d^1 & d^2 & d^3 & e^- & e^+ & \hat{d}^3 & \hat{d}^2 & \hat{d}^1 \end{array} \right)^{\dot{\gamma} \to L} ,
$$
$$
J \to \left(\begin{array}{cccc|cccc} 1 & 1 & 1 & -3 & 5 & 1 & 1 & 1 \\ 1 & 1 & 1 & -3 & 1 & -3 & -3 & -3 \end{array} \right) . \tag{25}
$$

The assignment of $\hat{J}$-charges in Eq. (25) follows from the fermionic oscillator representation of the spin $2n$ associated Γ algebra through n such oscillators and the associated embedding spin10 $\supset$ SU5 (Ref. 6) for $n = 5$ here[16]

$$
\{a_s, a_t^\dagger\} = \delta_{st} \, ; \quad s, t = 1, 2, \ldots, n \, ; \quad \{a_s, a_t\} = 0 = \{a_s^\dagger, a_t^\dagger\} ,
$$
$$
J_n = \sum_{s=1}^{n} \begin{pmatrix} a_s^\dagger a_s \\ -a_s a_s^\dagger \end{pmatrix} = 2\hat{n} - n \P_{2^n \times 2^n} \, ; \quad \hat{n} = \sum_{s=1}^{n} a_s^\dagger a_s \, . \tag{26}
$$

The eigenvalues (X) *and* multiplicities $(\#)$ of J_n

$$
\begin{array}{c|cccccc} (X) & n & n-2 & n-4 & \cdots & -n+2 & -n \\ \hline (\#) & \binom{n}{0} & \binom{n}{1} & \binom{n}{2} & \cdots & \binom{n}{n-1} & \binom{n}{n} \end{array} \tag{27}
$$

The orthogonal series for n even $\leftrightarrow$ real (spin8, spin12, ...) has another decomposition within the associated Γ algebra, than the one with n odd $\leftrightarrow$ complex (spin10, spin14, ...). We give here the explicit numbers according to Eq. (27) for $n = 5$, i.e. spin10

$$\begin{array}{c|cccccc} (X) & 5 & 3 & 1 & -1 & -3 & -5 \\ \hline (\#) & \binom{5}{0} & \binom{5}{1} & \binom{5}{2} & \binom{5}{3} & \binom{5}{4} & \binom{5}{5} \\ \text{SU5} & \{1\} & \{5\} & \{10\} & \{\overline{10}\} & \{\bar{5}\} & \{\bar{1}\} \end{array} \tag{28}$$

The subset of states in Eq. (28) $(X) = \{5, 1, -3\}$ forms the 16 representations of spin10, while $(X) = \{3, -1, -5\}$ the complex conjugate $\overline{16}$.

This opens the "path" of linking the "tilt to the left" with a substructure based on the primary breakdown of the local gauged chargelike symmetry associated with

$$J = -4I_{3R} + 3(B - L)\,. \tag{29}$$

J as defined through integer eigenvalues (X) given in Eqs. (25) and (28) is normalized differently from the other Cartan subalgebra charges I_{3L}, I_{3R}, $B - L$

$$\begin{aligned} |Q_C|^2 &= \sum_{\{16\}} (Q_C(f))^2\,, \quad |I_{3L}|^2 = 2\,, \quad |I_{3R}|^2 = 2\,, \\ |B - L|^2 &= \frac{16}{3}\,, \quad |J|^2 = 80\,. \end{aligned} \tag{30}$$

The consequence as far as neutrino-mass and mixing is concerned follows from identifying the J direction with a major axis of primary spontaneous gauge-symmetry breaking, bringing about the "tilt to the left" from Eq. (45) below

$$\begin{aligned} &j_\varrho(B - L)|_{3\times 15} \\ &= \sum_{fm} \Big[\frac{1}{3}((u^*)^{\alpha\dot{c}}(\sigma_\mu)_{\alpha\dot{\gamma}}(u)^{\dot{\gamma}c} - (\hat{u}^*)^{\alpha c}(\sigma_\mu)_{\alpha\dot{\gamma}}(\hat{u})^{\dot{\gamma}\dot{c}}) \\ &\quad + \frac{1}{3}((d^*)^{\alpha\dot{c}}(\sigma_\mu)_{\alpha\dot{\gamma}}(d)^{\dot{\gamma}c} - (\hat{d}^*)^{\alpha c}(\sigma_\mu)_{\alpha\dot{\gamma}}(\hat{d})^{\dot{\gamma}\dot{c}}) \\ &\quad - (e^-)^{*\alpha}(\sigma_\mu)_{\alpha\dot{\gamma}}(e^-)^{\dot{\gamma}} + (e^+)^{*\alpha}(\sigma_\mu)_{\alpha\dot{\gamma}}(e^+)^{\dot{\gamma}} - (\nu)^{*\alpha}(\sigma_\mu)_{\alpha\dot{\gamma}}(\nu)^{\dot{\gamma}} \Big] e^\mu_\varrho\,, \end{aligned} \tag{31}$$

$$g_{\varrho\tau} = e^\mu_\varrho \eta_{\mu\nu} e^\nu_\tau : \text{ metric}; \quad e^\mu_\varrho : \text{ vierbein};$$

$$* : \text{ Hermitian operator conjugation};$$

$$(u^*)^{\alpha\dot{c}} \equiv (u^{\dot{\alpha}c})^*\,; \quad \eta_{\mu\nu} = \text{diag}(1, -1, -1, -1) : \text{ tangent space metric};$$

$$^{c}\,(^{\dot{c}}) : \text{ color and anticolor}; \quad c = 1, 2, 3\,.$$

The contribution of charged fermion (pairs) q, $\hat{q}$; $e^\mp$ can be combined to vector currents $\bar{q}\gamma_\mu q$; $\bar{e}\gamma_\mu e$ with $q \to u$, d, c, s, t, b; $e \to e^-$, μ^-, τ^-.

The anomalous Ward identity for the $B-L$ current(-density) defined in Eq. (31) takes the form

$$\begin{aligned}
& d^4x \sqrt{|g|} D^{\varrho} j_{\varrho}(B-L)\big|_{3\times 15} = 3\hat{A}_1(X)\,, \\
& \hat{A}_1(X) = -\frac{1}{24}\,\mathrm{tr}\,X^2\,; \quad (X)^a{}_b = \frac{1}{2\pi}\frac{1}{2}\,dx^{x\varrho}\wedge dx^{\tau}(R^a{}_b)_{\varrho\tau}\,, \\
& (R^a{}_b)_{\varrho\tau} : \begin{cases} \text{Riemann curvature tensor} \\ \textbf{mixed components: } {}^a{}_b \rightarrow \textbf{tangent space} \\ \qquad\qquad\qquad\qquad {}_{\mu\nu} \rightarrow \textbf{covariant space} \end{cases}, \\
& D^{\varrho} j_{\varrho}(B-L)|_{3\times(16)} = 0\,.
\end{aligned} \tag{32}$$

Before discussing the extension $j_{\varrho}(B-L)|_{3\times(15)} \rightarrow j_{\varrho}(B-L)|_{3\times(16)}$ which renders the latter current conserved, let us define the quantities appearing in Eq. (32):

$$\begin{aligned}
(R^a{}_b)_{\varrho\tau} &= e^a_{\mu} e_{b\nu}(R^{\mu}{}_{\nu})_{\varrho\tau}\,; e_{b\nu} = \ \eta_{bb'} e^{b'}_{\nu}\,, \\
(R^{\mu}{}_{\nu})_{\varrho\tau} &= (\partial_{\varrho}\Gamma_{\tau} - \partial_{\tau}\Gamma_{\varrho} + \Gamma_{\varrho}\Gamma_{\tau} - \Gamma_{\tau}\Gamma_{\varrho})^{\mu}{}_{\nu}\,, \\
(\Gamma^{\mu}{}_{\nu})_{\tau}&: \ \text{matrix valued } (GL(4,\mathbb{R})) \text{ connection minimal here}\,,
\end{aligned} \tag{33}$$

$$\begin{aligned}
& d^4x \sqrt{|g|} D^{\varrho} j_{\varrho}(B-L)|_{3\times 15} = 3\hat{A}_1(X)\,, \\
& \hat{A}_1(X) = -\frac{1}{24}\,\mathrm{tr}\,X^2\,; \quad (X)^a{}_b = \frac{1}{2\pi}\frac{1}{2}dx^{\varrho}\wedge dx^{\tau}(R^a{}_b)_{\varrho\tau}\,, \\
& (R^a{}_b)_{\varrho\tau} : \begin{cases} \text{Riemann curvature tensor} \\ \textbf{mixed components: } {}^a{}_b \rightarrow \textbf{tangent space} \\ \qquad\qquad\qquad\qquad {}_{\mu\nu} \rightarrow \textbf{covariant space} \end{cases}, \\
& D^{\varrho} j_{\varrho}(B-L)|_{3\times(16)} = 0\,.
\end{aligned} \tag{34}$$

In Eq. (32) $\hat{A}(X \rightarrow \lambda) = \frac{1}{2}\lambda/\sinh\left(\frac{1}{2}\lambda\right)$ denotes the Atiyah–Hirzebruch character or $\hat{A}$-genus[17] with its integral over a compact, Euclidean signatured closed manifold M_4, capable of carrying an SO4-spin structure, becomes the index of the associated *elliptic* Dirac equation

$$\int \hat{A}(X_E) = n_R - n_L = \text{integer}\,. \tag{35}$$

In Eq. (35) $n_{R,L}$ denote the numbers of right- and left-chiral solutions of the Dirac equation on M_4. The index ${}_E \rightarrow X_E$ shall indicate the Euclidean transposed curvature 2-form, and is *adapted* here to physical curved and uncurved space–time. For the latter case the first relation in Eq. (32) yields the integrated form, in the limit of infinitely heavy $\mathcal{N}_F$ (Eq. (3)),

$$\begin{aligned}
\Delta_{R-L} n_{\nu} &= \int d^4x \sqrt{|g|} D^{\mu} j_{\mu}^{B-L(15)} = 3\Delta n(\hat{A})\,, \\
3 &= \text{number of families} = \text{odd}\,; \quad m_{\nu_F} \rightarrow 0\,.
\end{aligned} \tag{36}$$

In Eq. (36) $\Delta_{R-L} n_\nu$ denotes the difference of right-chiral $(\hat{\nu})$ [a] and left-chiral (ν) flavors between times $t \to \pm\infty$.

Here a subtlety arises *precisely* because the number of families on the level of G_{SM} is odd, and the light neutrino flavors are not "Dirac-doubled," which according to Eq. (36) could potentially lead to a change in fermion number being odd, which violates the rotation by 2π symmetry, equivalent to $\hat{\Theta}^2$ (CPT^2), *unless*[b]

$$\Delta n(\hat{A}) = \text{even} \quad (\surd \text{ for dim} = 4 \bmod 8)\,. \tag{37}$$

We turn to the SO10 inspired cancellation of the gravity induced anomaly, giving rise to the completion of neutrino flavors to 3 families of 16-plets, also called "right-handed" neutrino flavors, denoted $\mathcal{N}$ in the left-chiral basis in Eq. (3) (for a more complete account of left- *and* right-chiral bases see also Ref. 18)

$$\left(\begin{array}{cccc|cccc} u^1 & u^2 & u^3 & \nu & \mathcal{N} & \hat{u}^3 & \hat{u}^2 & \hat{u}^1 \\ d^1 & d^2 & d^3 & e^- & e^+ & \hat{d}^3 & \hat{d}^2 & \hat{d}^1 \end{array} \right)^{\dot{\gamma} \to L} = (f)^{\dot{\gamma}}\,, \tag{38}$$

$$j_\varrho(B-L)|_{3\times 15} \to j_\varrho(B-L)|_{3\times 16}\,, \tag{39}$$

$$\begin{gathered} d^4x\,\sqrt{|g|}D^\varrho j_\varrho(B-L)|_{3\times 15} = 3\hat{A}_1(X)\,, \\ \hat{A}_1(X) = -\frac{1}{24}\,\mathrm{tr}\,X^2\,; \quad (X)^a_{\ b} = \frac{1}{2\pi}\frac{1}{2}dx^\varrho \wedge dx^\tau (R^a_{\ b})_{\varrho\tau}\,, \\ (R^a_{\ b})_{\varrho\tau} : \left\{ \begin{array}{l} \text{Riemann curvature tensor} \\ \textbf{mixed components: } {}^a_{\ b} \to \textbf{tangent space} \\ \qquad\qquad\qquad\quad {}_{\mu\nu} \to \textbf{covariant space} \end{array} \right.\,, \\ D^\varrho j_\varrho(B-L)|_{3\times(16)} = 0\,, \end{gathered} \tag{40}$$

$$\begin{aligned} j_\varrho(B-L)|_{3\times 15} &\to j_\varrho(B-L)|_{3\times 16} \\ &= \sum_{fm} \left[\frac{1}{3}\left((u^*)^{\alpha\dot{c}}(\sigma_\mu)_{\alpha\dot{\gamma}}(u)^{\dot{\gamma}c} - (\hat{u}^*)^{\alpha c}(\sigma_\mu)_{\alpha\dot{\gamma}}(\hat{u})^{\dot{\gamma}\dot{c}}\right) \right. \\ &\quad + \frac{1}{3}\left((d^*)^{\alpha\dot{c}}(\sigma_\mu)_{\alpha\dot{\gamma}}(d)^{\dot{\gamma}c} - (\hat{d}^*)^{\alpha c}(\sigma_\mu)_{\alpha\dot{\gamma}}(\hat{d})^{\dot{\gamma}\dot{c}}\right) \\ &\quad - (e^-)^{*\alpha}(\sigma_\mu)_{\alpha\dot{\gamma}}(e^-)^{\dot{\gamma}} + (e^+)^{*\alpha}(\sigma_\mu)_{\alpha\dot{\gamma}}(e^+)^{\dot{\gamma}} \\ &\quad \left. - (\nu)^{*\alpha}(\sigma_\mu)_{\alpha\dot{\gamma}}(\nu)^{\dot{\gamma}} + (\mathcal{N})^{*\alpha}(\sigma_\mu)_{\alpha\dot{\gamma}}(\mathcal{N})^{\dot{\gamma}} \right] e^\mu_\varrho\,, \end{aligned} \tag{41}$$

---

[a] $\hat{\nu}_\alpha \equiv \varepsilon_{\alpha\beta}(\nu^*)^\gamma$; $\varepsilon = i\sigma_2$; (2nd Pauli matrix) stands for the left-chiral neutrino fields transformed to the right-chiral basis.

[b] The obviously nontrivial relation between the compact Euclidean — and noncompact asymptotic and locality restricted form of the index theorem involves not clearly formulated *boundary conditions*.

$$g_{\varrho\tau} = e^{\mu}_{\varrho}\eta_{\mu\nu}e^{\nu}_{\tau} : \text{ metric}; \quad e^{\mu}_{\varrho} : \text{ vierbein};$$

$$* : \text{ Hermitian operator conjugation}; \quad (u^*)^{\alpha\dot{c}} \equiv (u^{\dot{\alpha}c})^* ;$$

$$\eta_{\mu\nu} = \text{diag}(1, -1, -1, -1) : \text{ tangent space metric};$$

$$^{c}\,(^{\dot{c}}) : \text{ color and anticolor}; \quad c = 1, 2, 3; \quad D^{\varrho} j_{\varrho}(B-L)|_{3\times(16)} = 0 \rightarrow$$

Let me illustrate the triple-doubling inherent in the elimination of the anomaly in the covariant divergence of $j_{\varrho}(B-L)|_{3\times 15}$ in Eq. (31) as seen through the left-chiral basis, repeating only the ν, $\mathcal{N}$ components of the $B-L$ current in Eq. (41)

$$j_{\varrho}(B-L)|_{3\times 16} = \sum_{fm} \left[\begin{array}{c} \cdots \\ -(\nu)^{*\alpha}(\sigma_{\mu})_{\alpha\dot{\gamma}}(\nu)^{\dot{\gamma}} + \underbrace{(\mathcal{N})^{*\alpha}(\sigma_{\mu})_{\alpha\dot{\gamma}}(\mathcal{N})^{\dot{\gamma}}} \end{array} \right], \tag{42}$$

	$\nu^{\dot{\gamma}}_F$	$\mathcal{N}^{\dot{\gamma}}_F$
$B-L$	-1	$+1$

, $F = 1, 2, 3$ family.

The Majorana representation of $\{\boldsymbol{\Gamma}_{p,q}; \mathbb{R}\}$ ***for*** $p = 1,\ q = 3$

$$\Gamma^{(\chi)}_{\mu} = \begin{pmatrix} 0 & i\sigma_{\mu} \\ i\tilde{\sigma}_{\mu} & 0 \end{pmatrix}; \quad \sigma_{\mu} = (\sigma_0; \sigma_k), \quad \tilde{\sigma}_{\mu} = (\sigma_0; -\sigma_k); \quad k = 1, 2, 3,$$

$$\sigma_0 = \P_{2\times 2}; \quad \sigma_1 = \begin{pmatrix} 0 & 1 \\ 1 & 0 \end{pmatrix}; \quad \sigma_2 = \begin{pmatrix} 0 & -i \\ i & 0 \end{pmatrix}; \quad \sigma_3 = \begin{pmatrix} 1 & 0 \\ 0 & -1 \end{pmatrix}, \tag{43}$$

$$\Gamma_{\mu} \equiv \eta_{\mu\nu}\Gamma^{\nu}, \quad \Gamma_5 = \Gamma_0\Gamma_1\Gamma_2\Gamma_3, \quad \Gamma_5^2 = -\P_{4\times 4}|_{p=1q=3}$$

in any basis.

In the chiral basis we have

$$\Gamma^{(\chi)}_5 = i \begin{pmatrix} \P & 0 \\ 0 & -\P \end{pmatrix} = i\gamma_{5R} = -i\gamma_{5L}\,. \tag{44}$$

***Some conclusions from Subsecs.* 1.1 *and* 1.1.2**

(C1) The oscillation phenomena indicate clearly, that a *genuinely chiral* extension of $B-L$ to a conserved, global symmetry, generating a *continuous* U1-group of transformations, is not involved.

(C2) On the other hand the binary code of a (minimally) supposed unifying gauge group SO10 or spin10 could, if $B-L$ is *not* gauged, equivalently generate a global symmetry of the vectorlike nature. The latter however would allow neutrino mass through the (electroweak doublet–singlet) pairing

$$-\mathcal{L}_{\mathcal{M}} = \mu_{FG}\mathcal{N}^F_{\dot{\gamma}}\nu^{\dot{\gamma}G} + \text{h.c.}, \quad F, G = 1, 2, 3 \text{ family} \tag{45}$$

without symmetry restrictions on the mass matrix μ_{FG} in Eq. (45).

(C3) Then however the question arises, why the mass matrix μ, involving the scalar doublet(s) within the electroweak gauge group, also generating masses of charged spin $\frac{1}{2}$ fermions, gives rise to very small physical neutrino masses. Thus we follow the *hypothesis* that SO10 *is* gauged and that it is the *large* mass scale of the gauge boson associated with $B - L$ in particular, which distinguishes neutrino flavors.[4,11,12]

We turn to the so induced structure of the $\mathcal{N}$, ν mass term

$$\begin{aligned} \mathcal{H}_{\mathcal{M}} &= \mu_{FG} \mathcal{N}^F_{\dot\gamma} \nu^{\dot\gamma G} + \text{h.c.} + \mathcal{H}_M\,, \\ \mathcal{H}_M &= \frac{1}{2} M_{FG} \mathcal{N}^F_{\dot\gamma} \mathcal{N}^{\dot\gamma G} + \text{h.c.}\,, \quad F, G = 1,2,3\,, \\ M_{FG} &= M_{GF}\text{: complex arbitrary otherwise}\,; \quad |M| \gg |\mu|\,. \end{aligned} \tag{46}$$

It is the primary breakdown along the direction of $\hat{J}$ which contrary to all "mirror complexes" brings on the level of (pseudo-)scalar fields to the foreground the complex bosonic 126 and $\overline{126}$ representations of SO10

$$\begin{aligned} &\mathcal{H}_M \leftarrow \left(\Phi^{\,\overline{126}\,FG}\right)^{\bar\xi} (f_{a16F})_{\dot\gamma} (f_{b16G})^{\dot\gamma} C \begin{pmatrix} 126 & 16 & 16 \\ \xi & a & b \end{pmatrix} + \text{h.c.}\,, \\ &(\Phi^{\,\overline{126}\,FG})^{\bar\xi}\text{: (pseudo-)scalar fields in the } \overline{126} \text{ representation of SO10}\,. \end{aligned} \tag{47}$$

In Eq. (47), $C\begin{pmatrix} 126 & 16 & 16 \\ \xi & a & b \end{pmatrix}$ denotes the coupling coefficients, projecting the symmetric product of two 16-representation of spin10 on the 126 one of SO10. The 126 *complex* representations of SO10 exhibits the value of $J\ 10 = 2 \times 5\mathcal{N}\mathcal{N}$. The relatively complex conjugate representations $126 \oplus \overline{126}$ are contained in the *real, reducible* fivefold antisymmetric tensor representation of SO10 decomposing into the irreducible pair upon the duality conditions

$$\begin{aligned} &t^{[A_1 A_2 \cdots A_5]}\,; \quad A_{1\cdots 5} = 1,2,\ldots,10\,, \\ &t^{[A_{\pi_1} A_{\pi_2} \cdots A_{\pi_5}]} = \text{sgn}\begin{pmatrix} 1 & 2 & \cdots & 5 \\ \pi_1 & \pi_2 & \cdots & \pi_5 \end{pmatrix} t^{[A_1 A_2 \cdots A_5]}\,, \\ &\frac{1}{5!} \varepsilon_{A_1 \cdots A_5 B_1 \cdots B_5} t_{\pm}^{[B_1 B_2 \cdots B_5]} = (\pm i) t_{\pm}^{[A_1 A_2 \cdots A_5]}\,, \\ &\varepsilon_{A_1 \cdots A_5 A_6 \cdots A_{10}} = \text{sgn}\begin{pmatrix} 1 & 2 & \cdots & 10 \\ \pi_1 & \pi_2 & \cdots & \pi_{10} \end{pmatrix} \varepsilon_{A_{\pi_1} \cdots A_{\pi_5} A_{\pi_6} \cdots A_{\pi_{10}}}\,, \quad \varepsilon_{12 \cdots 10} = 1\,. \end{aligned} \tag{48}$$

Within the complex spin ($2\nu = 4\tau + 2$), $\tau = 2, 3, \ldots$ series — $\tau = 2 \leftrightarrow$ spin10 — the relatively complex conjugate spinorial pair of representations with dimension $4^\tau \leftarrow 16(64, \ldots)$ and the complex self-dual–antiself-dual pair of representations with dimension $\frac{1}{2}\binom{4\tau+2}{2\tau+1} \leftarrow 126$ ($11.12.13 = 1716, \ldots$) are intrinsically related for $\tau = 2, 3, 4, \ldots$.

***More conclusions and questions from Subsec.* 1.1.2**

(Q1) Is it enough to consider the primary breakdown and its characteristic, the "tilt to the left" concerning 3 families, as due *essentially* to spin10, which is the *lowest* simple spin group along the complex *orthogonal* chain?

It has been argued interestingly by Feza Gursey and collaborators,[19] that it is the chain of exceptional groups which encode intrinsically the number 3, which in turn underlies the 3 as the number of (left-chiral) families as well as the strong interaction gauge group $SU3_c$.

(A1) I think the answer is to the affirmative, since all higher gauge groups, including the exceptional chain and especially E8, but also spin14, spin18 do *not* explain the #3 of families, rather generate together with even the apparently correct 3 families — for E8 — also mirror families — 3 for E8, and powers of 2 for the orthogonal chain with $\tau \geq 3$.

The tentative conclusion remains, that the structure of families has to be explained outside spin10 and *also* larger unifying gauge groups containing spin10, whereas the origin of neutrino mass is laid out by the lowest member of the complex orthogonal chain → spin10. Maybe a systematic study of discrete groups can bring new insights: e.g. following Ref. 20.

(C4) The two apparently different phenomena of (a) "tilt to the left" and (b) baryon number violation, are intrinsically associated with the *unusual* sequence of (pseudo)scalar fields generating primary breakdown (Eq. (24))

$$
\begin{aligned}
&\text{spin10} \to \text{SU5} \times \text{U1}_J \to \text{SU3}_c \times \text{SU2}_L \times \text{U1}_{\mathcal{Y}} = \text{G}_{\text{SM}} \\
&[16] = \{1\}_{+5} \qquad + \{10\}_{+1} \qquad + \{\bar{5}\}_{-3} \\
&[\overline{16}] = \{1\}_{-5} \qquad + \{\overline{10}\}_{-1} \qquad + \{5\}_{+3} \\
&\{\bar{5}\}_{-3} = \left](\bar{3},1)_{+\frac{1}{3}}\right[_{-3} + \left](1,2)_{-\frac{1}{2}}\right[_{-3} \,.
\end{aligned}
\tag{49}
$$

3. Recent Deliberation Justifying the Special Properties of SO10 as Unifying Gauge Theory

(1) Quantized e.m. charges
The cancellation of the gravitation induced anomaly in SO10, contrary to the fractional representation structure in SU5, as described in Eq. (42) has an essential consequence for the stability of quantized e.m. charges, through the large scale *spontaneous* gauge breaking at or near the unification scale $M_U \sim 10^{16\pm 1}$ GeV in Eq. (20).

(2) An argument in favor of SO10 unification over SU5
This is an argument in favor of SO10 unification, quite independently of the parallel structure of "mass by mixing" or "seesaw" for 3 light and 3 heavy neutrino flavors.

(3) Mass and mixing parameter estimates, unchanged since 2005.[8]

3.1. *"Mass from mixing" in vacuo or "seesaw"*

(1) Neutrinos oscillate

Like neutral kaons (yes, but how?)[c] Here let me continue the "flow of thought" embedding neutrino masses in SO10.

The special feature, pertinent to (electrically neutral) neutrinos is, that the ν-extending degrees of freedom $\mathcal{N}$ are singlets under the whole SM gauge group $\mathcal{G}_{\text{SM}} = SU3_c \otimes SU2_L \otimes U1_{\mathcal{Y}}$, in fact remain singlets under the larger gauge group $SU5 \supset \mathcal{G}_{\text{SM}}$. This allows an arbitrary (Majorana-)mass term, involving the bilinears formed from two $\mathcal{N}$'s.

In the present setup (minimal ν-extended SM) the full neutrino mass term is thus of the form

$$\mathcal{H}_{\mathcal{M}} = \frac{1}{2}[\nu \mathcal{N}]\mathcal{M}\begin{bmatrix}\nu \\ \mathcal{N}\end{bmatrix} + \text{h.c.}, \quad \mathcal{M} = \begin{pmatrix}0 & \mu^T \\ \mu & M\end{pmatrix}, \quad \mathcal{M} = \mathcal{M}^T \to M = M^T. \tag{50}$$

Again within primary SO10 breakdown the full $\mathcal{M}$ extends the scalar sector to the representations $(10) \oplus (120) \oplus (126)$.[d]

Especially the 0 entry needs explanation. It is an exclusive property of the minimal ν-extension assumed here. Since the "active" flavors ν_F all carry $I_{3w} = \frac{1}{2}$ terms of the form

$$\frac{1}{2}{}_{F'}\nu\chi_{F'F}\nu_F = \frac{1}{2}\nu^T\chi\nu, \quad \chi = \chi^T \tag{51}$$

cannot arise as Lagrangian masses, except induced by an I_w-triplet of scalars, developing a vacuum expected value independent from the doublet(s).

"seesaw" $\swarrow\!\!\nearrow$

The relative "size" of μ and M shall define the "mass from mixing" situation and segregates 3 heavy neutrino flavors from the 3 light ones:

$$\swarrow \; ||\mu|| \ll ||M|| \; \nearrow \,. \tag{52}$$

(2) Mass from mixing

Here we introduce the arithmetic mean measure for 3×3 matrices A, not to be confused with the norms $||\cdot||$ defined in Eq. (54)

$$|A| = |\,\text{Det}\,A|^{1/3}, \tag{53}$$

$$||\mu||^2 = \text{tr}\,\mu\mu^\dagger, \quad ||M||^{-2} = \text{tr}\,M^{-1}M^{-1\dagger}; \quad \vartheta = ||\mu||/||M|| \ll 1, \tag{54}$$

$$\mathcal{M}_1 = -t\mathcal{M}_2 t^T \tag{55}$$

---

[c] Bruno Pontecorvo, "Mesonium and antimesonium," *JETP* (*USSR*) **33**, 549 (1957), English translation: *Soviet Physics, JETP* **6**, 429 (1958).

[d] It is from here where, to the best of my knowledge, the discussion of the *necessarily nonvanishing nature* and of the magnitude of the light neutrino masses (re-)started in 1974.[11,12] The structure in Eq. (50) is reserved for the minimal $SU2_L \times U1_{\mathcal{Y}}$ case "tilted to the left."[13]

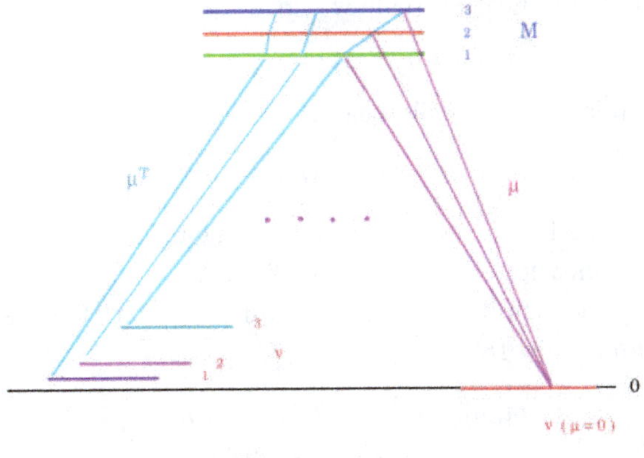

Fig. 2. Key questions → which is the scale of M? $O(10^{10})$ GeV → is there any evidence for this scale today? hardly! → and what about susy?

Eq. (55) then implies

$$\begin{aligned} |\mathcal{M}_1|/|\mathcal{M}_2| &= |t|^2\,, \\ |\mathcal{M}_1| = |m_{\text{diag}}| &= (m_1 m_2 m_3)^{1/3}\,, \\ |\mathcal{M}_2| = |M_{\text{diag}}| &= (M_1 M_2 M_3)^{1/3}\,. \end{aligned} \tag{56}$$

We consider the arithmetic mean of the light and heavy neutrino masses and the corresponding "would be" masses if μ and μ^T would be the only parts of the full 6×6 mass matrix $\mathcal{M}$

$$\begin{aligned} \bar{m} &= (m_1 m_2 m_3)^{1/3}\,, & \bar{M} &= (M_1 M_2 M_3)^{1/3}\,, \\ \mu &= u_\mu \mu_{\text{diag}}(\mu_1, \mu_2, \mu_3) v_\mu^{-1}\,, & \bar{\mu} &= (\mu_1, \mu_2, \mu_3)^{1/3}\,. \end{aligned} \tag{57}$$

Then beyond Eq. (56) there is one more (exact) relation[e]

$$\begin{aligned} \hat{t} = (\tan a_1 \tan a_2 \tan a_3)^{1/3} = |t|\,, \quad |\mu|^2 = |\mathcal{M}_1|\,|\mathcal{M}_2| \rightarrow \bar{m}/\bar{\mu} = \hat{t}\,, \\ \bar{m}/\bar{M} = \hat{t}^2 \quad \text{or equivalently}\,, \quad \bar{m} = \hat{t}\bar{\mu} \swarrow\!\!\nearrow \bar{M} = \hat{t}^{-1}\bar{\mu} \quad \text{seesaw (of type I)}\,. \end{aligned} \tag{58}$$

The estimates below are based on the assumption that the scalar doublets (2) are part of a complex 10-representation of SO10 with Yukawa couplings of the form

$$\mathcal{H}_Y = \lambda_{F'F} \begin{pmatrix} 16 & 16 & 10 \\ B & A & D \end{pmatrix}_{F'} f_B f_{AF} + \text{h.c.} \rightarrow \lambda_{F'F} = \lambda_{FF'}\,. \tag{59}$$

[e]For MSSM inspired seesaw of type II realizations see e.g. Ref. 22.

It follows that at the unification scale we have[f]

$$\mu = \mu^T = \mu_u \,. \tag{60}$$

We shall use the relation at a scale near 100 GeV

$$\mu \sim \frac{1}{3}(\mu_u) \,. \tag{61}$$

The factor $\frac{1}{3}$ accounts for the color rescaling reducing the (colored) up-quark mass matrix from the unification scale down to 100 GeV.

It follows using the definitions in Eq. (57) and the quark masses $m_u \sim 5.25$ MeV, $m_c \sim 1.25$ GeV and $m_t \sim 180$ GeV

$$\bar{\mu}_u = (m_u m_c m_t)^{1/3} \sim 1 \text{ GeV} \to \bar{\mu} \sim \frac{1}{3} \text{ GeV} \,. \tag{62}$$

Further lets approximate the mass square differences obtained from the combined neutrino oscillation measurements by

$$\Delta m_{12}^2 \sim 10^{-4} \text{ eV}^2 \,, \quad \Delta m_{23}^2 \sim 2.5 \cdot 10^{-3} \text{ eV}^2 \,. \tag{63}$$

Finally "pour fixer les idées" I set the lowest light neutrino mass ~1 meV and assume hierarchical (123) light masses. This implies

$$m_1 \sim 1 \text{ meV} \,, \quad m_2 \sim 10 \text{ meV} \,, \quad m_3 \sim 50 \text{ meV} \to \bar{m} \sim 8 \text{ meV} \,. \tag{64}$$

It follows from Eq. (58)

$$\begin{aligned} \hat{t} &= \bar{m}/\bar{\mu} \sim 2.4 \cdot 10^{-11} \,, \\ \bar{M} &= \bar{\mu}/\hat{t} \sim 1.4 \cdot 10^{10} \text{ GeV} \,, \\ \hat{t}^2 &\sim 5.8 \cdot 10^{-22} \,. \end{aligned} \tag{65}$$

Light neutrino masses are indeed small.[g]

4. Mass Ranges and Limits on Heavy-Light Neutrino Flavor Mixings from a CMS Experiment[24]

Recently the CMS collaboration obtained new results in Ref. 23. Following Tao Han and Bin Zhang,[24] we consider the production mechanism shown below. The amplitudes described by the Feynman diagrams in Fig. 3 depend on the mass and mixing parameters pertaining to neutrino masses, light and heavy. The range of heavy "right-handed" neutrino masses is sensitive to values

$$40 \text{ GeV} \le M_{\mathcal{N}} \le 500 \text{ GeV} \,. \tag{66}$$

The heavy-light individual mixing elements $V_{\mu M}$ satisfy the constraints depending on the heavy neutrino flavor mass.

[f] In order to obtain a general (not a symmetric) heavy-light mass matrix μ a combination of SO10 representations (120)⊕(10) is needed, which however would "destroy" the mass relation in Eq. (59). Key question → is this relevant? Estimate shall be estimate.

[g] Puzzling questions → is susy bringing down in "small steps" the $B - L$ protecting mass scale $\bar{M} =\sim 1.4 \cdot 10^7$ TeV to 1 TeV? — or is $\bar{M} =\sim 1.4 \cdot 10^7$ TeV in view of seesaw type II too small? $\mu \to e\gamma$ at a rate of?

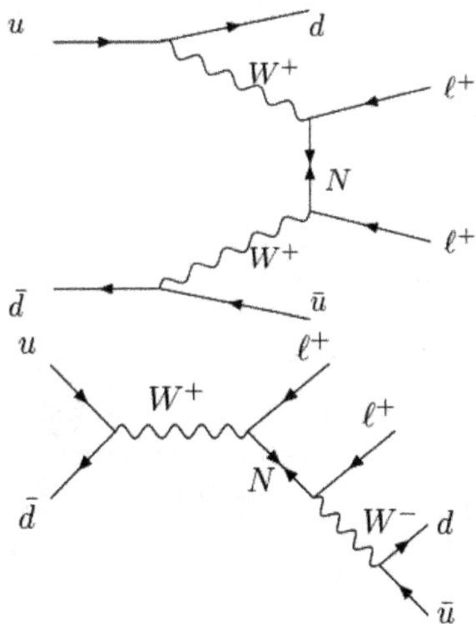

Fig. 3. Feynman diagrams for $\Delta L = 2$ processes induced by a Majorana neutrino N in $q\bar{q}'$ collisions.

For $m_{\mathcal{N}} = 90$ GeV the experiment finds

$$|V_{\mu M}|^2 < 0.00470\,. \tag{67}$$

At $m_{\mathcal{N}} = 200$ GeV the limit is

$$|V_{\mu M}|^2 < 0.0123$$

and at $m_{\mathcal{N}} = 500$ GeV becomes

$$|V_{\mu M}|^2 < 0.583\,.$$

We compare the above limits with the estimate on the deviation from 3×3 unitarity of the light 3 neutrino mass matrix in Eq. (65)

$$\hat{t}^2 \sim 5.8 \cdot 10^{-22}\,.$$

5. Concluding Remarks

Apparent paradox of unification of forces and stability of *large, primary* breakdown scales of order $\bar{M} \sim 10^{10}$ GeV – $m_{\text{Pl}} \sim 10^{19}$ GeV.

In a *concluding remark* (not a *conclusion*), let me point out that the example of the large scale inherent in heavy neutrino masses is not only responsible for the small masses of observed neutrino flavors, but at the same time the stability of $\bar{M} \sim 10^{10}$ GeV — a key feature unexplained so far — serves as *protecting scale* for the *approximative* conservation of leptonic as well as baryonic numbers at electroweak scales much below $\bar{M}$. As a consequence even unified charge-like gauge groups: SO10, E6, ... act first through a primary breakdown of local gauge invariance at scales $M_{\text{unif}} \sim 10^{16}$ GeV, which produces essential deviations from unified symmetries, yet no anomalies in the so broken charge like gauge theories! This is related to the quantization of electromagnetic charge.

6. Note Added in Proof

Dark matter is formed by the 3 predominantly active light neutrino–antineutrino flavors, proof by elimination of alternatives.

In correcting the original slidefile: singanu2015.pdf, which contains my contribution to the Conference on Massive Neutrinos, 9–13 February 2015, IAS, Nanyang Technological University, Singapore, entitled "Massive Neutrinos and SO (10) Unification," I realized that combining its content with related papers of the author, not discussed there, an important consequence for the nature of Dark Matter can be reached by elimination of alternatives. — This Note is devoted to this topic.

The first alternative concerns the "Strong CP Problem," which is rejected in its very existence in the paper by the author from 1978, forming Ref. 25.

The second alternative concerns a rigid supersymmetry structure enriching an exact and broken gauge theory containing the Standard Model, as well as extension of this supersymmetric structure to open and closed superstrings. Hereto let me cite a recent paper by Ashoke Sen, Ref. 26.

A candidate for Dark Matter in the above framework, being the Lowest Supersymmetric Partner Particle (LSP) being stable or sufficiently long-lived, is compromised by the trance anomaly in uncurved $3+1$ dimensions in Ref. 27. The latter investigation was prompted by underpinning the universal scale of the slope of Regge trajectories, whence interpreted as oscillatory modes of u, d, s quark and antiquark flavors in QCD, published for mesons in Ref. 28 and for baryons in Ref. 29.

Later we investigated and established in my group of collaborators at the ITP in Bern direct proofs for special cases of supersymmetric gauge theories. I cite here 3 papers in ascending order of time: 1 with Markus Leibundgut Ref. 30 and 2 with Luzi Bergamin Refs. 31 and 32. — This eliminates Alternative 2.

Qod erat demonstrandum	What was to be demonstrated
Ibi tertium non datur	In this outline a 3d Alternative is not given

What remains to be done, in the hopefully not too distant future, is to show how the electromagnetic long-range interactions of the mainly active 3 neutrino and antineutrino flavors develop a vacuum condensate and as a consequence their equilibrium density and pressure do *not* correspond to (almost) massless free states at a corresponding temperature of $\sim$2K. Here the original framework was laid out in a paper with Raul Horvat and Josip Trampetic: Ref. 33.

References

1. F. Englert, Nobel lecture, The BEH mechanism and its scalar boson, http://www.nobelprize.org/nobel_prizes/physics/laureates/2013/englert-lecture.html, in file: englert-lecture-slides.pdf.

2. P. Higgs, Nobel lecture, Evading the Goldstone theorem, http://www.nobelprize.org/nobel_prizes/physics/laureates/2013/higgs-lecture.html, in file: higgs-lecture.pdf.
3. Particle Data Group (K. A. Olive *et al.*), *Chin. Phys. C* **38**, 090001 (2014).
4. H. Fritzsch and P. Minkowski, *Ann. Phys.* **93**, 193 (1975); H. Georgi, *AIP Conf. Proc.* **23**, 575 (1975).
5. J. C. Pati and A. Salam, *Phys. Rev. D* **10**, 275 (1974) [Erratum: *ibid.* **11**, 703 (1975)].
6. H. Georgi and S. L. Glashow, *Phys. Rev. Lett.* **32**, 438 (1974), DOI: 10.1103/PhysRevLett.32.438.
7. H. Georgi, H. R. Quinn and S. Weinberg, *Phys. Rev. Lett.* **33**, 451 (1974), DOI: 10.1103/PhysRevLett.33.451.
8. P. Minkowski, Neutrino oscillations: A historical overview and its projection, May 2005, Contributed to conference: C05-02-22.1, pp. 7–27, Proceedings, arXiv:hep-ph/0505049, URL: http://www.mink.itp.unibe.ch, under "Lectures and Talks" in the file venice3.pdf.
9. P. Minkowski, *J. Phys. Conf. Ser.* **171**, 012016 (2009), DOI: 10.1088/1742-6596/171/1/012016.
10. W. Buchmüller, C. Greub and P. Minkowski, *Phys. Lett. B* **267**, 395 (1991), DOI: 10.1016/0370-2693(91)90952-M.
11. H. Fritzsch, M. Gell-Mann and P. Minkowski, *Phys. Lett. B* **59**, 256 (1975).
12. H. Fritzsch and P. Minkowski, *Phys. Lett. B* **62**, 72 (1976).
13. P. Minkowski, *Phys. Lett. B* **67**, 421 (1977).
14. M. Gell-Mann, P. Ramond and R. Slansky, Complex spinors and unified theories, in *Supergravity*, eds. P. van Nieuwenhuizen and D. Z. Freedman (North Holland Publ. Co., 1979), Stony Brook Wkshp.1979:0315 (QC178:S8:1979); T. Yanagida, Horizontal symmetry and masses of neutrinos, in *Proceedings of the Workshop on the Baryon Number of the Universe and Unified Theories*, eds. O. Sawada and A. Sugamoto, Tsukuba, Japan, 13–14 Feb. 1979), and in (QCD161:W69:1979); S. Glashow, Quarks and leptons, in *Proceedings of the Cargèse Lectures*, ed. M. Lévy (Plenum Press, New York, 1980); R. Mohapatra and G. Senjanovič, *Phys. Rev. Lett.* **44**, 912 (1980).
15. R. Coquereaux, *Phys. Lett. B* **115**, 389 (1982).
16. P. Minkowski, Fermionic oscillators and the embedding SO(2n) ⊃ SUn, Seminar at the Enrico Fermi Summer School of Physics, Varenna, Italy, Jul 21–Aug 2, 1980 (unpublished).
17. F. Hirzebruch, *Topological Methods in Algebraic Geometry*, Die Grundlehren der mathematischen Wissenschaften, Band 131 (Springer Verlag, Berlin, 1966).
18. P. Minkowski, Neutrino flavors light and heavy — how heavy?, in nus-albufeira2007.pdf, URL: http://www.mink.itp.unibe.ch/, and references cited therein.
19. F. Gursey, P. Ramond and P. Sikivie, *Phys. Lett. B* **60**, 177 (1976).
20. Y. BenTov, Fermion masses without symmetry breaking in two spacetime dimensions, arXiv:1412.0154 [cond-mat.str-el].
21. B. Pontecorvo, *JETP* (*USSR*) **33**, 549 (1957) [English translation: *Soviet Physics JETP* **6**, 429 (1958)].
22. H. S. Goh, R. N. Mohapatra and S. Nasri, *Phys. Rev. D* **70**, 075022 (2004), arXiv:hep-ph/0408139.
23. CMS Collab. (V. Khachatryan *et al.*), Search for heavy Majorana neutrinos in $\mu^\pm\mu^\pm +$ jets events inproton-proton collisions at $\sqrt{s} = 8$ TeV, CMS-EXO-12-057, CERN-PH-EP-2015-001, arXiv:1501.05566 [hep-ex].
24. T. Han and B. Zhang, *Phys. Rev. Lett.* **97**, 171804 (2006), DOI: 10.1103/PhysRevLett.97.171804, arXiv:hep-ph/0604064.

25. P. Minkowski, The mass of (958), CP and P conservation in QCD, manifestations of a Josephson effect, submitted to: *Phys. Lett. B*, MPI-PAE-PTh-37-78, scanned version available on Inspirehep.net through Link to Fulltext.
26. A. Sen, Gauge invariant 1PI effective superstring field theory: Inclusion of the Ramond sector, arXiv:1501.00988 [hep-th].
27. P. Minkowski, On the anomalous divergence of the dilatation current in gauge theories, PRINT-76-0813 (BERN), scanned version available on Inspirehep.net through Link to Fulltext.
28. P. Minkowski, *NATO Sci. Ser. B* **49**, 315 (1979), DOI: 10.1007/978-1-4684-7665-1_10, Conference: C78-09-04 Proceedings, scanned version available on Inspirehep.net through Link to Fulltext.
29. P. Minkowski, *Nucl. Phys. B* **174**, 258 (1980), DOI: 10.1016/0550-3213(80)90202-3.
30. M. Leibundgut and P. Minkowski, *Nucl. Phys. B* **531**, 95 (1998), DOI: 10.1016/S0550-3213(98)00439-8, arXiv:hep-th/9708061.
31. L. Bergamin and P. Minkowski, SUSY glue balls, dynamical symmetry breaking and nonholomorphic potentials, TUW-03-03, BUTP-2003-01, arXiv:hep-th/0301155.
32. L. Bergamin and P. Minkowski, No supersymmetry without supergravity: Induced supersymmetry representations on composite effective superfields, TUW-03-36, arXiv:hep-th/0312034.
33. R. Horvat, P. Minkowski and J. Trampetic, *Phys. Lett. B* **671**, 51 (2009), DOI: 10.1016/j.physletb.2008.11.055, arXiv:0809.0582 [hep-ph].

Relating Small Neutrino Masses and Mixing

Soumita Pramanick* and Amitava Raychaudhuri†
Department of Physics, University of Calcutta,
92 Acharya Prafulla Chandra Road, Kolkata 700009, India
**soumitapramanick5@gmail.com*
†*palitprof@gmail.com*

Experiments on neutrino oscillations have uncovered several small parameters, θ_{13} being a prominent one. Others are the solar mass splitting *vis-à-vis* the atmospheric one and the deviation of θ_{23} from maximal mixing. In this talk, we elaborate on a neutrino mass model based on the see-saw mechanism in which the mixing angles to start with are either vanishing (θ_{13} and θ_{12}) or $\pi/4$ (θ_{23}). The atmospheric mass splitting is taken as a part of this initial structure but the solar splitting is absent. A perturbative contribution, originating from a Type-I see-saw, results in nonzero values of θ_{13}, θ_{12}, $\Delta m^2_{\rm solar}$, and shifts θ_{23} slightly from $\pi/4$, interrelating them all. The model incorporates CP-violation, the phase δ being close to $3\pi/2$ for (a) quasidegeneracy or (b) inverted mass ordering. It will be put to test as the neutrino parameters get better determined.

Keywords: Neutrino mixing; θ_{13}; leptonic CP-violation; neutrino mass ordering; perturbation.

1. Introduction

Neutrino mass and mixing have been subjects of intensive exploration as they shed light on the physics beyond the standard model. Atmospheric and solar neutrinos indicate two very different scales of neutrino mass splitting — $\Delta m^2_{\rm solar}/|\Delta m^2_{\rm atm}| \sim 10^{-2}$ — which are confirmed in accelerator and reactor experiments. The lepton mixing is captured in the PMNS matrix.[a] The global fits to the data[2,3] from atmospheric, solar, accelerator, and reactor experiments indicate θ_{13} to be small[b] ($\sin\theta_{13} \sim 0.1$) and θ_{23} to be near maximal ($\sim \pi/4$). Here we discuss a model in which the atmospheric mass splitting with maximal mixing in this sector, $\theta_{23} = \pi/4$, follows from a zero-order mass matrix which sets the scale of the problem. There

[a]We use the PDG[1] parametrization of the PMNS matrix.

[b]For the present status of θ_{13}, see presentations from Double Chooz, RENO, Daya Bay, MINOS/MINOS+ and T2K at Neutrino 2014.
https://indico.fnal.gov/conferenceOtherViews.py?view=standard&confId=8022.

is at this stage no solar splitting and the other two mixing angles are also absent.[c] θ_{13} and a small shift to θ_{23} arise from a Type-I see-saw[4–8] contribution which also results in the solar mass splitting, acting as a perturbation. A nonzero θ_{12} is also produced and due to the degeneracy of masses it is not small. The three nonzero mixing angles open the possibility of CP-violation in the lepton sector. This model accommodates a CP-phase δ which must be close to maximal ($\delta \sim \pi/2, 3\pi/2$) if the neutrinos have an inverted mass ordering or if they are quasidegenerate.[9] Previous work[d] partially address similar issues, but to our knowledge this is the first time that *all* the small parameters have been shown to have the same perturbative origin and are consistent with the latest data.

2. The Model

The starting choice of the mixing angles imply the following form of the mixing matrix, the columns of it being the unperturbed flavor basis:[e]

$$U^0 = \begin{pmatrix} 1 & 0 & 0 \\ 0 & \sqrt{\frac{1}{2}} & \sqrt{\frac{1}{2}} \\ 0 & -\sqrt{\frac{1}{2}} & \sqrt{\frac{1}{2}} \end{pmatrix}. \tag{1}$$

Solar splitting is absent at this stage causing the first two mass eigenvalues to be degenerate.[f] Thus the unperturbed neutrino mass matrix is $M^0 = \text{diag}\{m_1^{(0)}, m_1^{(0)}, m_3^{(0)}\}$ in the mass basis. The unperturbed mass eigenvalues are made real and positive by suitable choices of the Majorana phases. The atmospheric splitting is given by $\Delta m^2_{\text{atm}} = \left(m_3^{(0)}\right)^2 - \left(m_1^{(0)}\right)^2$. It is useful to define $m^\pm = m_3^{(0)} \pm m_1^{(0)}$ and express the unperturbed mass matrix in flavor basis as,

$$(M^0)^{\text{flavor}} = U^0 M^0 U^{0T} = \frac{1}{2}\begin{pmatrix} 2m_1^{(0)} & 0 & 0 \\ 0 & m^+ & m^- \\ 0 & m^- & m^+ \end{pmatrix}. \tag{2}$$

As already hinted, the perturbation can originate from a Type-I see-saw. In order to reduce the number of independent parameters the Dirac mass term is taken to be proportional to the identity, i.e. $M_D = m_D \mathbb{I}$, in the flavor basis. This choice completely determines the right-handed flavor basis although the form of M_R^{flavor} can be chosen at will to suit our purpose. In the interest of minimality we seek symmetric matrices with the fewest nonzero entries. Five texture zero matrices fail

[c]One mixing angle being $\pi/4$ and the other two zero can be a manifestation of some underlying symmetry.

[d]Earlier work on neutrino mass models in which a few elements dominate over others can be traced to Ref. 10. Models with somewhat similar points of view as those espoused here are Refs. 11 and 12. For more recent work after the determination of θ_{13} see, for example, Refs. 13–21.

[e]In this flavor basis the charged lepton mass matrix is taken to be diagonal.

[f]Due to this degeneracy θ_{12} is arbitrary and can be chosen to be zero as done here.

the invertibility criterion[g] and therefore are not pursued. Next we try four texture zero options. By examining the different alternatives it can be seen that all the perturbation goals that we have set for ourselves could be achieved by only two such candidates out of which one is scripted below:[h]

$$M_R^{\text{flavor}} = m_R \begin{pmatrix} 0 & xe^{-i\phi_1} & 0 \\ xe^{-i\phi_1} & 0 & 0 \\ 0 & 0 & ye^{-i\phi_2} \end{pmatrix}, \tag{3}$$

where x, y are dimensionless constants of $\mathcal{O}(1)$. The Dirac mass is kept real without any loss of generality.

3. Real M_R ($\phi_1 = 0$ or π, $\phi_2 = 0$ or π)

For notational simplicity in this section the phases are not written explicitly, instead x (y) is taken as positive or negative depending on whether ϕ_1 (ϕ_2) is 0 or π.

Employing Type-I see-saw one can write,

$$\begin{aligned} M'^{\text{mass}} &= U^{0T}\left[M_D^T (M_R^{\text{flavor}})^{-1} M_D\right] U^0 \\ &= \frac{m_D^2}{\sqrt{2}xym_R} \begin{pmatrix} 0 & y & y \\ y & \dfrac{x}{\sqrt{2}} & -\dfrac{x}{\sqrt{2}} \\ y & -\dfrac{x}{\sqrt{2}} & \dfrac{x}{\sqrt{2}} \end{pmatrix}. \end{aligned} \tag{4}$$

The changes in the solar sector are determined by the 2×2 submatrix of M'^{mass},

$$M'^{\text{mass}}_{2\times 2} = \frac{m_D^2}{\sqrt{2}xym_R} \begin{pmatrix} 0 & y \\ y & x/\sqrt{2} \end{pmatrix}. \tag{5}$$

From the above one has

$$\tan 2\theta_{12} = 2\sqrt{2}\left(\frac{y}{x}\right). \tag{6}$$

The tribimaximal mixing value of θ_{12}, which is disallowed by the 1σ data but is allowed in the 3σ range,[i] is obtained if $y/x = 1$. From the data, $\tan 2\theta_{12} > 0$ always, forcing x and y to have the same sign. Thus it can be inferred that either $\phi_1 = 0 = \phi_2$ or $\phi_1 = \pi = \phi_2$. The global fits of θ_{12} provide a bound on this ratio as,

$$0.682 < \frac{y}{x} < 1.075 \quad \text{at} \quad 3\sigma. \tag{7}$$

From Eq. (5),

$$\Delta m^2_{\text{solar}} = \frac{m_D^2}{xym_R} m_1^{(0)} \sqrt{x^2 + 8y^2}. \tag{8}$$

---

[g] Existence of the inverse of M_R is an essential condition for the see-saw mechanism.

[h] The other alternative is a mere $2 \leftrightarrow 3$ exchange of this configuration and the corresponding results vary only up to a relative sign.

[i] From Ref. 2 we use $7.03 \le \Delta m^2_{21}/10^{-5}\ \text{eV}^2 \le 8.03$ and $31.30° \le \theta_{12} \le 35.90°$ at 3σ.

The first-order corrected third wave function $|\psi_3\rangle$ is:

$$|\psi_3\rangle = \begin{pmatrix} \kappa \\ \dfrac{1}{\sqrt{2}}\left(1 - \dfrac{\kappa}{\sqrt{2}}\dfrac{x}{y}\right) \\ \dfrac{1}{\sqrt{2}}\left(1 + \dfrac{\kappa}{\sqrt{2}}\dfrac{x}{y}\right) \end{pmatrix}, \tag{9}$$

where

$$\kappa \equiv \frac{m_D^2}{\sqrt{2} x m_R m^-}. \tag{10}$$

If $x > 0$ the sign of m^- determines that of κ. Comparing Eq. (9) with the third column of the PMNS matrix, we write,

$$\sin\theta_{13}\cos\delta = \kappa = \frac{m_D^2}{\sqrt{2} x m_R m^-}. \tag{11}$$

For normal mass ordering (NO), $\delta = 0$ while for inverted mass ordering (IO) $\delta = \pi$ if $x > 0$, both being CP conserving.[j] For $x < 0$ NO (IO) corresponds to $\delta = \pi(0)$. From Eqs. (11), (6), and (8) we get,

$$\Delta m^2_{\text{solar}} = \text{sgn}(x) m^- m_1^{(0)} \frac{4\sin\theta_{13}\cos\delta}{\sin 2\theta_{12}}, \tag{12}$$

which relates the solar sector with θ_{13}. The requirement $\Delta m^2_{\text{solar}} > 0$ is ensured by $\text{sgn}(x) m^- \sin\theta_{13}\cos\delta > 0$ from Eq. (11). If the neutrino mass splittings, θ_{12}, and θ_{13} are given, Eq. (12) determines the lightest neutrino mass, m_0.

Inverted ordering is excluded by Eq. (12) as we now show. If $z \equiv m^- m_1^{(0)}/\Delta m^2_{\text{atm}}$ and $m_0/\sqrt{|\Delta m^2_{\text{atm}}|} \equiv \tan\xi$, then

$$\begin{aligned} z &= \sin\xi/(1+\sin\xi) \quad \text{(normal ordering)}, \\ z &= 1/(1+\sin\xi) \quad\quad \text{(inverted ordering)}. \end{aligned} \tag{13}$$

It is seen that $0 \leq z \leq 1/2$ for NO and $1/2 \leq z \leq 1$ for IO and as $z \to 1/2$ one approaches quasidegeneracy, i.e. $m_0 \to$ large, in both cases. From Eq. (12)

$$z = \left(\frac{\Delta m^2_{\text{solar}}}{|\Delta m^2_{\text{atm}}|}\right)\left(\frac{\sin 2\theta_{12}}{4\sin\theta_{13}|\cos\delta|}\right), \tag{14}$$

where $|\cos\delta| = 1$ for real M_R. For the observed ranges of the oscillation parameters $z \sim 10^{-2}$, as a result of which inverted mass ordering is disallowed.

[j] The usual convention of all the mixing angles θ_{ij} belonging to the the first quadrant is followed.

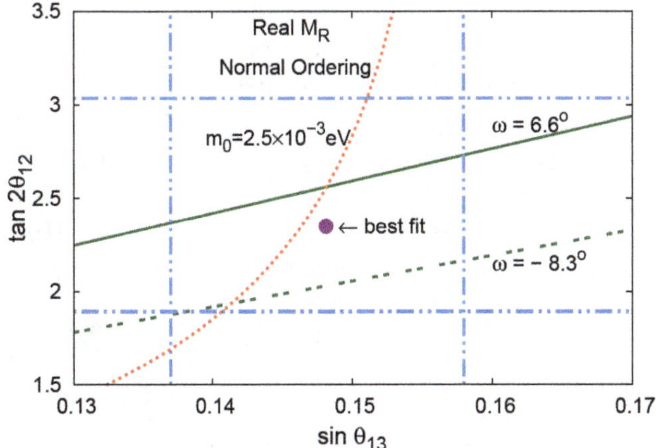

Fig. 1. (Color online) The 3σ range of $\sin\theta_{13}$ and $\tan 2\theta_{12}$ from global fits is represented by the blue dot-dashed box with the best-fit shown as a violet dot. When the best-fit values of the two mass-splittings are used, Eq. (12) gives the red dotted curve for $m_0 = 2.5$ meV. From Eq. (16) for the first (second) octant the portion below the green solid (dashed) straight line is excluded by θ_{23} at 3σ. Equation (12) does not allow inverted ordering for real M_R.

From Eq. (9),

$$\tan\theta_{23} \equiv \tan(\pi/4 - \omega) = \frac{1 - \frac{\kappa}{\sqrt{2}}\frac{x}{y}}{1 + \frac{\kappa}{\sqrt{2}}\frac{x}{y}}\,. \tag{15}$$

Using Eqs. (6) and (11) in the above we get,

$$\tan\omega = \frac{2\sin\theta_{13}\cos\delta}{\tan 2\theta_{12}}\,. \tag{16}$$

The octant of θ_{23} is dictated by the sign of ω which in its turn is determined by δ. θ_{23} lies in the first (second) octant, when ω is positive (negative), i.e. $\delta = 0$ (π). For NO, the only allowed option, this corresponds to $x > 0$ ($x < 0$).

Our results for the real perturbation are shown in Fig. 1. The 3σ global-fit range of $\sin\theta_{13}$ and $\tan 2\theta_{12}$ is marked by the blue dot-dashed box and the best-fit value is indicated by a violet dot in Fig 1. For any point in this region, if the two mass splittings are specified, the z (or equivalently m_0) that produces the correct solar splitting is determined by Eq. (14).

From the 3σ data $\omega_{\rm min} = 0$ for both octants and $\omega_{\rm max} = 6.6°$ ($-8.3°$) for the first (second) octant.[2] In this model in case of real M_R we get $|\omega| \geq 5.14°$ for both octants using Eq. (16), as $|\cos\delta| = 1$. This limits the range in which θ_{23} can be obtained.[k] Equation (16) for $\omega_{\rm max}$ for the first (second) octant is denoted by the green solid (dashed) straight lines below which the model does not hold in each case. Needless to mention that the best-fit point is allowed only if θ_{23} is in the second octant.

[k]This range is excluded at 1σ for the first octant.

One obtains $z_{\max} = 6.03 \times 10^{-2}$ using the 3σ limits of θ_{13} and θ_{12} in Eq. (14), implying $(m_0)_{\max} = 3.10$ meV. The consistency of Eq. (14) with Eq. (16) at $\omega_{\max}$ sets $z_{\min} = 4.01 \times 10^{-2}$ (3.88×10^{-2}) for the first (second) octant corresponding to $(m_0)_{\min} = 2.13$ (2.06) meV. For example, if $m_0 = 2.5$ meV and if the best-fit values of the solar and atmospheric mass splittings are used then Eq. (12) yields the red dotted curve in Fig. 1.

The free parameters here are m_0, m_D^2/xm_R and y for real M_R with which the solar mass splitting, θ_{12}, θ_{13}, θ_{23} are obtained for normal mass ordering. Inverted ordering is not allowed so long as the perturbation is real.

4. Complex M_R

For the more general complex M_R in the mass basis, one gets in place of Eq. (4):

$$M'^{\text{mass}} = \frac{m_D^2}{\sqrt{2}xym_R}\begin{pmatrix} 0 & ye^{i\phi_1} & ye^{i\phi_1} \\ ye^{i\phi_1} & \dfrac{xe^{i\phi_2}}{\sqrt{2}} & \dfrac{-xe^{i\phi_2}}{\sqrt{2}} \\ ye^{i\phi_1} & \dfrac{-xe^{i\phi_2}}{\sqrt{2}} & \dfrac{xe^{i\phi_2}}{\sqrt{2}} \end{pmatrix}. \tag{17}$$

Here x and y are positive. One observes that M' in Eq. (17) is not Hermitian. To proceed, one chooses the Hermitian combination $(M^0 + M')^\dagger(M^0 + M')$ treating $M^{0\dagger}M^0$ as the zeroth order term and $(M^{0\dagger}M' + M'^\dagger M^0)$ as the lowest-order perturbation. The unperturbed eigenvalues now are $(m_i^{(0)})^2$ and the perturbation matrix, which is Hermitian by construction, is

$$(M^{0\dagger}M' + M'^\dagger M^0)^{\text{mass}}$$
$$= \frac{m_D^2}{\sqrt{2}xym_R}\begin{pmatrix} 0 & 2m_1^{(0)}y\cos\phi_1 & yf(\phi_1) \\ 2m_1^{(0)}y\cos\phi_1 & \dfrac{2}{\sqrt{2}}m_1^{(0)}x\cos\phi_2 & -\dfrac{1}{\sqrt{2}}xf(\phi_2) \\ yf^*(\phi_1) & -\dfrac{1}{\sqrt{2}}xf^*(\phi_2) & \dfrac{2}{\sqrt{2}}m_3^{(0)}x\cos\phi_2 \end{pmatrix}, \tag{18}$$

with

$$f(\xi) = m^+\cos\xi - im^-\sin\xi\,. \tag{19}$$

Beyond this point steps similar to those for real M_R are followed.

From Eq. (18), the solar mixing angle now is

$$\tan 2\theta_{12} = 2\sqrt{2}\frac{y}{x}\frac{\cos\phi_1}{\cos\phi_2}\,. \tag{20}$$

Thus, $(\cos\phi_1/\cos\phi_2)$ has to be positive. The limits given in Eq. (7) will now apply on the combination $(y/x)(\cos\phi_1/\cos\phi_2)$.

In the complex M_R case, including first-order corrections, $|\psi_3\rangle$ becomes

$$|\psi_3\rangle = \begin{pmatrix} \kappa f(\phi_1)/m^+ \\ \frac{1}{\sqrt{2}}\left(1 - \frac{\kappa}{\sqrt{2}}\frac{x}{y} f(\phi_2)/m^+\right) \\ \frac{1}{\sqrt{2}}\left(1 + \frac{\kappa}{\sqrt{2}}\frac{x}{y} f(\phi_2)/m^+\right) \end{pmatrix}. \tag{21}$$

Now κ is positive (negative) for NO (IO) always. Equation (21) implies

$$\sin\theta_{13}\cos\delta = \kappa\cos\phi_1\,, \quad \sin\theta_{13}\sin\delta = \kappa\,\frac{m^-}{m^+}\sin\phi_1\,. \tag{22}$$

So, $\cos\delta$ has the same sign (opposite sign) as that of $\cos\phi_1$ for NO (IO). Further, the product $\sin\theta_{13}\sin\delta$ in the CP-violation Jarlskog parameter, J, is dependent on $\sin\phi_1$. The phase ϕ_2 has no affect on δ.

For normal ordering — $\kappa > 0$ — the quadrants of δ and ϕ_1 are the same while for inverted ordering — $\kappa < 0$ — δ has to be in the first (third) quadrant when ϕ_1 happens to be in the second (fourth) quadrant and *vice versa*. So, a near-maximal $\delta = 3\pi/2 - \epsilon$ can be obtained if $\phi_1 \sim 3\pi/2 - \epsilon$ $(3\pi/2 + \epsilon)$ for normal (inverted) ordering.

From Eq. (21),

$$\tan\theta_{23} = \frac{1 - \frac{\kappa}{\sqrt{2}}\frac{x}{y}\cos\phi_2}{1 + \frac{\kappa}{\sqrt{2}}\frac{x}{y}\cos\phi_2}\,. \tag{23}$$

Using Eqs. (20) and (22),

$$\tan\omega = \frac{2\sin\theta_{13}\cos\delta}{\tan 2\theta_{12}}\,. \tag{24}$$

The corresponding result for real M_R — Eq. (16) — is recovered if $\cos\delta = \pm 1$. From Eq. (24), θ_{23} is in the first octant if δ lies in the first or the fourth quadrant — which result in opposite signs of J — otherwise it is in the second octant. This correlation does not depend on the mass ordering. Thus the first (second) octant goes with $\delta = 3\pi/2 + \epsilon(3\pi/2 - \epsilon)$ if δ is near $3\pi/2$.

If m_D and m_R are expressed in terms of $\sin\theta_{13}\cos\delta$, one finds

$$\Delta m^2_{\text{solar}} = \text{sgn}(\cos\phi_2) m^- m_1^{(0)} \frac{4\sin\theta_{13}\cos\delta}{\sin 2\theta_{12}}\,, \tag{25}$$

which is of very similar form as Eq. (12) for real M_R. Obviously, Eqs. (13) and (14) still apply. If one notes the factors which determine the sign of $\cos\delta$ one can conclude that the positivity of $\Delta m^2_{\text{solar}}$ is ensured if $\text{sgn}(\cos\phi_1\cos\phi_2)$ is positive for both mass orderings. Therefore, the sign of the solar mass splitting accommodates both octants of θ_{23} irrespective of the mass ordering. The admissible range of δ can be identified by reexpressing Eq. (14) as:

$$|\cos\delta| = \left(\frac{\Delta m^2_{\text{solar}}}{|\Delta m^2_{\text{atm}}|}\right)\left(\frac{\sin 2\theta_{12}}{4\sin\theta_{13} z}\right). \tag{26}$$

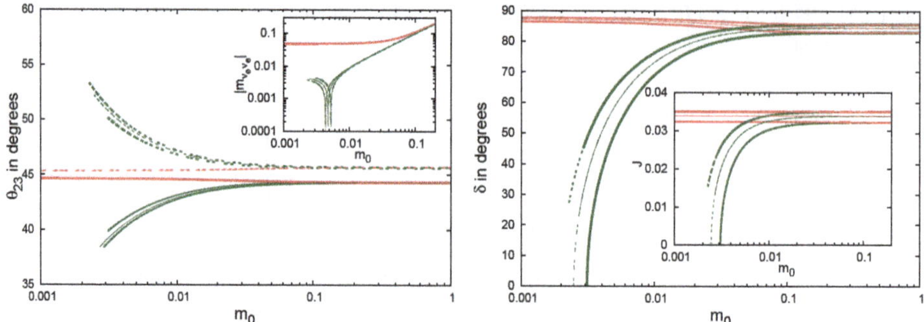

Fig. 2. (Color online) θ_{23}, $|m_{\nu_e\nu_e}|$ (in eV), δ, and J as a function of the lightest neutrino mass m_0 (in eV). The green (pink) curves are for NO (IO). The 3σ allowed region is between the thick curves of each type while the thin curves are for the best-fit input values. The solid (dashed) curves are for the first (second) octant of θ_{23}. Left: The variation of θ_{23}. The inset shows $|m_{\nu_e\nu_e}|$ (in eV), the effective mass controlling $0\nu2\beta$ processes. Right: CP-phase δ. The inset exhibits the CP-violation measure J.

In the analysis which we report m_0, θ_{13}, and θ_{12} are the inputs. We get δ and θ_{23} from Eqs. (26) and (24). The CP-violation measure, J, and the combination $|m_{\nu_e\nu_e}|$ which contributes to $0\nu2\beta$ are then easily obtained.

The results for complex M_R are presented in Fig. 2. The left panel (thick curves) shows the variation of θ_{23} with m_0 when the neutrino mass square splittings and the angles θ_{13} and θ_{12} cover their 3σ ranges. The thin curves are for the best-fit values. In this figure the green (pink) curves are always for NO (IO) while solid (dashed) curves are for θ_{23} in the first (second) octant. For IO the thick and thin curves are too close for distinction in this panel. Note that $\theta_{23} = \pi/4$ is not consistent with the 3σ predictions from this model. It is seen that θ_{23} is symmetrically distributed about $\pi/4$, which is expected from Eq. (24). For IO the obtained range is outside 1σ but are admissible at 3σ. When θ_{23} is better measured one of the mass orderings will be eliminated unless the neutrinos are in the quasidegenerate regime.

The 3σ limits of θ_{23} in the two octants determine the minimum permitted value of m_0 for NO. For IO Eq. (25) allows m_0 to be arbitrarily small (see below). In the inset of this panel $|m_{\nu_e\nu_e}|$ has been plotted. Direct neutrino mass measurements[22] are expected to be sensitive to masses up to 200 meV. Planned $0\nu2\beta$ experiments will access m_0 in the quasidegenerate regime.[23] Figure 2 indicates that to distinguish the alternate mass ordering possibilities an order of magnitude improvement in their sensitivity will be needed. Large atmospheric neutrino detectors such as INO or long-baseline experiments are alternate avenues for determining the mass ordering.

In the right panel of Fig. 2 the variation of δ with m_0 for both mass orderings is shown. The dependence of J appears in the inset. The two panels of Fig. 2 use the same conventions. Since the three mixing angles are kept in the first quadrant, J is positive if $0 \leq \delta \leq \pi$ and is negative otherwise. As mentioned before, the quadrant of δ can be altered by choosing the quadrant of ϕ_1 suitably.[l]

[l]From Eq. (22), $\delta \to \pi + \delta$ if $\phi_1 \to \pi + \phi_1$.

However, for a particular mass ordering from Eq. (26) the dependence of $|\cos\delta|$ on m_0 is the same for the different alternatives, which are $\pm\delta$ and $(\pi \pm \delta)$. Keeping this in mind, in Fig. 2 (right panel) δ has been plotted in the first quadrant and J has been taken as positive.

As θ_{23} is symmetric around $\pi/4$ in this model and J is proportional to $\sin 2\theta_{23}$ so it is independent of the octant. For inverted mass ordering both δ and J remain nearly unaffected by variations of m_0.

If m_0 is smaller than 10 meV, then the CP-phase δ is much larger for inverted ordering. Once the mass ordering is known and CP-violation in the neutrino sector is measured this could provide a clear test of this model. Consistent with Sec. 3, the limit of real M_R is admissible only for NO, and that too for only a portion of the 3σ range.

As discussed, one has $0 \leq z \leq 1/2$ for NO and $1/2 \leq z \leq 1$ for IO. It is seen from Eq. (26) that as a consequence of this the allowed δ are complementary for the two mass orderings tending towards a common value in the quasidegenerate limit, which sets in from around $m_0 = 100$ meV. Unlike the real M_R case, in Eq. (14) by taking $\cos\delta$ small one can make $z \equiv m^- m_1^{(0)}/\Delta m^2_{\rm atm} \sim 1$ so that solutions exist for m_0 for IO corresponding to even m_0 arbitrarily small unlike for NO where the lower limit of m_0 is set by $\cos\delta = 1$, i.e. real M_R.

5. Conclusions

Summarizing, a neutrino mass model is presented in which the observed solar mass splitting, θ_{12}, θ_{13}, and $\omega = \pi/4 - \theta_{23}$ all originated from a single perturbation (derived out of a Type-I see-saw mechanism) and are thereby related to each other. The atmospheric mass splitting preferred by the data and maximal mixing in this sector play the role of the unperturbed framework. In order to restrict free parameters to a minimum the Dirac term in the see-saw is taken as proportional to the identity matrix and the right-handed neutrino mass matrix, M_R, has a four-zero texture in the flavor basis. Requiring that the mixing angles and solar mass splitting identified by the global fits be reproduced, for a real M_R a narrow range of the lightest neutrino mass ($m_0 \sim$ a few meV) is permitted for normal ordering. It leaves the option open for θ_{23} to belong to the first or the second octant. Such a CP conserving real perturbation forbids inverted ordering. The more general complex M_R enables a considerable enlargement of the range of m_0 and determines in its terms the CP-phase δ and the octant of θ_{23} as well, while accommodating both mass orderings. In the quasidegenerate limit and in case of inverted ordering $\delta \sim 3\pi/2$ is a natural prediction. Future improved measurements of δ, θ_{23}, $0\nu2\beta$ and determination of the neutrino mass ordering will test the model from various angles.

Acknowledgments

A. Raychaudhuri thanks the organizers for arranging a very stimulating meeting on neutrino physics. S. Pramanick acknowledges a Senior Research Fellowship from CSIR, India. A. Raychaudhuri is partially funded by the Department of Science and Technology Grant No. SR/S2/JCB-14/2009.

References

1. Particle Data Group (K. A. Olive *et al.*), *Chin. Phys. C* **38**, 090001 (2014).
2. M. C. Gonzalez-Garcia, M. Maltoni, J. Salvado and T. Schwetz, *J. High Energy Phys.* **1212**, 123 (2012), arXiv:1209.3023v3 [hep-ph], NuFIT 1.3 (2014).
3. D. V. Forero, M. Tortola and J. W. F. Valle, *Phys. Rev. D* **86**, 073012 (2012), arXiv:1205.4018 [hep-ph].
4. P. Minkowski, *Phys. Lett. B* **67**, 421 (1977).
5. M. Gell-Mann, P. Ramond and R. Slansky, *Supergravity*, eds. F. van Nieuwenhuizen and D. Freedman (North Holland, Amsterdam, 1979), p. 315.
6. T. Yanagida, *Proc. Workshop on Unified Theory and the Baryon Number of the Universe* (KEK, Japan, 1979).
7. S. L. Glashow, *NATO Sci. Ser. B* **59**, 687 (1980).
8. R. N. Mohapatra and G. Senjanović, *Phys. Rev. D* **23**, 165 (1981).
9. S. Pramanick and A. Raychaudhuri, arXiv:1411.0320 [hep-ph].
10. F. Vissani, *J. High Energy Phys.* **9811**, 025 (1998), arXiv:hep-ph/9810435.
11. E. K. Akhmedov, *Phys. Lett. B* **467**, 95 (1999), arXiv:hep-ph/9909217.
12. M. Lindner and W. Rodejohann, *J. High Energy Phys.* **0705**, 089 (2007), arXiv:hep-ph/0703171.
13. B. Brahmachari and A. Raychaudhuri, *Phys. Rev. D* **86**, 051302 (2012), arXiv:1204.5619 [hep-ph].
14. B. Adhikary, A. Ghosal and P. Roy, *Int. J. Mod. Phys. A* **28**, 1350118 (2013), arXiv:1210.5328 [hep-ph].
15. D. Aristizabal Sierra, I. de Medeiros Varzielas and E. Houet, *Phys. Rev. D* **87**, 093009 (2013), arXiv:1302.6499 [hep-ph].
16. R. Dutta, U. Ch, A. K. Giri and N. Sahu, *Int. J. Mod. Phys. A* **29**, 1450113 (2014), arXiv:1303.3357 [hep-ph].
17. L. J. Hall and G. G. Ross, *J. High Energy Phys.* **1311**, 091 (2013), arXiv:1303.6962 [hep-ph].
18. T. Araki, *PTEP* **2013**, 103B02 (2013), arXiv:1305.0248 [hep-ph].
19. M.-C. Chen, J. Huang, K. T. Mahanthappa and A. M. Wijangco, *J. High Energy Phys.* **1310**, 112 (2013), arXiv:1307.7711 [hep-ph].
20. S. Pramanick and A. Raychaudhuri, *Phys. Rev. D* **88**, 093009 (2013), arXiv:1308.1445 [hep-ph].
21. B. Brahmachari and P. Roy, *J. High Energy Phys.* **15**, 135 (2015).
22. KATRIN Collab. (M. Haag), *PoS* (EPS-HEP2013) 518 (2013).
23. W. Rodejohann, *Int. J. Mod. Phys. E* **20**, 1833 (2011), arXiv:1106.1334 [hep-ph].

Predictions for the Dirac CP Violation Phase in the Neutrino Mixing Matrix*

S. T. Petcov,[†,‡,§] I. Girardi[†] and A. V. Titov[†]

[†] *SISSA/INFN, Trieste, Italy*

[‡] *Kavli IPMU (WPI), University of Tokyo, Tokyo, Japan*

[§] *petcov@sissa.it*

Using the fact that the neutrino mixing matrix $U = U_e^\dagger U_\nu$, where U_e and U_ν result from the diagonalization of the charged lepton and neutrino mass matrices, we analyze the predictions based on the sum rules which the Dirac phase δ present in U satisfies when U_ν has a form dictated by, or associated with, discrete flavor symmetries and U_e has a "minimal" form (in terms of angles and phases it contains) that can provide the requisite corrections to U_ν, so that the reactor, atmospheric and solar neutrino mixing angles θ_{13}, θ_{23} and θ_{12} have values compatible with the current data.

Keywords: Neutrino mixing; Dirac leptonic CP violation; flavor symmetries; sum rules.

1. Introduction

One of the major goals of the future experimental studies in neutrino physics is the searches for CP violation (CPV) effects in neutrino oscillations (see, e.g. Refs. 1–4). It is part of a more general and ambitious program of research aiming to determine the status of the CP symmetry in the lepton sector.

In the case of the reference 3-neutrino mixing scheme, CPV effects in the flavor neutrino oscillations, i.e. a difference between the probabilities of $\nu_l \to \nu_{l'}$ and $\bar{\nu}_l \to \bar{\nu}_{l'}$ oscillations in vacuum,[5,6] $P(\nu_l \to \nu_{l'})$ and $P(\bar{\nu}_l \to \bar{\nu}_{l'})$, $l \neq l' = e, \mu, \tau$, can be caused, as is well known, by the Dirac phase present in the Pontecorvo, Maki, Nakagawa and Sakata (PMNS) neutrino mixing matrix U. If the neutrinos with definite masses ν_i, $i = 1, 2, 3$, are Majorana particles, the 3-neutrino mixing matrix contains two additional Majorana CPV phases.[6] However, the flavor neutrino oscillation probabilities $P(\nu_l \to \nu_{l'})$ and $P(\bar{\nu}_l \to \bar{\nu}_{l'})$, l, $l' = e, \mu, \tau$, do not depend on the Majorana phases.[6,7] Our interest in the CPV phases present in the neutrino mixing matrix is stimulated also by the intriguing possibility that the Dirac phase and/or the Majorana phases in U can provide the CP violation necessary for

*Presented by S. T. Petcov.

the generation of the observed baryon asymmetry of the Universe[8,9] (see also, e.g. Refs. 10 and 11).

In the framework of the reference 3-flavor neutrino mixing we will consider, the PMNS neutrino mixing matrix is always given by $U = U_e^\dagger U_\nu$, where U_e and U_ν are 3×3 unitary matrices originating from the diagonalization of the charged lepton and the neutrino (Majorana) mass terms. We will suppose in what follows that U_ν has a form which is dictated by, or associated with, symmetries (see, e.g. Refs. 12 and 13). In the present article, we consider the following symmetry forms of U_ν: (i) tri-bimaximal (TBM),[14–16] (ii) bimaximal (BM) (or corresponding to the conservation of the lepton charge $L' = L_e - L_\mu - L_\tau$ (LC)),[17–20] (iii) golden ratio type A (GRA),[21,22] (iv) golden ratio type B (GRB),[23] and (v) hexagonal (HG).[24] For all these symmetry forms U_ν can be written as

$$U_\nu = \Psi_1 \tilde{U}_\nu Q_0 = \Psi_1 R_{23}(\theta^\nu_{23}) R_{12}(\theta^\nu_{12}) Q_0\,, \tag{1}$$

where $R_{23}(\theta^\nu_{23})$ and $R_{12}(\theta^\nu_{12})$ are orthogonal matrices describing rotations in the 2–3 and 1–2 planes, respectively, and Ψ_1 and Q_0 are diagonal phase matrices each containing two phases. The phases in the matrix Q_0 give contribution to the Majorana phases in the PMNS matrix. The symmetry forms of $\tilde{U}_\nu$ of interest, TBM, BM (LC), GRA, GRB and HG, are characterized by the same values of the angles $\theta^\nu_{13} = 0$ and $\theta^\nu_{23} = -\pi/4$, but correspond to different fixed values of the angle θ^ν_{12} and thus of $\sin^2\theta^\nu_{12}$, namely, to (i) $\sin^2\theta^\nu_{12} = 1/3$ (TBM), (ii) $\sin^2\theta^\nu_{12} = 1/2$ (BM (LC)), (iii) $\sin^2\theta^\nu_{12} = (2+r)^{-1} \cong 0.276$ (GRA), r being the golden ratio, $r = (1+\sqrt{5})/2$, (iv) $\sin^2\theta^\nu_{12} = (3-r)/4 \cong 0.345$ (GRB), and (v) $\sin^2\theta^\nu_{12} = 1/4$ (HG). The best fit values (b.f.v.) and 1σ errors of the three corresponding neutrino mixing parameters in the standard parametrization of the PMNS matrix,[1] which we will employ, read:[25]

$$\sin^2\theta_{12} = 0.308^{+0.017}_{-0.017}\,, \tag{2}$$

$$\sin^2\theta_{13} = 0.0234^{+0.0020}_{-0.0019}\,, \tag{3}$$

$$\sin^2\theta_{23} = 0.437^{+0.033}_{-0.023}\,, \tag{4}$$

where the quoted values correspond to neutrino mass spectrum with normal ordering (NO); the values for spectrum with inverted ordering (IO) found in Ref. 25 differ insignificantly. The minimal form of U_e of interest that can provide the requisite corrections to U_ν, so that the neutrino mixing angles θ_{13}, θ_{23} and θ_{12} have values compatible with the current data, including a possible sizeable deviation of θ_{23} from $\pi/4$, includes a product of two orthogonal matrices describing rotations in the 2–3 and 1–2 planes,[26] $R_{23}(\theta^e_{23})$ and $R_{12}(\theta^e_{12})$, θ^e_{23} and θ^e_{12} being two (real) angles.[a] This leads to the following parametrization of the PMNS matrix U:

$$U = R_{12}(\theta^e_{12}) R_{23}(\theta^e_{23}) \Psi R_{23}(\theta^\nu_{23}) R_{12}(\theta^\nu_{12}) Q_0\,, \tag{5}$$

[a] For a detailed discussion of alternative possibilities see Ref. 27.

where $\Psi = \mathrm{diag}(1, e^{-i\psi}, e^{-i\omega})$, and $\theta^\nu_{23} = -\pi/4$. Equation (5) can be recast in the form:[26]

$$U = R_{12}(\theta^e_{12})\Phi(\phi)R_{23}(\hat{\theta}_{23})R_{12}(\theta^\nu_{12})\hat{Q}\,, \tag{6}$$

where we have defined $\Phi = \mathrm{diag}(1, e^{i\phi}, 1)$, ϕ being a CPV phase, $\hat{\theta}_{23}$ is a function of θ^e_{23}, $\sin^2\hat{\theta}_{23} = 1/2 - \sin\theta^e_{23}\cos\theta^e_{23}\cos(\omega - \psi)$, and $\hat{Q}$ is a diagonal phase matrix. The phases in $\hat{Q}$ give contributions to the Majorana phases in the PMNS matrix. The angle $\hat{\theta}_{23}$, however, can be expressed in terms of the angles θ_{23} and θ_{13} of the PMNS matrix:

$$\sin^2\theta_{23} = \frac{|U_{\mu 3}|^2}{1 - |U_{e3}|^2} = \frac{\sin^2\hat{\theta}_{23} - \sin^2\theta_{13}}{1 - \sin^2\theta_{13}}\,, \tag{7}$$

and the value of $\hat{\theta}_{23}$ is fixed by the values of θ_{23} and θ_{13}.

2. Predicting the Dirac Phase in the PMNS Matrix

In the scheme under discussion, the four observables θ_{12}, θ_{23}, θ_{13} and the Dirac phase δ in the PMNS matrix are functions of three parameters θ^e_{12}, $\hat{\theta}_{23}$ and ϕ. As a consequence, the Dirac phase δ can be expressed as a function of the three PMNS angles θ_{12}, θ_{23} and θ_{13}, leading to a new "sum rule" relating δ and θ_{12}, θ_{23} and θ_{13}. Within the approach employed this sum rule is exact. Its explicit form depends on the symmetry form of the matrix $\tilde{U}_\nu$, i.e. on the value of the angle θ^ν_{12}. For arbitrary fixed value of θ^ν_{12} the sum rule of interest reads:[28]

$$\cos\delta = \frac{\tan\theta_{23}}{\sin 2\theta_{12}\sin\theta_{13}}\left[\cos 2\theta^\nu_{12} + \left(\sin^2\theta_{12} - \cos^2\theta^\nu_{12}\right)\left(1 - \cot^2\theta_{23}\sin^2\theta_{13}\right)\right]. \tag{8}$$

A similar sum rule can be derived for the phase ϕ.[28]

In Refs. 28 and 29 we have derived predictions for $\cos\delta$, δ and the rephasing invariant J_{CP}, which controls the magnitude of the CPV effects in neutrino oscillations,[30] using the sum rule in Eq. (8) and the measured values of the lepton mixing angles θ_{12}, θ_{13} and θ_{23}. In the present article, we first summarize the predictions for these observables obtained in Refs. 28 and 29 in a simplified analysis employing the best fit values (b.f.v.) and the 3σ allowed ranges of the three relevant neutrino mixing parameters $\sin^2\theta_{12}$, $\sin^2\theta_{13}$ and $\sin^2\theta_{23}$. This is followed by a summary of the results of the statistical analysis of the predictions performed in Ref. 29, which is based on (i) the current, and most importantly, (ii) the prospective, uncertainties in the measured values of $\sin^2\theta_{12}$, $\sin^2\theta_{13}$ and $\sin^2\theta_{23}$.

We note first that the predicted values of $\cos\delta$ vary significantly with the symmetry form of $\tilde{U}_\nu$.[28] For the best fit values of $\sin^2\theta_{12} = 0.308$, $\sin^2\theta_{13} = 0.0234$ and $\sin^2\theta_{23} = 0.437$ found in Ref. 25, for instance, we get $\cos\delta = (-0.0906), (-1.16), 0.275, (-0.169)$ and 0.445 for the TBM, BM (LC), GRA, GRB and HG forms, respectively. For the TBM, GRA, GRB and HG forms these values correspond to $\delta = \pm 95.2°, \pm 74.0°, \pm 99.7°, \pm 63.6°$, respectively. The unphysical value of $\cos\delta$ in the BM (LC) case is a reflection of the fact that the scheme under discussion with BM (LC) form of the matrix $\tilde{U}_\nu$ does not provide a good description of the current data on θ_{12}, θ_{23} and θ_{13}.[26] Physical values of $\cos\delta$ can be obtained, for instance, for the b.f.v. of $\sin^2\theta_{13}$ and $\sin^2\theta_{23}$ if $\sin^2\theta_{12}$ has a larger value:[29] for, e.g. $\sin^2\theta_{12} = 0.34$ allowed at 2σ by the current data, we have $\cos\delta = -0.943$, corresponding to $\delta = \pm 160.6°$. Similarly, for $\sin^2\theta_{12} = 0.32$, $\sin^2\theta_{23} = 0.41$ and $\sin\theta_{13} = 0.158$ we have:[28] $\cos\delta = -0.978$, $\delta = \pm 168.1°$.

The results quoted above imply[28] that the measurement of $\cos\delta$ can allow to distinguish between the different symmetry forms of $\tilde{U}_\nu$, provided θ_{12}, θ_{13} and θ_{23} are known with a sufficiently good precision. Even determining the sign of $\cos\delta$ will be sufficient to eliminate some of the possible symmetry forms of $\tilde{U}_\nu$.

It was also found in Ref. 29 that the sum rule predictions for $\cos\delta$ exhibit strong dependence on the value of $\sin^2\theta_{12}$ when the latter is varied in its 3σ experimentally allowed range[25] (0.259–0.359). The predictions for $\cos\delta$ change significantly not only in magnitude, but also the sign of $\cos\delta$ changes in the TBM, GRA and GRB cases.[29] In the case of $\theta^e_{23} = 0$, for instance, we get for the TBM form of $\tilde{U}_\nu$ for the three values of $\sin^2\theta_{12} = 0.308$, 0.259 and 0.359: $\cos\delta = (-0.114), (-0.469)$ and 0.221, thus $\cos\delta = 0$ is allowed for a certain value of $\sin^2\theta_{12}$. For the GRA and GRB forms of $\tilde{U}_\nu$ we have, respectively, $\cos\delta = 0.289, (-0.044), 0.609$, and $\cos\delta = (-0.200), (-0.559), 0.138$. Similarly, for the HG form we find for the three values of $\sin^2\theta_{12}$: $\cos\delta = 0.476, 0.153, 0.789$.

In what concerns the dependence of the sum rule predictions for $\cos\delta$ when $\sin^2\theta_{23}$ is varied in its 3σ allowed interval, $0.374 \leq \sin^2\theta_{23} \leq 0.626$, the results we obtained for $\sin^2\theta_{23} = 0.374$ and $\sin^2\theta_{23} = 0.437$, setting $\sin^2\theta_{12}$ and $\sin^2\theta_{13}$ to their best fit values, do not differ significantly. However, the differences between the predictions for $\cos\delta$ obtained for $\sin^2\theta_{23} = 0.437$ and for $\sin^2\theta_{23} = 0.626$ are rather large[29] — they differ by the factors of 2.05, 1.25, 1.77 and 1.32 in the TBM, GRA, GRB and HG cases, respectively.

Similar analysis can be performed for the predictions for the cosine of the phase[29] ϕ which in many theoretical models serves as a "source" for the Dirac phase δ. The phase ϕ is related to, but does not coincide with, the Dirac phase δ. This leads to the confusing identification of ϕ with δ: the sum rules satisfied by $\cos\phi$ and $\cos\delta$ differ significantly.[28] Correspondingly, the predicted values of $\cos\phi$ and $\cos\delta$ in the cases of the TBM, GRA, GRB and HG symmetry forms of $\tilde{U}_\nu$ considered by us also differ significantly. This conclusion is not valid for the BM (LC) form: for this form the sum rules predictions for $\cos\phi$ and $\cos\delta$ are rather similar.[28]

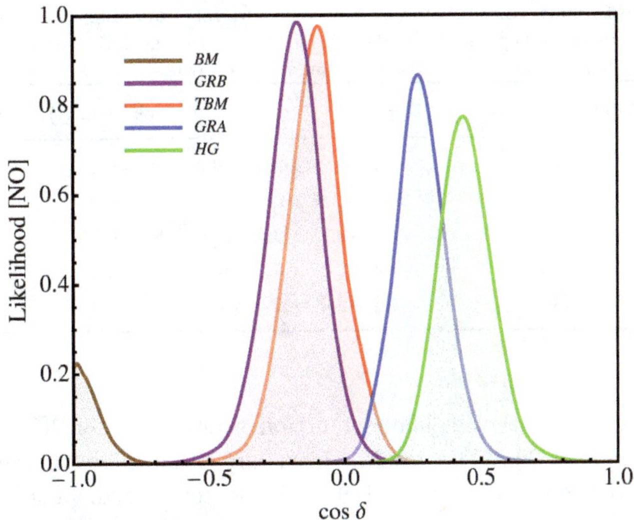

Fig. 1. The likelihood function versus $\cos\delta$ for NO neutrino mass spectrum after marginalizing over $\sin^2\theta_{13}$ and $\sin^2\theta_{23}$, for the TBM, BM (LC), GRA, GRB and HG symmetry forms of the mixing matrix $\tilde{U}_\nu$ (see text for further details). (From Ref. 29.)

We next present results of the statistical analysis of the predictions for δ, $\cos\delta$ and the rephasing invariant $J_{\rm CP}$ performed in Ref. 29 in the cases of the TBM, BM (LC), GRA, GRB and HG symmetry forms of the matrix $\tilde{U}_\nu$. In this analysis the latest results on $\sin^2\theta_{12}$, $\sin^2\theta_{13}$, $\sin^2\theta_{23}$ and δ, obtained in the global analysis of the neutrino oscillation data performed in Ref. 25 were used as input. The aim was to derive the allowed ranges for $\cos\delta$ and $J_{\rm CP}$, predicted on the basis of the current data on the neutrino mixing parameters for each of the symmetry forms of $\tilde{U}_\nu$ considered. For this purpose the χ^2 function was constructed in the following way:[29] $\chi^2(\{x_i\}) = \sum_i \chi_i^2(x_i)$, with $x_i = \{\sin^2\theta_{12}, \sin^2\theta_{13}, \sin^2\theta_{23}, \delta\}$. The functions χ_i^2 have been extracted from the one-dimensional projections given in Ref. 25 and, thus, the correlations between the oscillation parameters have been neglected. This approximation is sufficiently precise since it allows to reproduce the contours in the planes $(\sin^2\theta_{23}, \delta)$, $(\sin^2\theta_{13}, \delta)$ and $(\sin^2\theta_{23}, \sin^2\theta_{13})$, given in Ref. 25, with a rather high accuracy. We calculated $\chi^2(\cos\delta)$ by marginalizing $\chi^2(\{x_i\})$ over $\sin^2\theta_{13}$ and $\sin^2\theta_{23}$ for a fixed value of $\cos\delta$. Given the global fit results, the likelihood function,

$$L(\cos\delta) \propto \exp\left(-\frac{\chi^2(\cos\delta)}{2}\right), \tag{9}$$

represents the most probable value of $\cos\delta$ for each of the considered symmetry forms of $\tilde{U}_\nu$. The $n\sigma$ confidence level (C.L.) region corresponds to the interval of values of $\cos\delta$ in which $L(\cos\delta) \geq L(\chi^2 = \chi^2_{\rm min}) \cdot L(\chi^2 = n^2)$, where $\chi^2_{\rm min}$ is the value of χ^2 in the minimum.

Table 1. Best fit values (b.f.v.) of $J_{\rm CP}$ and $\cos\delta$ and corresponding 3σ ranges (found fixing $\chi^2 - \chi^2_{\rm min} = 9$) in our setup using the data from Ref. 25 for the NO neutrino mass spectrum. (From Ref. 29, where results for the IO spectrum are also given.)

Scheme	$J_{\rm CP}/10^{-2}$ (b.f.v.)	$J_{\rm CP}/10^{-2}$ (3σ range)	$\cos\delta$ (b.f.v.)	$\cos\delta$ (3σ range)
TBM	−3.4	$[-3.8, -2.8] \cup [3.1, 3.6]$	−0.07	$[-0.47,\ 0.21]$
BM (LC)	−0.5	$[-2.6, 2.1]$	−0.99	$[-1.00, -0.72]$
GRA	−3.3	$[-3.7, -2.7] \cup [3.0, 3.5]$	0.25	$[-0.08,\ 0.69]$
GRB	−3.4	$[-3.9, -2.6] \cup [3.1, 3.6]$	−0.15	$[-0.57,\ 0.13]$
HG	−3.1	$[-3.5, -2.0] \cup [2.6, 3.4]$	0.47	$[\ 0.16,\ 0.80]$

In Fig. 1 we show the likelihood function versus $\cos\delta$ for NO neutrino mass spectrum from Ref. 29. The results shown are obtained by marginalizing over all the other relevant parameters of the scheme considered. The dependence of the likelihood function on $\cos\delta$ in the case of IO neutrino mass spectrum differs little from that shown in Fig. 1. As can be observed in Fig. 1, a rather precise measurement of $\cos\delta$ would allow to distinguish between the different symmetry forms of $\tilde{U}_\nu$ considered by us. For the TBM and GRB forms there is a significant overlap of the corresponding likelihood functions. The same observation is valid for the GRA and HG forms. However, the overlap of the likelihood functions of these two groups of symmetry forms occurs only at 3σ level in a very small interval of values of $\cos\delta$. This implies that in order to distinguish between TBM/GRB, GRA/HG and BM (LC) symmetry forms, a not very demanding measurement (in terms of accuracy) of $\cos\delta$ might be sufficient. The value of the non-normalized likelihood function at the maximum in Fig. 1 is equal to $\exp(-\chi^2_{\rm min}/2)$, which allows us to make conclusions about the compatibility of the symmetry schemes considered with the current global data. The results of this analysis for $\cos\delta$ are summarized in Table 1.

We have also performed in Ref. 29 a similar statistical analysis of the predictions for the rephasing invariant $J_{\rm CP}$ in the cases of the TBM, BM (LC), GRA, GRB and HG symmetry forms of the matrix $\tilde{U}_\nu$ considered. In this analysis we used as input the latest results on $\sin^2\theta_{12}$, $\sin^2\theta_{13}$, $\sin^2\theta_{23}$ and δ, obtained in the global analysis of the neutrino oscillation data performed in Ref. 25, and minimized χ^2 for a fixed value of $J_{\rm CP}$. The obtained b.f.v. and 3σ ranges are given in Table 1. We have found, in particular, that the CP-conserving value of $J_{\rm CP} = 0$ is excluded in the cases of the TBM, GRA, GRB and HG neutrino mixing symmetry forms, respectively, at approximately 5σ, 4σ, 4σ and 3σ C.L. with respect to the C.L. of the corresponding best fit value. These results reflect the predictions we have obtained for δ, more specifically, the C.L. at which the CP-conserving values of $\delta = 0$ (2π), π, are excluded in the discussed cases. We found also that the 3σ allowed intervals of values of δ and $J_{\rm CP}$ are rather narrow for all the symmetry forms considered, except for the BM (LC) form. More specifically, for the TBM, GRA, GRB and

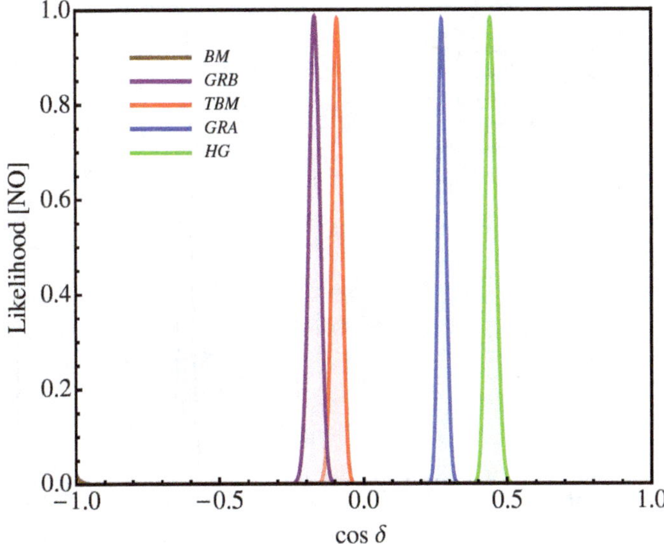

Fig. 2. The same as in Fig. 1, but using the prospective 1σ uncertainties in the determination of $\sin^2\theta_{12}$, $\sin^2\theta_{13}$ and $\sin^2\theta_{23}$ within the Gaussian approximation. The three neutrino mixing parameters are fixed to their current best fit values (i.e. $\sin^2\theta_{12} = 0.308$, etc.). See text for further details. (From Ref. 29.)

HG symmetry forms we have obtained at 3σ: $0.020 \leq |J_{\rm CP}| \leq 0.039$. For the b.f.v. of $J_{\rm CP}$ we have found, respectively: $J_{\rm CP} = (-0.034)$, (-0.033), (-0.034), and (-0.031). Our results indicate that distinguishing between the TBM, GRA, GRB and HG symmetry forms of the neutrino mixing would require extremely high precision measurement of the $J_{\rm CP}$ factor.[28]

In Fig. 2 we present the likelihood function versus $\cos\delta$ within the Gaussian approximation, i.e. using $\chi^2_{\rm G} = \sum_i (y_i - \bar{y}_i)^2/\sigma^2_{y_i}$, with $y_i = \{\sin^2\theta_{12}, \sin^2\theta_{13}, \sin^2\theta_{23}\}$, where we used the current b.f.v. ($\bar{y}_i$) of the mixing angles for NO neutrino mass spectrum given in Ref. 25 and the prospective 1σ uncertainties (σ_{y_i}) in the determination of $\sin^2\theta_{12}$ (0.7% from JUNO[31]), $\sin^2\theta_{13}$ (3% derived from an expected error on $\sin^2 2\theta_{13}$ of 3% from Daya Bay, see Refs. 4 and 32) and $\sin^2\theta_{23}$ (5% derived from the potential sensitivity of NOvA and T2K on $\sin^2 2\theta_{23}$ of 2%, see Ref. 4, this sensitivity can be also achieved in future neutrino facilities as T2HK[33]). The BM (LC) case is very sensitive to the b.f.v. of $\sin^2\theta_{12}$ and $\sin^2\theta_{23}$ and is disfavored at more than 2σ for the current b.f.v. found in Ref. 25. This case might turn out to be compatible with the data for larger (smaller) measured values of $\sin^2\theta_{12}$ ($\sin^2\theta_{23}$), as can be seen from Fig. 3, which was obtained for $\sin^2\theta_{12} = 0.332$ (the best fit values of the two other mixing angles being kept intact). With the increase of the value of $\sin^2\theta_{23}$ the BM (LC) form becomes increasingly disfavored, while the TBM/GRB (GRA/HG) predictions for $\cos\delta$ are shifted somewhat to the left (right) with respect to those shown in Fig. 2. For, e.g. the best fit values of

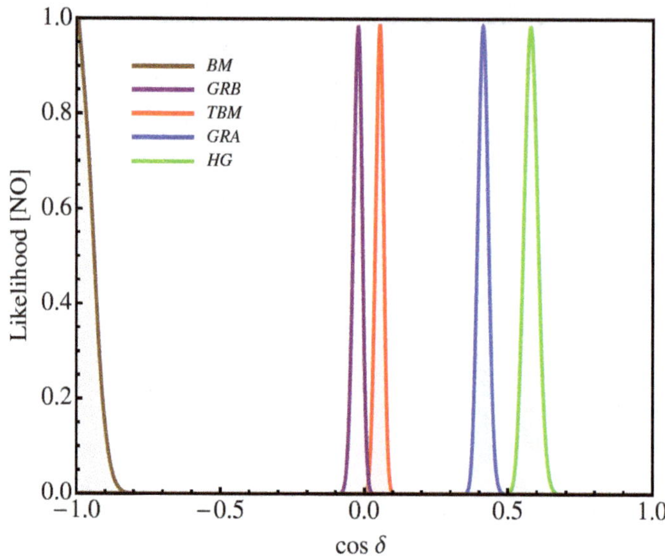

Fig. 3. The same as in Fig. 2, but using $\sin^2\theta_{12} = 0.332$. (From Ref. 29.)

$\sin^2\theta_{12} = 0.304$, $\sin^2\theta_{13} = 0.0219$ and $\sin^2\theta_{23} = 0.579$, found in Ref. 34 for the IO neutrino mass spectrum, these shifts in $\cos\delta$ are approximately by 0.1.

The measurement of $\sin^2\theta_{12}$, $\sin^2\theta_{13}$ and $\sin^2\theta_{23}$ with the quoted precision will open up the possibility to distinguish between the BM (LC), TBM/GRB, GRA and HG forms of $\tilde{U}_\nu$. Distinguishing between the TBM and GRB forms would require relatively high precision measurement of $\cos\delta$. Assuming that $|\cos\delta| < 0.93$, which means for 76% of values of δ, the error on δ, $\Delta\delta$, for an error on $\cos\delta$, $\Delta(\cos\delta) = 0.10$ (0.08), does not exceed $\Delta\delta \lesssim \Delta(\cos\delta)/\sqrt{1-0.93^2} = 16^\circ$ (12°). This accuracy is planned to be reached in the future neutrino experiments like T2HK (ESSνSB).[4] Therefore, a measurement of $\cos\delta$ in the quoted range will allow one to distinguish between the TBM/GRB, BM (LC) and GRA/HG forms at approximately 3σ C.L., if the precision achieved on $\sin^2\theta_{12}$, $\sin^2\theta_{13}$ and $\sin^2\theta_{23}$ is the same as in Figs. 2 and 3.

3. Summary and Conclusions

In conclusions, we have derived in Ref. 29 the ranges of the predicted values of $\cos\delta$ and $J_{\rm CP}$ for the TBM, BM (LC), GRA, GRB and HG symmetry forms of $\tilde{U}_\nu$, from a statistical analysis using the sum rule in Eq. (8) obtained in Ref. 28 and the current global neutrino oscillation data.[25] The results of this analysis are summarized in Table 1 and in Fig. 1. We found, in particular, that in the TBM, GRA, GRB and HG cases, the best fit values of $J_{\rm CP}$ lie in the narrow interval $(-0.034) \leq J_{\rm CP} \leq (-0.031)$, while at 3σ we have $0.020 \leq |J_{\rm CP}| \leq 0.039$. The predictions for δ, $\cos\delta$ and $J_{\rm CP}$ in the case of the BM (LC) symmetry form of $\tilde{U}_\nu$, as the results of the statistical analysis performed by us showed, differ significantly: the best fit value of $\delta \cong \pi$, and, correspondingly, of $J_{\rm CP} \cong 0$. For the 3σ range in the case of NO (IO) neutrino mass spectrum we find: -0.026 $(-0.025) \leq J_{\rm CP} \leq 0.021$ (0.023), i.e. it includes a subinterval of values centred on zero, which does not overlap with the 3σ allowed intervals of values of $J_{\rm CP}$, corresponding to the TBM, GRA, GRB and HG symmetry forms of $\tilde{U}_{\nu}$.

Finally, we have derived in Ref. 29 predictions for $\cos\delta$ using the prospective 1σ uncertainties in the determination of $\sin^2\theta_{12}$, $\sin^2\theta_{13}$ and $\sin^2\theta_{23}$ respectively in JUNO, Daya Bay and accelerator and atmospheric neutrino experiments (Figs. 2 and 3). The results thus obtained show that (i) the measurement of the sign of $\cos\delta$ will allow to distinguish between the TBM/GRB, BM (LC) and GRA/HG forms of $\tilde{U}_{\nu}$, (ii) for a best fit value of $\cos\delta = -1$ (-0.1) distinguishing at 3σ between the BM (TBM/GRB) and the other forms of $\tilde{U}_{\nu}$ would be possible if $\cos\delta$ is measured with 1σ uncertainty of 0.3 (0.1).

The results obtained in the studies performed in Refs. 28 and 29 show, in particular, that the experimental measurement of the Dirac phase δ of the PMNS neutrino mixing matrix in the future neutrino experiments, combined with the data on the neutrino mixing angles can provide unique information about the possible discrete symmetry origin of the observed pattern of neutrino mixing.

Acknowledgments

This work was supported in part by the European Union FP7 ITN INVISIBLES (Marie Curie Actions, PITN-GA-2011-289442-INVISIBLES), by the INFN program on Theoretical Astroparticle Physics (TASP), by the research grant 2012CPPYP7 (*Theoretical Astroparticle Physics*) under the program PRIN 2012 funded by the Italian Ministry of Education, University and Research (MIUR) and by the World Premier International Research Center Initiative (WPI Initiative), MEXT, Japan (STP).

References

1. Particle Data Group (K. Nakamura *et al.*), *Chin. Phys. C* **38**, 090001 (2014).
2. S. K. Agarwalla *et al.*, *J. High Energy Phys.* **1405**, 094 (2014).
3. C. Adams *et al.*, arXiv:1307.7335.
4. A. de Gouvea *et al.*, arXiv:1310.4340.
5. N. Cabibbo, *Phys. Lett. B* **72**, 333 (1978).
6. S. M. Bilenky, J. Hosek and S. T. Petcov, *Phys. Lett. B* **94**, 495 (1980).
7. P. Langacker, S. T. Petcov, G. Steigman and S. Toshev, *Nucl. Phys. B* **282**, 589 (1987).
8. S. Pascoli, S. T. Petcov and A. Riotto, *Phys. Rev. D* **75**, 083511 (2007).
9. S. Pascoli, S. T. Petcov and A. Riotto, *Nucl. Phys. B* **774**, 1 (2007).
10. S. Davidson, E. Nardi and Y. Nir, *Phys. Rep.* **466**, 105 (2008).
11. G. Branco, R. G. Felipe and F. Joaquim, *Rev. Mod. Phys.* **84**, 515 (2012).
12. G. Altarelli and F. Feruglio, *Rev. Mod. Phys.* **82**, 2701 (2010).

13. S. F. King and C. Luhn, *Rep. Prog. Phys.* **76**, 056201 (2013).
14. P. Harrison, D. Perkins and W. Scott, *Phys. Lett. B* **530**, 167 (2002).
15. Z. Z. Xing, *Phys. Lett. B* **533**, 85 (2002).
16. L. Wolfenstein, *Phys. Rev. D* **18**, 958 (1978).
17. S. T. Petcov, *Phys. Lett. B* **110**, 245 (1982).
18. F. Vissani, arXiv:hep-ph/9708483.
19. V. D. Barger, S. Pakvasa, T. J. Weiler and K. Whisnant, *Phys. Lett. B* **437**, 107 (1998).
20. A. J. Baltz, A. S. Goldhaber and M. Goldhaber, *Phys. Rev. Lett.* **81**, 5730 (1998).
21. L. L. Everett and A. J. Stuart, *Phys. Rev. D* **79**, 085005 (2009).
22. Y. Kajiyama, M. Raidal and A. Strumia, *Phys. Rev. D* **76**, 117301 (2007).
23. W. Rodejohann, *Phys. Lett. B* **671**, 267 (2009).
24. C. H. Albright, A. Dueck and W. Rodejohann, *Eur. Phys. J. C* **70**, 1099 (2010).
25. F. Capozzi *et al.*, *Phys. Rev. D* **89**, 093018 (2014).
26. D. Marzocca, S. T. Petcov, A. Romanino and M. C. Sevilla, *J. High Energy Phys.* **1305**, 073 (2013).
27. I. Girardi, S. T. Petcov and A. V. Titov, arXiv:1504.00658 [hep-ph].
28. S. T. Petcov, *Nucl. Phys. B* **892**, 400 (2015).
29. I. Girardi, S. T. Petcov and A. V. Titov, *Nucl. Phys. B* **894**, 733 (2015).
30. P. I. Krastev and S. T. Petcov, *Phys. Lett. B* **205**, 84 (1988).
31. Y. Wang, *PoS* (Neutel2013) 030 (2013).
32. C. Zhang *et al.*, arXiv:1501.04991.
33. P. Coloma, H. Minakata and S. J. Parke, *Phys. Rev. D* **90**, 093003 (2014).
34. M. C. Gonzalez-Garcia, M. Maltoni and T. Schwetz, *J. High Energy Phys.* **1411**, 052 (2014).

Sterile Neutrinos in E_6

Jonathan L. Rosner
Enrico Fermi Institute, University of Chicago, Chicago, IL 60637, USA
rosner@hep.uchicago.edu

The opportunity to accommodate three flavors of sterile neutrinos exists within the exceptional group E_6. Implications of this description are discussed.

Keywords: Sterile neutrinos; exceptional groups; E_6.

1. Introduction

Sterile neutrinos are weak isosinglet neutrinos, visible through mixing with one or more of the three "active" neutrinos ν_e, ν_μ, ν_τ. Several tentative indications exist that the three active neutrinos are not enough to fit all oscillation data; sterile neutrinos are one possibility. Present data prefer at least one sterile neutrino, but there are tensions even with two. In the grand unified group E_6 three sterile neutrinos are natural; here we explore some distinguishing features of such a description.[1]

We first review the shortcomings of a description with only three active neutrinos (Sec. 2); this topic has been covered in greater detail by Giunti.[2] We then discuss mass matrices in E_6 and its subgroups (Sec. 3), and their relevance for short-baseline neutrino oscillation experiments (Sec. 4). One of the three sterile neutrinos could play the role of a 7 keV dark matter candidate (Sec. 5). We conclude in Sec. 6.

2. Evidence for Sterile Neutrinos

An early conflict with the picture of three active neutrinos was seen by the LSND experiment at Los Alamos.[3] The MiniBooNE experiment at Fermilab confirmed this result (after a re-analysis of their data).[4–7] The signal is mainly at low energy, falling below an initial energy cut of 475 MeV. It is not clear whether the signal is $e^\pm$ or photons. A possible photon source would arise from a Z–ω–γ Wess–Zumino–Witten (WZW) coupling[8,9] giving rise to neutral-current coherent photon production [Fig. 1(a)] off a nuclear target.[10,11]

A claimed 6% deficit with respect to expectations in the flux of reactor neutrinos could be due to very-short-baseline neutrino oscillations.[16–18] A cautionary note[19]

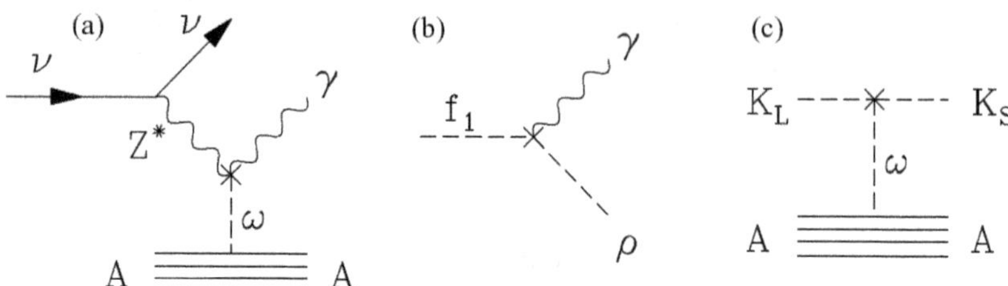

Fig. 1. A virtual Z^* transforms as a $J^{PC} = 1^{++}$ a_1 meson. The decays $a_1^0(1260) \to \omega\gamma$ and the related decay $a_1^{\pm}(1260) \to \rho^{\pm}\gamma$ are hard to look for. Possible evidence for a related WZW coupling comes from the decay $f_1(1285) \to \rho^0\gamma$ (b); the Mark III Collaboration at SPEAR observed this process[12] at a rate 12 times the non-WZW prediction.[13,14] A related term generates an $\omega K\bar{K}$ coupling giving rise to K_S regeneration off a nuclear target (c). One estimate of the contribution of the process (a)[15] gives a rate less than 1/4 that needed to explain the MiniBooNE result.

identifies an additional uncertainty associated with 30% of the flux coming from forbidden decays, whose intensity and energy spectra are hard to evaluate.

More evidence in favor of an anomaly comes from the use of ^{51}Cr and ^{37}Ar radioactive sources to calibrate the SAGE and Gallex solar neutrino detector,[20,21] finding an observed/predicted ratio of 0.84 ± 0.05.[22] Both the reactor and the gallium anomalies could be due to short-baseline neutrino oscillations with $\Delta m^2 = \mathcal{O}(\text{eV}^2)$. Such a large splitting cannot be accommodated with three active neutrinos, whose masses satisfy $\Delta m^2_{21} \simeq 7.6 \times 10^{-5}$ eV^2; $|\Delta m^2_{32}| \simeq 2.4 \times 10^{-3}$ eV^2. A full set of constraints, including ones from the BNL E-776, CDHSW, Daya Bay, ICARUS, KARMEN, MiniBooNE, MINOS, NOMAD, OPERA, SciBooNE, and T2K experiments, is discussed in Ref. 23 and by Giunti.[2] The absence of oscillations to one flavor ($N = 1$) of sterile neutrino is disfavored at the 6.3σ level.

Even with more than one sterile neutrino a basic tension remains between disappearance and appearance experiments. In one fit[24] these are compatible only at the level of 0.008%, mainly owing to a poor fit to the low-energy MiniBooNE "$e^{\pm}$" signal. For another fit, see Ref. 25. An initial incompatibility between neutrino and antineutrino fits, favoring $N > 1$, has been resolved with subsequent data, so there is no longer preference for more than one flavor of sterile neutrino.[2] Nevertheless, should such a need arise in the future, the E$_6$ scheme provides a natural home for three sterile flavors ($N = 3$).

3. Mass Matrices in E$_6$ and Subgroups

The group SU(5)[26] is the unique one of rank 4 containing the Standard Model (SM) group SU(3)$_{\text{color}}$ × SU(2)$_{\text{L}}$ × U(1). The quarks and leptons belong to $5^* + 10$ representations; there is no need for a right-handed neutrino. The group SO(10)[27] contains SU(5); its 16-plet spinor contains the SU(5) representations $5^* + 10 + 1$, where the SU(5) singlet is a right-handed neutrino N. If this state is given a large Majorana mass, the corresponding left-handed neutrino Majorana mass can

Table 1. U(1) charges of 27-plet fermions in E_6.

27 member (SO(10), SU(5))	$2\sqrt{6}\,Q_\psi$	$2\sqrt{10}\,Q_\chi$	$2\sqrt{10}\,Q_N$
$\nu_e(16, 5^*)$	-1	3	-2
(16, 10)	-1	-1	-1
$N_e^c(16, 1)$	-1	-5	0
$\nu_E(10, 5^*)$	2	-2	3
$N_E^c(10, 5)$	2	2	2
$n(1, 1)$	-4	0	-5

be made very small (the *seesaw* mechanism[28]). The rank-5 nature of SO(10) implies the existence of an extra U(1) and the possible observability of a Z' at the TeV scale or above.

The exceptional group[29] E_6 contains SO(10). Each fundamental 27-plet of E_6 contains 16+10+1 of SO(10). The 10 of SO(10) contains $5+5^*$ of SU(5), where the 5 contains a color-triplet weak isosinglet quark and a color-singlet weak isodoublet lepton. The singlet of SO(10) ("n") is a sterile neutrino candidate. Since one needs three 27's to account for three families of ordinary quarks and leptons, there are three sterile neutrinos in E_6. The rank-6 nature of E_6 implies the possibility of two extra U(1)'s or at least one linear combination surviving symmetry breaking down to LHC energies. The U(1) charges are defined as

$$E_6 \to SO(10) \times U(1)_\psi(Q_\psi)\,; \qquad SO(10) \to SU(5) \times U(1)_\chi(Q_\chi)\,.$$

A Z_θ can couple to $Q_\psi \cos\theta + Q_\chi \sin\theta$. The combination $Q_N \equiv -(1/4)Q_\chi + (\sqrt{15}/4)Q_\psi$ vanishes for the right-handed neutrino N. A large Majorana N mass is then permitted by Q_N conservation, enabling a seesaw mechanism with fermion masses generated by a 27-plet Higgs representation.[30] The U(1) charges for various members of a 27-plet are shown in Table 1.

The Z_N, coupling to Q_N, has characteristic branching fractions. Within a single

Table 2. Branching fractions of Z_N within a family.

Decay product	L	Helicity R	Sum	Percent of total
$e\bar{e}$	4/120	1/120	5/120	4.17
$\nu_e\bar{\nu}_e$	4/120	—	4/120	3.33
$u\bar{u}$	3/120	3/120	6/120	5.00
$d\bar{d}$	3/120	12/120	15/120	12.50
$E\bar{E}$	9/120	4/120	13/120	10.83
$\nu_E\bar{\nu}_E$	9/120	4/120	13/120	10.83
$h\bar{h}$	27/120	12/120	39/120	32.50
$n\bar{n}$	25/120	—	25/120	20.83

Table 3. U(1) charges (see text) in the product of two 27's of E$_6$.

	$\nu_e(-1,3,2)$	$N_e^c(-1,-5,0)$	$\nu_E(2,-2,3)$	$N_E^c(2,2,2)$	$n(-4,0,-5)$
$\nu_e(-1,3,2)$	—	$(-2,-2,-2)$	—	$(1,5,0)$	—
$N_e^c(-1,-5,0)$	$(-2,-2,-2)$	$\boxed{(-,-,0)}$	—	$(1,-3,2)$	—
$\nu_E(2,-2,3)$	—	—	—	$(4,0,5)$	$(-2,-2,-2)$
$N_E^c(2,2,2)$	$(1,5,0)$	$(1,-3,2)$	$(4,0,5)$	—	$(-2,2,-3)$
$n(-4,0,-5)$	—	—	$(-2,-2,-2)$	$(-2,2,-3)$	—

family, 25% of its decays are to ordinary fermions (above the middle line in Table 2) while 75% are to exotic fermions (below the middle line). These consist of a vector-like charged lepton E, its neutrino ν_E, an isosinglet quark h, and the sterile neutrino n. If the Z_N is found at the LHC, it is a potential source of exotic quarks and leptons. The differences between left-handed (L) and right-handed (R) couplings give rise to characteristic production and decay asymmetries.[31]

While $27 \times 27 = 27^* + 351 + 351'$, we wish to see what follows from assuming 27^* dominates, which was a popular assumption in the early days of string theory.[32–36] Some mass matrix elements will be absent as their (Q_ψ, Q_χ) values are not in 27^*. The U(1) charges for the product 27×27 are shown in Table 3, where we have listed values of $(2\sqrt{6}\,Q_\psi, 2\sqrt{10}\,Q_\chi, 2\sqrt{10}\,Q_N)$. Blank entries denote charges not found in a 27^*-plet, implying a zero entry in the mass matrix. The exception (in the box) is a Majorana mass for the right-handed neutrino N_e^c, which must be generated by a higher-dimension operator conserving Q_N.

For simplicity we make two further assumptions. First, we let ν_E pair up with N_E^c to obtain a large Dirac mass M_{34}. Second, we assume an approximate Z_2 symmetry to suppress vacuum expectation values stemming from SO(10) 16-plets in comparison with those from SO(10) 10's or singlets. The mass matrix in the basis $(\nu_e, N_e^c, \nu_E, N_E^c, n)$, where we have used small letters to denote entries with weak isospin $\Delta I = 1/2$ and large letters to denote entries with $\Delta I = 0$, is

$$\mathcal{M} = \begin{bmatrix} 0 & m_{12} & 0 & M_{14} & 0 \\ m_{12} & M_{22} & 0 & m_{24} & 0 \\ 0 & 0 & 0 & M_{34} & m_{35} \\ M_{14} & m_{24} & M_{34} & 0 & m_{45} \\ 0 & 0 & m_{35} & m_{45} & 0 \end{bmatrix} .$$

It is convenient to diagonalize this matrix with respect to the large entry M_{34}, leading to

$$\mathcal{M}' = \begin{bmatrix} 0 & m_{12} & M_{14}/\sqrt{2} & M_{14}/\sqrt{2} & 0 \\ m_{12} & M_{22} & m_{24}/\sqrt{2} & m_{24}/\sqrt{2} & 0 \\ M_{14}/\sqrt{2} & m_{24}/\sqrt{2} & M_{34} & 0 & (m_{35}+m_{45})/\sqrt{2} \\ M_{14}/\sqrt{2} & m_{24}/\sqrt{2} & 0 & -M_{34} & (m_{45}-m_{35})/\sqrt{2} \\ 0 & 0 & (m_{35}+m_{45})/\sqrt{2} & (m_{45}-m_{35})/\sqrt{2} & 0 \end{bmatrix}.$$

Now we can perturb about the three eigenvectors $[0,1,0,0,0]^T$, $[0,0,1,0,0]^T$, and $[0,0,0,1,0]^T$ corresponding to the large eigenvalues $M_{22}, M_{34}, -M_{34}$. For the small masses, the resulting 2×2 mass matrix in the (ν_e, n) basis is

$$\mathcal{S}_2 = \begin{bmatrix} -m_{12}^2/M_{22} & -M_{14}m_{35}/M_{34} \\ -M_{14}m_{35}/M_{34} & -2m_{35}m_{45}/M_{34} \end{bmatrix}.$$

We look for solutions with small mixing and $m_n > m_\nu$:

$$\nu = \begin{bmatrix} \cos\theta \\ \sin\theta \end{bmatrix}, \qquad n = \begin{bmatrix} -\sin\theta \\ \cos\theta \end{bmatrix}, \qquad t \equiv \tan\theta\,,$$

so we seek a small-t solution of a quadratic equation in t, which in its linearized form is

$$t \simeq \left(\frac{m_{12}^2 M_{34}}{M_{14}m_{35}M_{22}} - \frac{2m_{45}}{M_{14}} \right)^{-1}.$$

Barring accidental cancellations, after several steps we get $m_n > m_\nu$ with small mixing if $M_{14} \ll m_{45}$ and

$$\left| \frac{m_{35}m_{45}M_{22}}{M_{34}m_{12}^2} \right| > 1\,, \qquad \frac{m_{45}}{M_{14}} \gg 1\,.$$

The smallness of M_{14} is curious but achievable via the approximate Z_2 symmetry mentioned earlier.

The neutrino mass matrix can be related to those for charged fermions at a unification scale. Thus, m_{12} and m_{35} are related to masses of quarks of charge 2/3, while m_{24}, m_{45}, M_{14} and M_{34} are related to charge $-1/3$ quark and charged lepton masses. Specifically, for up-type quarks, the U(1) charges of masses are $(-2,-2,-2)$, corresponding to m_{21} and m_{35}. The relation of m_{12} to m_u is familiar from SO(10) unification. For down-type quarks and charged leptons, the correspondences are $(-2,2,-3) \sim m_{45}$, $(1,5,0) \sim M_{14}$, $(1,-3,2) \sim m_{24}$, and $(4,0,5) \sim M_{34}$. In the absence of mixing, m_{45} is related to Dirac masses of d and e, while M_{34} is related to Dirac masses of quarks and charged leptons in the 10 of SO(10). Weak universality suggests $|m_{24}| \ll |m_{45}|$ (because isosinglet impurities in left-handed charged leptons and down-type quarks should be small), while there is less of a constraint on M_{14} as it has $\Delta I = 0$.

4. Relevance for Short-Baseline Neutrino Oscillation Experiments

The present model has mixing only within single families. In order to explain the LSND and MiniBooNE electron appearance signals one needs both muon and electron neutrinos to mix with the same sterile neutrino. The freedom of setting a sterile neutrino mass and mixing for one family (the matrix $\mathcal{S}_2$ in the previous section) is encouraging for the case of three families (which must be represented by a 6×6 matrix). Furthermore, if data improve to the extent that three sterile neutrinos are needed to explain oscillations, E_6 is available.

5. One Neutrino as a Possible Dark Matter Candidate

Another possible use of a third sterile ν is as a warm dark matter candidate at the keV scale, as suggested some time ago.[37,38] For more recent reviews see Refs. 23 and 39. In contrast to many schemes, the present one distinguishes between right-hand neutrinos (usually taken very heavy, at the seesaw scale) and the n's (one of which can easily have keV-scale mass).

There have been two claims for observation of an X-ray line near 3.5 keV.[40,41] These signals could arise from a 7 keV "neutrino" decaying to a photon and a much lighter "neutrino." A corresponding signal is *not* seen, however, in the Milky Way.[42]

There are some special features of E_6 concerning a 7 keV dark matter candidate. The Higgs vacuum expectation values considered here correspond to the five neutral complex scalar bosons in the 27^* representation of E_6. The masses of these bosons are free parameters; two of the five are those of the minimal supersymmetric standard model or SO(10). Exchanges of these bosons can produce the states n; for example, in the processes

$$d_l + h_L^c \to n_L + N_{EL}^c\,, \qquad e^- + E_L^+ \to n_L + N_{EL}^c\,.$$

A TeV-scale Z_N produced in the early universe would have appreciable branching ratio into nn^c pairs, so n are candidates for early overproduction unless their abundance is diluted by subsequent entropy production.[43]

6. Summary

None of the various hints of sterile neutrinos rises to the level of a conclusive observation so it is crucial to strengthen or refute them. Some effects may be due to interesting non-ν physics: for example, if the MiniBooNE low-energy signal is photons and not electrons.

The grand unified group E_6 [the next step up from SO(10)] naturally incorporates three candidates for neutrinos with neither left-handed nor right-handed weak isospin. E_6 breaking to the standard model times a particular $U(1)_N$ allows a large Majorana mass for the right-handed neutrino N and hence the standard seesaw mechanism can proceed without constraints.[30,44] Masses and mixings of three sterile neutrinos are at one's disposal to fit oscillation data, assuming present anomalies are really due to sterile ν and not something else.

If at most two sterile neutrinos are needed to fit anomalies successfully, a third is left over as a dark matter candidate. The E_6 scheme appears to have enough free parameters to allow such a scenario to successfully navigate a number of constraints.

Acknowledgments

I am grateful to H. Fritzsch for the invitation to attend the conference on Massive Neutrinos, and to Louis Lim and K. K. Phua for generous hospitality. I thank Janet Conrad, P. S. Bhupal Dev, Mariana Frank and Rabi Mohapatra for pointing me to some helpful literature, and Kevork Abazajian, Joshua Frieman, Richard Hill, Lauren Hsu, Hitoshi Murayama, and Robert Shrock for useful discussions. This work was supported in part by the U.S. Department of energy under Grant No. DE–FG02-13ER41598 and in part by funds from the Physics Department of the University of Chicago.

References

1. J. L. Rosner, *Phys. Rev. D* **90**, 035005 (2014).
2. C. Giunti, *Mod. Phys. Lett. A* **30**, (2015); updated version of C. Giunti, M. Laveder, Y. F. Li and H. W. Long, *Phys. Rev. D* **88**, 073008 (2013).
3. LSND Collab. (A. Aguilar-Arevalo *et al.*), *Phys. Rev. D* **64**, 112007 (2001) and earlier references therein.
4. MiniBooNE Collab. (A. A. Aguilar-Arevalo *et al.*), *Phys. Rev. Lett.* **98**, 231801 (2007).
5. MiniBooNE Collab. (A. A. Aguilar-Arevalo *et al.*), *Phys. Rev. Lett.* **102**, 101802 (2009).
6. MiniBooNE Collab. (A. A. Aguilar-Arevalo *et al.*), *Phys. Rev. Lett.* **105**, 181801 (2010).
7. MiniBooNE Collab. (A. A. Aguilar-Arevalo *et al.*), *Phys. Rev. Lett.* **110**, 161801 (2013).
8. J. Wess and B. Zumino, *Phys. Lett. B* **37**, 95 (1971).
9. E. Witten, *Nucl. Phys. B* **223**, 422 (1983).
10. J. A. Harvey, C. T. Hill and R. J. Hill, *Phys. Rev. Lett.* **99**, 261601 (2007).
11. J. A. Harvey, C. T. Hill and R. J. Hill, *Phys. Rev. D* **77**, 085017 (2008).
12. MARK-III Collab. (D. Coffman *et al.*), *Phys. Rev. D* **41**, 1410 (1990).
13. J. Babcock and J. L. Rosner, *Phys. Rev. D* **14**, 1286 (1976).
14. J. L. Rosner, *Phys. Rev. D* **23**, 1127 (1981).
15. J. L. Rosner, arXiv:1502.01704.
16. T. A. Mueller *et al.*, *Phys. Rev. C* **83**, 054615 (2011).
17. G. Mention *et al.*, *Phys. Rev. D* **83**, 073006 (2011).
18. P. Huber, *Phys. Rev. C* **84**, 024617 (2011) [Erratum-*ibid.* **85**, 029901 (2012)].
19. A. C. Hayes, J. L. Friar, G. T. Garvey, G. Jungman and G. Jonkmans, *Phys. Rev. Lett.* **112**, 202501 (2014).
20. SAGE Collab. (J. N. Abdurashitov *et al.*), *Phys. Rev. C* **80**, 015807 (2009) and earlier references therein.

21. F. Kaether, W. Hampel, G. Heusser, J. Kiko and T. Kirsten, *Phys. Lett. B* **685**, 47 (2010) and earlier references therein.
22. C. Giunti, M. Laveder, Y. F. Li, Q. Y. Liu and H. W. Long, *Phys. Rev. D* **86**, 113014 (2012).
23. K. N. Abazajian *et al.*, arXiv:1204.5379.
24. J. M. Conrad, C. M. Ignarra, G. Karagiorgi, M. H. Shaevitz and J. Spitz, *Adv. High Energy Phys.* **2013**, 163897 (2013).
25. J. Kopp, P. A. N. Machado, M. Maltoni and T. Schwetz, *JHEP* **1305**, 050 (2013).
26. H. Georgi and S. L. Glashow, *Phys. Rev. Lett.* **32**, 438 (1974).
27. H. Fritzsch and P. Minkowski, *Ann. Phys.* (*N. Y.*) **93**, 193 (1975). For historical context see P. Minkowski, in *International Conference on Massive Neutrinos* (World Scientific), to appear.
28. P. Minkowski, *Phys. Lett. B* **67**, 421 (1977); M. Gell-Mann, P. Ramond and R. Slansky, in *Supergravity*, eds. D. Freedman and P. Van Nieuwenhuizen (North-Holland, 1979), pp. 315–321; T. Yanagida, *Prog. Theor. Phys.* **64**, 1104 (1980); R. N. Mohapatra and G. Senjanovic, *Phys. Rev. Lett.* **44**, 912 (1980).
29. F. Gursey, P. Ramond and P. Sikivie, *Phys. Lett. B* **60**, 177 (1976).
30. E. Ma, *Phys. Lett. B* **380**, 286 (1996).
31. P. Langacker, R. W. Robinett and J. L. Rosner, *Phys. Rev. D* **30**, 1470 (1984).
32. M. Dine, V. Kaplunovsky, M. L. Mangano, C. Nappi and N. Seiberg, *Nucl. Phys. B* **259**, 549 (1985).
33. J. D. Breit, B. A. Ovrut and G. C. Segre, *Phys. Lett. B* **158**, 33 (1985).
34. S. Cecotti, J. P. Derendinger, S. Ferrara and M. Roncadelli, *Phys. Lett. B* **156**, 318 (1985).
35. J. L. Rosner, *Comments Nucl. Part. Phys.* **15**, 195 (1986).
36. S. Nandi and U. Sarkar, *Phys. Rev. Lett.* **56**, 564 (1986).
37. S. Dodelson and L. M. Widrow, *Phys. Rev. Lett.* **72**, 17 (1994).
38. X. D. Shi and G. M. Fuller, *Phys. Rev. Lett.* **82**, 2832 (1999).
39. A. Kusenko, *Phys. Rep.* **481**, 1 (2009).
40. E. Bulbul, M. Markevitch, A. Foster, R. K. Smith, M. Loewenstein and S. W. Randall, *Astrophys. J.* **789**, 13 (2014).
41. A. Boyarsky, O. Ruchayskiy, D. Iakubovskyi and J. Franse, *Phys. Rev. Lett.* **113**, 251301 (2014).
42. S. Riemer-Sorensen, arXiv:1405.7943.
43. R. J. Scherrer and M. S. Turner, *Phys. Rev. D* **33**, 1585 (1986) [Erratum-*ibid.* **34**, 3263 (1986)].
44. J. C. Callaghan, S. F. King and G. K. Leontaris, *JHEP* **1312**, 037 (2013) and references therein.

Phenomenology of Light Sterile Neutrinos

Carlo Giunti
INFN, Sezione di Torino, Via P. Giuria 1, I–10125 Torino, Italy
carlo.giunti@to.infn.it

We consider the extension of standard three-neutrino mixing with the addition of one or two light sterile neutrinos which can explain the anomalies found in short-baseline neutrino oscillation experiments. We review the results of the global analyses of short-baseline neutrino oscillation data in $3 + 1$, $3 + 2$ and $3 + 1 + 1$ neutrino mixing schemes.

Keywords: Neutrino masses and mixing; sterile neutrinos; neutrino oscillations.

1. Introduction

Neutrino oscillations have been measured with high accuracy in solar, atmospheric and long-baseline neutrino oscillation experiments (see Refs. 1–3). Hence, we know that neutrinos are massive and mixed particles (see Refs. 4 and 5) and that there are two independent squared-mass differences: the solar $\Delta m^2_{\rm SOL} \simeq 7.5 \times 10^{-5}$ eV^2 and the atmospheric $\Delta m^2_{\rm ATM} \simeq 2.3 \times 10^{-3}$ eV^2. This is in agreement with the standard three-neutrino mixing paradigm, in which the three active neutrinos ν_e, ν_μ, ν_τ are superpositions of three massive neutrinos ν_1, ν_2, ν_3 with respective masses m_1, m_2, m_3. The two measured squared-mass differences can be interpreted as $\Delta m^2_{\rm SOL} = \Delta m^2_{21}$ and $\Delta m^2_{\rm ATM} = |\Delta m^2_{31}| \simeq |\Delta m^2_{32}|$, with $\Delta m^2_{kj} = m^2_k - m^2_j$.

2. Beyond Three-Neutrino Mixing: Sterile Neutrinos

The completeness of the three-neutrino mixing paradigm has been challenged by the following indications in favor of short-baseline neutrino oscillations, which require the existence of at least one additional squared-mass difference, $\Delta m^2_{\rm SBL}$, which is much larger than $\Delta m^2_{\rm SOL}$ and $\Delta m^2_{\rm ATM}$:

1. The reactor antineutrino anomaly,[6] which is a deficit of the rate of $\bar{\nu}_e$ observed in several short-baseline reactor neutrino experiments in comparison with that expected from a new calculation of the reactor neutrino fluxes.[7,8] The statistical significance is about 2.8σ.

2. The Gallium neutrino anomaly,[9–13] consisting in a short-baseline disappearance of ν_e measured in the Gallium radioactive source experiments GALLEX[14] and SAGE[15] with a statistical significance of about 2.9σ.
3. The LSND experiment, in which a signal of short-baseline $\bar{\nu}_\mu \to \bar{\nu}_e$ oscillations has been observed with a statistical significance of about 3.8σ.[16,17]

In this review, we consider 3 + 1,[18–21] 3 + 2[22–25] and 3 + 1 + 1[26–29] neutrino mixing schemes in which there are one or two additional massive neutrinos at the eV scale and the masses of the three standard massive neutrinos are much smaller. Since from the LEP measurement of the invisible width of the Z boson we know that there are only three active neutrinos (see Ref. 4), in the flavor basis the additional massive neutrinos correspond to sterile neutrinos,[30] which do not have standard weak interactions.

The possible existence of sterile neutrinos is very interesting, because they are new particles which could give us precious information on the physics beyond the Standard Model (see Refs. 31 and 32). The existence of light sterile neutrinos is also very important for astrophysics (see Ref. 33) and cosmology (see Refs. 34–37).

In the 3 + 1 scheme, the effective probability of $\overset{(-)}{\nu}_\alpha \to \overset{(-)}{\nu}_\beta$ transitions in short-baseline experiments has the two-neutrino-like form[19]

$$P_{\overset{(-)}{\nu}_\alpha \to \overset{(-)}{\nu}_\beta} = \delta_{\alpha\beta} - 4|U_{\alpha 4}|^2(\delta_{\alpha\beta} - |U_{\beta 4}|^2) \sin^2\left(\frac{\Delta m^2_{41} L}{4E}\right), \tag{1}$$

where U is the mixing matrix, L is the source-detector distance, E is the neutrino energy and $\Delta m^2_{41} = m^2_4 - m^2_1 = \Delta m^2_{\text{SBL}} \sim 1\ \text{eV}^2$. The electron and muon neutrino and antineutrino appearance and disappearance in short-baseline experiments depend on $|U_{e4}|^2$ and $|U_{\mu 4}|^2$, which determine the amplitude $\sin^2 2\vartheta_{e\mu} = 4|U_{e4}|^2|U_{\mu 4}|^2$ of $\overset{(-)}{\nu}_\mu \to \overset{(-)}{\nu}_e$ transitions, the amplitude $\sin^2 2\vartheta_{ee} = 4|U_{e4}|^2(1 - |U_{e4}|^2)$ of $\overset{(-)}{\nu}_e$ disappearance, and the amplitude $\sin^2 2\vartheta_{\mu\mu} = 4|U_{\mu 4}|^2(1 - |U_{\mu 4}|^2)$ of $\overset{(-)}{\nu}_\mu$ disappearance.

Since the oscillation probabilities of neutrinos and antineutrinos are related by a complex conjugation of the elements of the mixing matrix (see Ref. 4), the effective probabilities of short-baseline $\nu_\mu \to \nu_e$ and $\bar{\nu}_\mu \to \bar{\nu}_e$ transitions are equal. Hence, the 3 + 1 scheme cannot explain a possible CP-violating difference of $\nu_\mu \to \nu_e$ and $\bar{\nu}_\mu \to \bar{\nu}_e$ transitions in short-baseline experiments. In order to allow this possibility, one must consider a 3 + 2 scheme, in which, there are four additional effective mixing parameters in short-baseline experiments: $\Delta m^2_{51} \geq \Delta m^2_{41}$, $|U_{e5}|^2$, $|U_{\mu 5}|^2$ and $\eta = \arg[U^*_{e4} U_{\mu 4} U_{e5} U^*_{\mu 5}]$ (see Refs. 38 and 39). Since this complex phase appears with different signs in the effective 3 + 2 probabilities of short-baseline $\nu_\mu \to \nu_e$ and $\bar{\nu}_\mu \to \bar{\nu}_e$ transitions, it can generate measurable CP violations.

A puzzling feature of the 3 + 2 scheme is that it needs the existence of two sterile neutrinos with masses at the eV scale. We think that it may be considered as more plausible that sterile neutrinos have a hierarchy of masses. Hence, it is interesting to consider also the 3 + 1 + 1 scheme,[26–29] in which m_5 is much heavier than 1 eV and the oscillations due to Δm^2_{51} are averaged. Hence, in the analysis of short-baseline data in the $3+1+1$ scheme there is one effective parameter less than in the $3+2$ scheme (Δm^2_{51}), but CP violations generated by η are observable.

Updated global fits of short-baseline neutrino oscillation data have been presented in Refs. 40 and 41. These analyses take into account the final results of the MiniBooNE experiment, which was made in order to check the LSND signal with about one order of magnitude larger distance (L) and energy (E), but the same order of magnitude for the ratio L/E from which neutrino oscillations depend. Unfortunately, the results of the MiniBooNE experiment are ambiguous, because the LSND signal was not seen in neutrino mode ($\nu_\mu \to \nu_e$)[42] and the $\bar{\nu}_\mu \to \bar{\nu}_e$ signal observed in 2010[43] with the first half of the antineutrino data was not observed in the second half of the antineutrino data.[44] Moreover, the MiniBooNE data in both neutrino and antineutrino modes show an excess in the low-energy bins which is widely considered to be anomalous because it is at odds with neutrino oscillations.[45,46,a]

In the following we summarize the results of the analysis of short-baseline data presented in Ref. 41 of the following three groups of experiments:

(A) The $\overset{(-)}{\nu}_\mu \to \overset{(-)}{\nu}_e$ appearance data of the LSND,[17] MiniBooNE,[44] BNL-E776,[49] KARMEN,[50] NOMAD,[51] ICARUS[52] and OPERA[53] experiments.

(B) The $\overset{(-)}{\nu}_e$ disappearance data described in Ref. 13, which take into account the reactor[6–8] and Gallium[9–12,54] anomalies.

(C) The constraints on $\overset{(-)}{\nu}_\mu$ disappearance obtained from the data of the CDHSW experiment,[55] from the analysis[24] of the data of atmospheric neutrino oscillation experiments,[b] from the analysis[45] of the MINOS neutral-current data[58] and from the analysis of the SciBooNE-MiniBooNE neutrino[59] and antineutrino[60] data.

Table 1 summarizes the statistical results obtained in Ref. 41 from global fits of the data above in the $3+1$, $3+2$ and $3+1+1$ schemes. In the LOW fits all the MiniBooNE data are considered, including the anomalous low-energy bins, which are omitted in the HIG fits. There is also a $3+1$-noMB fit without MiniBooNE data and a $3+1$-noLSND fit without LSND data.

From Table 1, one can see that in all fits which include the LSND data the absence of short-baseline oscillations is disfavored by about 6σ, because the improvement of the χ^2 with short-baseline oscillations is much larger than the number of oscillation parameters.

In all the $3+1$, $3+2$ and $3+1+1$ schemes the goodness-of-fit in the LOW analysis is significantly worse than that in the HIG analysis and the appearance–disappearance parameter goodness-of-fit is much worse. This result confirms the fact that the MiniBooNE low-energy anomaly is incompatible with neutrino

[a]The interesting possibility of reconciling the low-energy anomalous data with neutrino oscillations through energy reconstruction effects proposed in Refs. 47 and 48 still needs a detailed study.
[b]The IceCube data, which could give a marginal contribution[56,57] have not been considered because the analysis is too complicated and subject to large uncertainties.

Table 1. Results of the fit of short-baseline data[41] taking into account all MiniBooNE data (LOW), only the MiniBooNE data above 475 MeV (HIG), without MiniBooNE data (noMB) and without LSND data (noLSND) in the $3+1$, $3+2$ and $3+1+1$ schemes. The first three lines give the minimum χ^2 (χ^2_{min}), the number of degrees of freedom (NDF) and the goodness-of-fit (GoF). The following five lines give the quantities relevant for the appearance–disappearance (APP–DIS) parameter goodness-of-fit (PG).[61] The last three lines give the difference between the χ^2 without short-baseline oscillations and χ^2_{min} ($\Delta\chi^2_{\text{NO}}$), the corresponding difference of number of degrees of freedom (NDF_{NO}) and the resulting number of σ's ($n\sigma_{\text{NO}}$) for which the absence of oscillations is disfavored.

	$3+1$ LOW	$3+1$ HIG	$3+1$ noMB	$3+1$ noLSND	$3+2$ LOW	$3+2$ HIG	$3+1+1$ LOW	$3+1+1$ HIG
χ^2_{min}	291.7	261.8	236.1	278.4	284.4	256.4	289.8	259.0
NDF	256	250	218	252	252	246	253	247
GoF	6%	29%	19%	12%	8%	31%	6%	29%
$(\chi^2_{\text{min}})_{\text{APP}}$	99.3	77.0	50.9	91.8	87.7	69.8	94.8	75.5
$(\chi^2_{\text{min}})_{\text{DIS}}$	180.1	180.1	180.1	180.1	179.1	179.1	180.1	180.1
$\Delta\chi^2_{\text{PG}}$	12.7	4.8	5.1	6.4	17.7	7.5	14.9	3.4
NDF_{PG}	2	2	2	2	4	4	3	3
GoF_{PG}	0.2%	9%	8%	4%	0.1%	11%	0.2%	34%
$\Delta\chi^2_{\text{NO}}$	47.5	46.2	47.1	8.3	54.8	51.6	49.4	49.1
NDF_{NO}	3	3	3	3	7	7	6	6
$n\sigma_{\text{NO}}$	6.3σ	6.2σ	6.3σ	2.1σ	6.0σ	5.8σ	5.8σ	5.8σ

oscillations, because it would require a small value of Δm^2_{41} and a large value of $\sin^2 2\vartheta_{e\mu}$,[45,46] which are excluded by the data of other experiments (see Ref. 41 for further details).[c] Note that the appearance–disappearance tension in the $3+2$-LOW fit is even worse than that in the $3+1$-LOW fit, since the $\Delta\chi^2_{\text{PG}}$ is so much larger that it cannot be compensated by the additional degrees of freedom (this behavior has been explained in Ref. 62). Therefore, we think that it is very likely that the MiniBooNE low-energy anomaly has an explanation which is different from neutrino oscillations and the HIG fits are more reliable than the LOW fits.

The $3+2$ mixing scheme was considered to be interesting in 2010 when the MiniBooNE neutrino[42] and antineutrino[43] data showed a CP-violating tension, but this tension almost disappeared in the final MiniBooNE data.[44] In fact, from Table 1 one can see that there is little improvement of the $3+2$-HIG fit with respect to the $3+1$-HIG fit, in spite of the four additional parameters and the additional possibility of CP violation. Moreover, since the p-value obtained by restricting the $3+2$ scheme to $3+1$ disfavors the $3+1$ scheme only at 1.2σ,[41] we think that considering the larger complexity of the $3+2$ scheme is not justified by the data.[d]

[c]One could fit the three anomalous MiniBooNE low-energy bins in a $3+2$ scheme[39] by considering the appearance data without the ICARUS[52] and OPERA[53] constraints, but the required large transition probability is excluded by the disappearance data.

[d]See however the somewhat different conclusions reached in Ref. 40.

The results of the 3+1+1-HIG fit presented in Table 1 show that the appearance–disappearance parameter goodness-of-fit is remarkably good, with a $\Delta\chi^2_{\rm PG}$ that is smaller than those in the $3+1$-HIG and $3+2$-HIG fits. However, the $\chi^2_{\rm min}$ in the $3+1+1$-HIG is only slightly smaller than that in the $3+1$-HIG fit and the p-value obtained by restricting the $3+1+1$ scheme to $3+1$ disfavors the $3+1$ scheme only at 0.8σ.[41] Therefore, there is no compelling reason to prefer the more complex $3+1+1$ to the simpler $3+1$ scheme.

Figure 1 shows the allowed regions in the $\sin^2 2\vartheta_{e\mu}$–Δm^2_{41}, $\sin^2 2\vartheta_{ee}$–Δm^2_{41} and $\sin^2 2\vartheta_{\mu\mu}$–$\Delta m^2_{41}$ planes obtained in the $3+1$-HIG fit of Ref. 41. These regions are relevant, respectively, for $\overset{(-)}{\nu}_\mu \to \overset{(-)}{\nu}_e$ appearance, $\overset{(-)}{\nu}_e$ disappearance and $\overset{(-)}{\nu}_\mu$ disappearance searches. The corresponding marginal allowed intervals of the oscillation parameters are given in Table 2. Figure 1 shows also the region allowed by $\overset{(-)}{\nu}_\mu \to \overset{(-)}{\nu}_e$ appearance data and the constraints from $\overset{(-)}{\nu}_e$ disappearance and $\overset{(-)}{\nu}_\mu$ disappearance

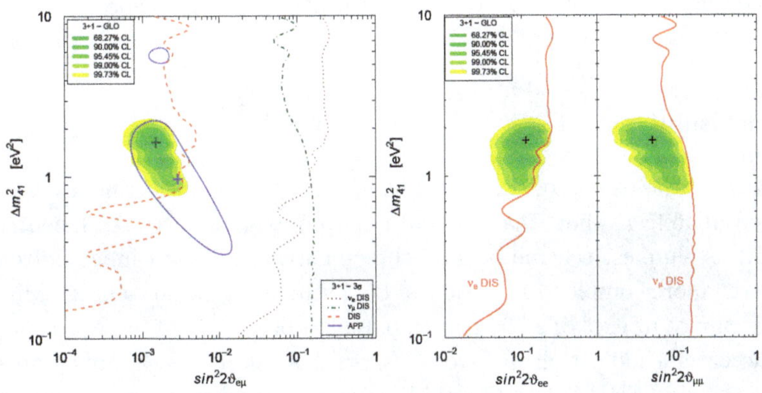

Fig. 1. Allowed regions in the $\sin^2 2\vartheta_{e\mu}$–Δm^2_{41}, $\sin^2 2\vartheta_{ee}$–Δm^2_{41} and $\sin^2 2\vartheta_{\mu\mu}$–$\Delta m^2_{41}$ planes obtained in the global (GLO) $3+1$-HIG fit[41] of short-baseline neutrino oscillation data compared with the 3σ allowed regions obtained from $\overset{(-)}{\nu}_\mu \to \overset{(-)}{\nu}_e$ short-baseline appearance data (APP) and the 3σ constraints obtained from $\overset{(-)}{\nu}_e$ short-baseline disappearance data (ν_e DIS), $\overset{(-)}{\nu}_\mu$ short-baseline disappearance data (ν_μ DIS) and the combined short-baseline disappearance data (DIS). The best-fit points of the GLO and APP fits are indicated by crosses.

Table 2. Marginal allowed intervals of the oscillation parameters obtained in the global $3+1$-HIG fit of short-baseline neutrino oscillation data.[41]

CL	Δm^2_{41} [eV^2]	$\sin^2 2\vartheta_{e\mu}$	$\sin^2 2\vartheta_{ee}$	$\sin^2 2\vartheta_{\mu\mu}$
68.27%	1.55–1.72	0.0012–0.0018	0.089–0.15	0.036–0.065
90.00%	1.19–1.91	0.001–0.0022	0.072–0.17	0.03–0.085
95.00%	1.15–1.97	0.00093–0.0023	0.066–0.18	0.028–0.095
95.45%	1.14–1.97	0.00091–0.0024	0.065–0.18	0.027–0.095
99.00%	0.87–2.09	0.00078–0.003	0.054–0.2	0.022–0.12
99.73%	0.82–2.19	0.00066–0.0034	0.047–0.22	0.019–0.14

data. One can see that the combined disappearance constraint in the $\sin^2 2\vartheta_{e\mu}$–Δm^2_{41} plane excludes a large part of the region allowed by $\overset{(-)}{\nu}_{\mu} \to \overset{(-)}{\nu}_{e}$ appearance data, leading to the well-known appearance-disappearance tension[39,40,45,46,62–65] quantified by the parameter goodness-of-fit in Table 1.

It is interesting to investigate what is the impact of the MiniBooNE experiment on the global analysis of short-baseline neutrino oscillation data. With this aim, the authors of Ref. 41 performed two additional $3+1$fits: a $3+1$-noMB fit without MiniBooNE data and a $3+1$-noLSND fit without LSND data. From Table 1 one can see that the results of the $3+1$-noMB fit are similar to those of the $3+1$-HIG fit and the exclusion of the case of no-oscillations remains at the level of 6σ. On the other hand, in the $3+1$-noLSND fit, without LSND data, the exclusion of the case of no-oscillations drops dramatically to 2.1σ. In fact, in this case the main indication in favor of short-baseline oscillations is given by the reactor and Gallium anomalies which have a similar statistical significance.[13] Therefore, it is clear that the LSND experiment is still crucial for the indication in favor of short-baseline $\bar{\nu}_{\mu} \to \bar{\nu}_{e}$ transitions and the MiniBooNE experiment has been rather inconclusive.

3. Conclusions

In conclusion, the results of the global fit of short-baseline neutrino oscillation data presented in Ref. 41 show that the data can be explained by $3+1$ neutrino mixing and this simplest scheme beyond three-neutrino mixing cannot be rejected in favor of the more complex $3+2$ and $3+1+1$ schemes. The low-energy MiniBooNE anomaly cannot be explained by neutrino oscillations in any of these schemes. Moreover, the crucial indication in favor of short-baseline $\bar{\nu}_{\mu} \to \bar{\nu}_{e}$ appearance is still given by the old LSND data and the MiniBooNE experiment has been inconclusive. Hence new experiments are needed in order to check this signal.[66–73]

References

1. D. Forero, M. Tortola and J. Valle, *Phys. Rev. D* **86**, 073012 (2012).
2. G. Fogli *et al.*, *Phys. Rev. D* **86**, 013012 (2012).
3. M. Gonzalez-Garcia, M. Maltoni, J. Salvado and T. Schwetz, *JHEP* **12**, 123 (2012).
4. C. Giunti and C. W. Kim, *Fundamentals of Neutrino Physics and Astrophysics* (Oxford Univ. Press, 2007).
5. M. C. Gonzalez-Garcia and M. Maltoni, *Phys. Rep.* **460**, 1 (2008).
6. G. Mention *et al.*, *Phys. Rev. D* **83**, 073006 (2011).
7. T. A. Mueller *et al.*, *Phys. Rev. C* **83**, 054615 (2011).
8. P. Huber, *Phys. Rev. C* **84**, 024617 (2011).
9. J. N. Abdurashitov *et al.*, *Phys. Rev. C* **73**, 045805 (2006).
10. M. Laveder, *Nucl. Phys. Proc. Suppl.* **168**, 344 (2007).
11. C. Giunti and M. Laveder, *Mod. Phys. Lett. A* **22**, 2499 (2007).
12. C. Giunti and M. Laveder, *Phys. Rev. C* **83**, 065504 (2011).
13. C. Giunti, M. Laveder, Y. Li, Q. Liu and H. Long, *Phys. Rev. D* **86**, 113014 (2012).

14. F. Kaether, W. Hampel, G. Heusser, J. Kiko and T. Kirsten, *Phys. Lett. B* **685**, 47 (2010).
15. J. N. Abdurashitov *et al.*, *Phys. Rev. C* **80**, 015807 (2009).
16. C. Athanassopoulos *et al.*, *Phys. Rev. Lett.* **75**, 2650 (1995).
17. A. Aguilar *et al.*, *Phys. Rev. D* **64**, 112007 (2001).
18. N. Okada and O. Yasuda, *Int. J. Mod. Phys. A* **12**, 3669 (1997).
19. S. M. Bilenky, C. Giunti and W. Grimus, *Eur. Phys. J. C* **1**, 247 (1998).
20. S. M. Bilenky, C. Giunti, W. Grimus and T. Schwetz, *Phys. Rev. D* **60**, 073007 (1999).
21. M. Maltoni, T. Schwetz, M. Tortola and J. Valle, *New J. Phys.* **6**, 122 (2004).
22. M. Sorel, J. Conrad and M. Shaevitz, *Phys. Rev. D* **70**, 073004 (2004).
23. G. Karagiorgi *et al.*, *Phys. Rev. D* **75**, 013011 (2007).
24. M. Maltoni and T. Schwetz, *Phys. Rev. D* **76**, 093005 (2007).
25. G. Karagiorgi, Z. Djurcic, J. Conrad, M. H. Shaevitz and M. Sorel, *Phys. Rev. D* **80**, 073001 (2009).
26. A. E. Nelson, *Phys. Rev. D* **84**, 053001 (2011).
27. J. Fan and P. Langacker, *JHEP* **04**, 083 (2012).
28. E. Kuflik, S. D. McDermott and K. M. Zurek, *Phys. Rev. D* **86**, 033015 (2012).
29. J. Huang and A. E. Nelson, *Phys. Rev. D* **88**, 033016 (2013).
30. B. Pontecorvo, *Sov. Phys. JETP* **26**, 984 (1968).
31. R. R. Volkas, *Prog. Part. Nucl. Phys.* **48**, 161 (2002).
32. R. N. Mohapatra and A. Y. Smirnov, *Annu. Rev. Nucl. Part. Sci.* **56**, 569 (2006).
33. A. Diaferio and G. W. Angus, arXiv:1206.6231.
34. J. Lesgourgues, G. Mangano, G. Miele and S. Pastor, *Neutrino Cosmology* (Cambridge Univ. Press, 2013).
35. S. Riemer-Sorensen, D. Parkinson and T. M. Davis, *Publ. Astron. Soc. Austral.* **30**, e029 (2013).
36. M. Archidiacono, E. Giusarma, S. Hannestad and O. Mena, *Adv. High Energy Phys.* **2013**, 191047 (2013).
37. J. Lesgourgues and S. Pastor, *New J. Phys.* **16**, 065002 (2014).
38. M. C. Gonzalez-Garcia and M. Maltoni, *Phys. Rep.* **460**, 1 (2008).
39. J. Conrad, C. Ignarra, G. Karagiorgi, M. Shaevitz and J. Spitz, *Adv. High Energy Phys.* **2013**, 163897 (2013).
40. J. Kopp, P. A. N. Machado, M. Maltoni and T. Schwetz, *JHEP* **1305**, 050 (2013).
41. C. Giunti, M. Laveder, Y. Li and H. Long, *Phys. Rev. D* **88**, 073008 (2013).
42. A. A. Aguilar-Arevalo *et al.*, *Phys. Rev. Lett.* **102**, 101802 (2009).
43. A. A. Aguilar-Arevalo *et al.*, *Phys. Rev. Lett.* **105**, 181801 (2010).
44. A. Aguilar-Arevalo *et al.*, *Phys. Rev. Lett.* **110**, 161801 (2013).
45. C. Giunti and M. Laveder, *Phys. Rev. D* **84**, 093006 (2011).
46. C. Giunti and M. Laveder, *Phys. Lett. B* **706**, 200 (2011).
47. M. Martini, M. Ericson and G. Chanfray, *Phys. Rev. D* **85**, 093012 (2012).
48. M. Martini, M. Ericson and G. Chanfray, *Phys. Rev. D* **87**, 013009 (2013).
49. L. Borodovsky *et al.*, *Phys. Rev. Lett.* **68**, 274 (1992).
50. B. Armbruster *et al.*, *Phys. Rev. D* **65**, 112001 (2002).
51. P. Astier *et al.*, *Phys. Lett. B* **570**, 19 (2003).
52. M. Antonello *et al.*, *Eur. Phys. J. C* **73**, 2599 (2013).
53. N. Agafonova *et al.*, *JHEP* **1307**, 004 (2013).
54. M. A. Acero, C. Giunti and M. Laveder, *Phys. Rev. D* **78**, 073009 (2008).
55. F. Dydak *et al.*, *Phys. Lett. B* **134**, 281 (1984).
56. A. Esmaili, F. Halzen and O. L. G. Peres, *J. Cosmol. Astropart. Phys.* **1211**, 041 (2012).

57. A. Esmaili and A. Y. Smirnov, *JHEP* **1312**, 014 (2013).
58. P. Adamson *et al.*, *Phys. Rev. Lett.* **107**, 011802 (2011).
59. K. B. M. Mahn *et al.*, *Phys. Rev. D* **85**, 032007 (2012).
60. G. Cheng *et al.*, *Phys. Rev. D* **86**, 052009 (2012).
61. M. Maltoni and T. Schwetz, *Phys. Rev. D* **68**, 033020 (2003).
62. M. Archidiacono, N. Fornengo, C. Giunti, S. Hannestad and A. Melchiorri, *Phys. Rev. D* **87**, 125034 (2013).
63. J. Kopp, M. Maltoni and T. Schwetz, *Phys. Rev. Lett.* **107**, 091801 (2011).
64. C. Giunti and M. Laveder, *Phys. Rev. D* **84**, 073008 (2011).
65. M. Archidiacono, N. Fornengo, C. Giunti and A. Melchiorri, *Phys. Rev. D* **86**, 065028 (2012).
66. K. N. Abazajian *et al.*, arXiv:1204.5379.
67. C. Rubbia, A. Guglielmi, F. Pietropaolo and P. Sala, arXiv:1304.2047.
68. M. Elnimr *et al.*, arXiv:1307.7097.
69. J.-P. Delahaye *et al.*, arXiv:1308.0494.
70. B. Fleming, O. Palamara and D. Schmitz, LAr1-ND: Testing Neutrino Anomalies with Multiple LArTPC Detectors at Fermilab (2013).
71. M. Antonello *et al.*, arXiv:1312.7252.
72. D. Adey *et al.*, arXiv:1402.5250.
73. C. Rubbia, arXiv:1408.6431.

Neutrino–Antineutrino Mass Splitting in the Standard Model: Neutrino Oscillation and Baryogenesis

Kazuo Fujikawa* and Anca Tureanu†

*Quantum Hadron Physics Laboratory, RIKEN Nishina Center,
Wako 351-0198, Japan

†Department of Physics, University of Helsinki, P. O. Box 64,
FIN-00014 Helsinki, Finland

*k-fujikawa@riken.jp

By adding a neutrino mass term to the Standard Model, which is Lorentz and SU(2) × U(1) invariant but nonlocal to evade CPT theorem, it is shown that nonlocality within a distance scale of the Planck length, that may not be fatal to unitarity in generic effective theory, can generate the neutrino–antineutrino mass splitting of the order of observed neutrino mass differences, which is tested in oscillation experiments, and non-negligible baryon asymmetry depending on the estimate of sphaleron dynamics. The one-loop order induced electron–positron mass splitting in the Standard Model is shown to be finite and estimated at $\sim 10^{-20}$ eV, well below the experimental bound $< 10^{-2}$ eV. The induced CPT violation in the K-meson in the Standard Model is expected to be even smaller and well below the experimental bound $|m_K - m_{\bar{K}}| < 0.44 \times 10^{-18}$ GeV.

1. CPT Theorem and Its Possible Evasion

The CPT theorem formulated by Pauli and Lüders,[1] which is valid for any Lorentz invariant and local theory described by a hermitian Lagrangian with normal spin-statistics, implies the equality of the masses of the particle and antiparticle. Nevertheless, the possible breaking of CPT theorem has been discussed by many people in the past. To evade CPT theorem, one may consider, for example,

1. nonlocal theory,
2. Lorentz non-invariant theory.

We have recently discussed the possible mass splitting of the neutrino and antineutrino in the Standard Model and its physical implications on the basis of Lorentz invariant nonlocal theory.[2] This is a sequel to the analyses of fermion–antifermion mass splitting in a Lorentz invariant nonlocal theory.[3,4] This Lorentz invariant nonlocal scheme of CPT breaking itself has been revived by the authors in Ref. 5; the model considered by them is based on the T-breaking with preserved CP, and thus no particle–antiparticle mass splitting. It is sometimes stated in the literature that CPT breaking implies particle–antiparticle mass splitting, but it is not the case; CPT breaking is a necessary condition but not sufficient to generate the particle–antiparticle mass splitting. The conceptual aspect of this Lorentz invariant nonlocal scheme has also been clarified in Ref. 6, since it is sometimes *erroneously* claimed in the literature that CPT breaking inevitably implies Lorentz symmetry breaking. It was emphasized in Ref. 6 that the Lorentz invariant CPT breaking scheme, which we adopt in the present study, is a very natural logical possibility.

The neutrino mass is outside the conventional Standard Model and thus may provide a window to "brave New World". It may be interesting to incorporate CPT breaking in the neutrino mass sector of a minimal extension of the Standard Model. We incorporate,

(a) C, CP and CPT breaking,
(b) Lorentz invariance,
(c) SU(2) × U(1) gauge invariance,
(d) Nonlocality within a distance scale of the Planck length,

in our model of the neutrino–antineutrino mass splitting.[2]

We have shown that sizable neutrino–antineutrino mass splitting, which is readily tested by oscillation experiment, is possible, but the induced electron–positron mass splitting, for example, is negligibly small and in this sense our scheme of CPT breaking is consistent with the existing experimental data.[7]

2. The Model

The Standard Model Lagrangian relevant to our discussion of the electron sector is given by

$$
\begin{aligned}
\mathcal{L} = {} & i\bar{\psi}_L\gamma^\mu\left(\partial_\mu - igT^aW_\mu^a - i\frac{1}{2}g'Y_LB_\mu\right)\psi_L \\
& + i\bar{e}_R\gamma^\mu(\partial_\mu + ig'B_\mu)e_R + i\bar{\nu}_R\gamma^\mu\partial_\mu\nu_R \\
& + \left[-\frac{\sqrt{2}m_e}{v}\bar{e}_R\phi^\dagger\psi_L - \frac{\sqrt{2}m_D}{v}\bar{\nu}_R\phi_c^\dagger\psi_L - \frac{m_R}{2}\nu_R^TC\nu_R + \text{h.c.}\right], \qquad (1)
\end{aligned}
$$

with *assumed* right-handed component ν_R. We denote the Higgs doublet and its SU(2) conjugate by ϕ and $\phi_c \equiv i\tau_2\phi^\star$, respectively. We tentatively set $m_R = 0$ with enhanced lepton number symmetry, namely, a "Dirac neutrino". In short, we assume that every mass arises from the Higgs doublet, which has been discovered recently.

One may add a Hermitian *nonlocal* Higgs coupling with a real parameter μ to the above Lagrangian,[2]

$$\mathcal{L}_{\rm CPT}(x) = -i\frac{2\sqrt{2}\mu}{v}\int d^4y\,\Delta_l(x-y)\theta(x^0-y^0)$$
$$\times\{\bar{\nu}_R(x)(\phi_c^\dagger(y)\psi_L(y)) - (\bar{\psi}_L(y)\phi_c(y))\nu_R(x)\}\,, \tag{2}$$

without spoiling Lorentz invariance and $\mathrm{SU}(2)_L \times \mathrm{U}(1)$ gauge symmetry. Here we defined a "time-like nonlocal factor",

$$\Delta_l(x-y) \equiv \delta((x-y)^2 - l^2) - \delta((x-y)^2 - l'^2) \tag{3}$$

with l standing for fixed length scale and $l' = 0$, for simplicity.

In the unitary gauge, the neutrino mass term becomes

$$S_{\nu\,\rm mass} = \int d^4x\Bigg\{-m_D\bar{\nu}(x)\nu(x)\left(1+\frac{\varphi(x)}{v}\right)$$
$$-i\mu\int d^4y\,\Delta_l(x-y)[\theta(x^0-y^0)-\theta(y^0-x^0)]\bar{\nu}(x)\nu(y)$$
$$+i\mu\int d^4y\,\Delta_l(x-y)\bar{\nu}(x)\gamma_5\nu(y) - i\frac{\mu}{v}\int d^4y\,\Delta_l(x-y)\theta(x^0-y^0)$$
$$\times[\bar{\nu}(x)(1-\gamma_5)\nu(y) - \bar{\nu}(y)(1+\gamma_5)\nu(x)]\varphi(y)\Bigg\}. \tag{4}$$

The term

$$-i\mu\int d^4x\int d^4y\,\Delta_l(x-y)[\theta(x^0-y^0)-\theta(y^0-x^0)]\bar{\nu}(x)\nu(y) \tag{5}$$

in the action preserves T but has C = CP = CPT = -1 and thus gives rise to particle–antiparticle mass splitting.

The equation of motion for the free neutrino is given by

$$i\gamma^\mu\partial_\mu\nu(x) = m_D\nu(x) + i\mu\int d^4y\,\Delta_l(x-y)[\theta(x^0-y^0)-\theta(y^0-x^0)]\nu(y)$$
$$-i\mu\int d^4y\,\Delta_l(x-y)\gamma_5\nu(y)\,. \tag{6}$$

By inserting an Ansatz, $\nu(x) = e^{-ipx}U(p)$, we obtain

$$\not{p}U(p) = \{m + i[f_+(p) - f_-(p)] - ig(p^2)\gamma_5\}U(p)\,, \tag{7}$$

where

$$f_\pm(p) = \mu\int d^4z\,e^{\pm ipz}\theta(z^0)[\delta((z)^2-l^2)-\delta(z^2)]\,,$$
$$g(p^2) = \mu\int d^4z\,e^{ipz}[\delta((z)^2-l^2)-\delta((z)^2)]\,. \tag{8}$$

The last term is parity violating mass term, which is C and CPT preserving.

The factor $f_\pm(p)$ is mathematically related to the two-point Wightman function,

$$\langle 0|\phi(x)\phi(y)|0\rangle = \int d^4p\, e^{i(x-y)p}\theta(p^0)\delta(p^2-m^2)\,. \tag{9}$$

We know the properties of the Wightman function well, and they are useful in our analysis. For example, the Wightman function has a quadratic divergence for the short distance, which is independent of mass. This implies that our CPT breaking term in the Dirac equation is free of quadratic divergence in the infrared.

For time-like $p^2 > 0$, one may go to the frame where $\mathbf{p} = 0$,

$$p_0 = \gamma^0[m_D - f(p_0) - ig(p_0^2)\gamma_5]\,, \tag{10}$$

with

$$\begin{aligned} f(p_0) &\equiv -i[f_+(p_0) - f_-(p_0)] \\ &= 4\mu\pi\int_0^\infty dz\left\{\frac{z^2\sin\left[p_0\sqrt{z^2+l^2}\,\right]}{\sqrt{z^2+l^2}} - \frac{z^2\sin\left[p_0\sqrt{z^2}\,\right]}{\sqrt{z^2}}\right\}, \\ g(p_0^2) &= 4\mu\pi\int_0^\infty dz\left\{\frac{z^2\cos\left[p_0\sqrt{z^2+l^2}\,\right]}{\sqrt{z^2+l^2}} - \frac{z^2\cos\left[p_0\sqrt{z^2}\,\right]}{\sqrt{z^2}}\right\}. \end{aligned} \tag{11}$$

For space-like $p^2 < 0$, one can confirm that the CPT violating term vanishes, $f(p) = 0$, by choosing $p_\mu = (0, \mathbf{p})$.

Since we are assuming that the CPT breaking terms are small, we may solve the mass eigenvalue equations in (10) iteratively

$$m_\pm \simeq m_D - i\gamma_5 g(m_D^2) \pm f(m_D)\,. \tag{12}$$

The parity violating mass $-i\gamma_5 g(m_D^2)$ is now transformed away by a suitable global chiral transformation. In this way, the neutrino–antineutrino mass splitting is incorporated in the Standard Model by the Lorentz invariant nonlocal CPT breaking mechanism, without spoiling the $SU(2)_L \times U(1)$ gauge symmetry.

The Higgs particle φ itself has a tiny C, CP and CPT violating coupling.

3. Evaluation of Mass Splitting

The CPT violating term is evaluated as

$$\begin{aligned} f(p) = -4\pi\mu l^2[\theta(p_0) - \theta(-p_0)]\theta(p^2)&\left\{\int_1^\infty du\frac{1}{2u\left(\sqrt{u^2-1}+u\right)^2}\sin(|p_0|lu)\right. \\ &\left. -\frac{1}{2}\int_0^1 du\frac{\sin(|p_0|lu)}{u} + \int_0^1 du\, u\sin(|p_0|lu) + \frac{1}{2}\int_0^\infty du\frac{\sin(u)}{u}\right\}. \end{aligned} \tag{13}$$

Our CPT violating term is characterized by the quantity

$$\mu l^2\,, \tag{14}$$

which has the dimension of mass.

For $|p_0|l \ll 1$, which is generally expected since we are going to choose l at about the Planck length, we have Lorentz invariant,

$$f(p) \simeq -\pi^2 \mu l^2 [\theta(p_0) - \theta(-p_0)]\theta(p^2) . \tag{15}$$

Thus the neutrino–antineutrino mass splitting is given by

$$\Delta m \simeq 2\pi^2 \mu l^2 . \tag{16}$$

Our CPT violating term $f(p_0)$ is odd in p_0 and $f(\pm 0) = \mp \Delta m/2$.

As for the parity violating mass term, we have (in the frame with $\mathbf{p} = 0$ for $p^2 > 0$)

$$g(p^2) = -4\pi \mu l^2 \left\{ \int_1^\infty du \frac{1}{\sqrt{u^2-1}+u} \cos[p_0 l u] + \frac{\sin[p_0 l]}{p_0 l} + \frac{\cos[p_0 l] - 1}{(p_0 l)^2} \right\}. \tag{17}$$

This formula is again well-defined if $p_0 = 0$ is excluded.

4. Neutrino–Antineutrino Mass Splitting

The Lorentz invariant nonlocal factor

$$[\theta(x^0 - y^0) - \theta(y^0 - x^0)][\delta((x-y)^2 - l^2) - \delta((x-y)^2)] , \tag{18}$$

mostly cancels out the infinite time-like volume effect and eliminates the quadratic infrared divergence completely. In effect, nonlocality is limited within the fluctuation around the tip of the light-cone characterized by the length scale l, which we choose to be the Planck length.

By setting

$$l = 1/M_{\rm P} , \qquad \mu = M^3 , \tag{19}$$

the neutrino–antineutrino mass splitting is given by

$$\Delta m = 2\pi^2 \mu l^2 = 2\pi^2 M (M/M_{\rm P})^2 , \tag{20}$$

which may be regarded as a gravitational effect due to the Newton constant $G_{\rm N} = 1/M_{\rm P}^2$. If one chooses $M \sim 10^9$ GeV, the neutrino–antineutrino mass splitting becomes of the order of the observed neutrino mass (difference) ~ 0.1 eV. The possible neutrino–antineutrino mass splitting has been discussed in the past in connection with neutrino oscillation phenomenology.[8–10] The neutrino–antineutrino mass splitting

$$\Delta m = 10^{-1} \sim 10^{-2} \text{ eV} , \tag{21}$$

which is intended to be of the order of $m_D/5$, is generated by $M \simeq 10^8$–10^9 GeV and appears to be allowed by the presently available experimental data such as MINOS.[11]

4.1. *Baryogenesis*

A neutrino–antineutrino mass difference would result in a leptonic matter–antimatter asymmetry proportional to the mass difference. This asymmetry is transmitted to the baryon sector through the chiral anomaly and sphaleron processes which preserve $B - L$ but violate $B + L$.

This "kinematical" picture implies the asymmetry in the neutrino and antineutrino of the order[12]

$$(n_\nu - n_{\bar{\nu}})/n_\nu \simeq m_D \Delta m / T^2 \,, \tag{22}$$

which is, however, too small at the electroweak energy scale to generate the baryon asymmetry via sphaleron processes in our case with $\Delta m = 10^{-1}$–10^{-2} eV. Besides, this initial asymmetry requires the lepton number non-conservation, while the lepton number is conserved in our model without sphaleron effects.

Thus the lepton and quark sectors need to be treated simultaneously in the presence of sphalerons. Barenboim, Borissov, Lykken and Smirnov discuss a rather elaborate sphaleron dynamics and conclude at weak scale[10] M_W

$$\frac{n_B}{n_\gamma} \sim \frac{\Delta m}{M_W} \,. \tag{23}$$

This estimate in the present case with $\Delta m = 10^{-1}$–10^{-2} eV, namely $n_B/n_\gamma \sim 10^{-12}$–$10^{-13}$, is smaller than the observed value $n_B/n_\gamma \simeq 10^{-10}$, but it still gives an interesting number.

This *equilibrium electroweak baryogenesis* does not need CP violation other than for the purpose of producing neutrino–antineutrino mass splitting. This mechanism differs from the more conventional baryogenesis[13,14] or leptogenesis.[15]

5. Higher Order Effects

The propagator of the neutrino in path integral, which is based on Schwinger's action principle,[16] is given by,[3]

$$\langle T^\star \nu(x) \bar{\nu}(y) \rangle = \int \frac{d^4p}{(2\pi)^4} e^{-ip(x-y)} \frac{i}{\not{p} - m_D + i\epsilon + i\gamma_5 g(p^2) + f(p)} \,. \tag{24}$$

We can show

$$f(p) = -i[f_+(p) - f_-(p)] \to 0 \,, \tag{25}$$

$$g(p^2) \to 0 \tag{26}$$

for $p \to \infty$ in Minkowski space, which is an analogue of the cluster property of the Wightman function in (9). The propagator for Minkowski momentum is thus well behaved and the effects of nonlocality are mild and limited. One may thus be tempted to replace $T^\star$ product by the canonical T product[16] in (24).

In the analysis of the renormalization, however, it is customary to consider the Euclidean amplitude obtained from the Minkowski amplitude by Wick rotation.

Our propagator, which contains trigonometric functions, has undesirable behavior under the Wick rotation such as

$$\sin p_0 z \to i \sinh p_4 \tag{27}$$

and exponentially divergent behavior is generally induced and the effects of non-locality become significant.

One might still argue that higher order effects in field theory defined on Minkowski space are in principle analyzed in Minkowski space and, if that is the case, our propagator suggests the ordinary renormalizable behavior. This issue is left for the future study.

5.1. *Induced electron–positron mass splitting*

Our Lorentz invariant CPT violating term is effectively replaced by

$$f(p) = -\pi^2 \mu l^2 [\theta(p_0) - \theta(-p_0)]\theta(p^2)\,, \tag{28}$$

which is similar to a constant mass term except for the CPT violating factor $[\theta(p_0) - \theta(-p_0)]\theta(p^2)$. When this term is inserted into a neutrino line in Feynman diagrams of the Standard Model, those Feynman diagrams are expected to show ordinary high energy behavior for a mass insertion, if the naive power counting works. Also, perturbative unitarity may not be spoiled since $f(p)$ is essentially constant in momentum space.

We examine the electron–positron mass splitting induced by the above factor $f(p)$, when inserted into a neutrino line in one-loop self-energy diagrams of the electron in the Standard Model. The W-boson contribution is then given by

$$g^2 \int \frac{d^4p}{(2\pi)^4} \left[\gamma^\alpha \frac{(1-\gamma_5)}{2} \frac{\not{p} + m_D}{p^2 - m_D^2 + i\epsilon} f(p) \frac{\not{p} + m_D}{p^2 - m_D^2 + i\epsilon} \gamma_\alpha \frac{(1-\gamma_5)}{2} \right] \times \frac{1}{(k-p)^2 - M_W^2 + i\epsilon}\,. \tag{29}$$

We thus obtain a finite result,

$$\sim \alpha [m_D \not{k} / M_W^2](\mu l^2)[\theta(k^0) - \theta(-k^0)]\theta(k^2)$$

with α standing for the fine structure constant, which in fact gives the leading contribution.

The induced CPT violating effect on the electron–positron splitting is *finite* and it is estimated at the order,[2]

$$\alpha [m_D m_e / M_W^2](\mu l^2)[\theta(k^0) - \theta(-k^0)]\theta(k^2)\,, \tag{30}$$

which, for $\pi^2 \mu l^2 = 10^{-1}$–10^{-2} eV, is

$$|m_e - m_{\bar{e}}| \sim 10^{-20}\ \text{eV} \tag{31}$$

and thus well below the present experimental upper bound $\leq 10^{-2}$ eV.[7]

The induced CPT violation is expected to be smaller in the quark sector (as a two-loop effect) than in the charged leptons in the SU(2) × U(1) invariant theory, and thus much smaller than the well-known limit on the K-meson,[7]

$$|m_K - m_{\bar{K}}| < 0.44 \times 10^{-18}\ \text{GeV}. \tag{32}$$

6. Conclusion

Our proposed model of Lorentz invariant nonlocal CPT breaking allows sizable neutrino–antineutrino mass splitting, which can be *tested by oscillation experiments*, without inducing detectable undesirable effects in other sectors of the Standard Model. Also, it has a potentially interesting implication on baryogenesis. The Lorentz invariant nonlocal CPT breaking at the Planck scale thus suggests a promising CPT breaking scheme for an effective field theory, although a deeper analysis of basic issues such as unitarity is required. The origin of CPT breaking at the Planck scale itself remains to be clarified.

Note Added

A. Suzuki of KamLAND and A. Y. Smirnov informed us that the sun neutrino data and the KamLAND antineutrino data show a discrepancy in the neutrino mass of $\sim e^{-3}$ eV which is a 2σ effect. It is an interesting subject to examine this and other related neutrino oscillation experiments in connection with the test of CPT symmetry. We thank them for this interesting information.

References

1. W. Pauli, *Niels Bohr and the Development of Physics*, ed. W. Pauli (Pergamon Press, 1955); G. Lüders, *Mat. Fys. Medd. Dan. Vid. Selsk.* **28**, 1 (1954).
2. K. Fujikawa and A. Tureanu, *Phys. Lett. B* **743**, 39 (2015).
3. M. Chaichian, K. Fujikawa and A. Tureanu, *Phys. Lett. B* **718**, 178 (2012).
4. M. Chaichian, K. Fujikawa and A. Tureanu, *Phys. Lett. B* **712**, 115 (2012); *ibid.* **718**, 1500 (2013); *Eur. Phys. J. C* **73**, 2349 (2013).
5. M. Chaichian, A. D. Dolgov, V. A. Novikov and A. Tureanu, *Phys. Lett. B* **699**, 177 (2011).
6. M. Duetsch and J. M. Gracia-Bondia, *Phys. Lett. B* **711**, 428 (2012).
7. Particle Data Group (J. Beringer *et al.*), *Phys. Rev. D* **86**, 010001 (2012).
8. V. D. Barger, S. Pakvasa, T.J. Weiler and K. Whisnant, *Phys. Rev. Lett.* **85**, 5055 (2000).
9. H. Murayama and T. Yanagida, *Phys. Lett. B* **520**, 263 (2001); S. M. Bilenky, M. Freund, M. Lindner, T. Ohlsson and W. Winter, *Phys. Rev. D* **65**, 073024 (2002); G. Barenboim, L. Borissov and J. Lykken, *Phys. Lett. B* **534**, 106 (2002); G. Barenboim and J. D. Lykken, *Phys. Rev. D* **80**, 113008 (2009).
10. G. Barenboim, L. Borissov, J. D. Lykken and A. Y. Smirnov, *JHEP* **0210**, 001 (2002).
11. MINOS Collab. (P. Adamson *et al.*), *Phys. Rev. Lett.* **108**, 191801 (2012).
12. A. D. Dolgov and Ya. B. Zeldovich, *Rev. Mod. Phys.* **53**, 1 (1981).

13. A. D. Sakharov, *JETP Lett.* **5**, 24 (1967).
14. M. Yoshimura, *Phys. Rev. Lett.* **41**, 281 (1978); A. Yu. Ignatiev, N. V. Krasnikov, V. A. Kuzmin and A. N. Tavkhelidze, *Phys. Lett. B* **76**, 436 (1978).
15. M. Fukugita and T. Yanagida, *Phys. Lett. B* **174**, 45 (1986).
16. K. Fujikawa, *Phys. Rev. D* **70**, 085006 (2004). As for the Bjorken–Johnson–Low method, see also Appendix in K. Fujikawa and P. van Nieuwenhuizen, *Ann. Phys.* **308**, 78 (2003).

The Strong CP Problem and Discrete Symmetries

Martin Spinrath

Institut für Theoretische Teilchenphysik, Karlsruhe Institute of Technology, Engesserstraße 7, D-76131 Karlsruhe, Germany

martin.spinrath@kit.edu

We discuss a possible solution to the strong CP problem which is based on spontaneous CP violation and discrete symmetries. At the same time we predict in a simple way the almost right-angled quark unitarity triangle angle ($\alpha \simeq 90°$) by making the entries of the quark mass matrices either real or imaginary. To prove the viability of our strategy we present a toy flavour model for the quark sector.

Keywords: Strong CP problem; discrete symmetries; CP violation; flavour models.

1. Motivation

It is fair to say that quantum chromodynamics (QCD) has emerged as the well-established theory of strong interactions. However, there are still puzzles about the strong interactions. One of them is the smallness of CP violation. Already in the '70s it was realised that the QCD Lagrangian can violate CP due to instanton effects[1,2] which is described by the strong phase

$$\bar{\theta} = \theta + \arg\det(M_u M_d)\,, \tag{1}$$

where θ is the coefficient of $\alpha_s/(8\pi)\tilde{G}_{\mu\nu}G^{\mu\nu}$, $G_{\mu\nu}$ is the field strength tensor of QCD, $\tilde{G}_{\mu\nu}$ its dual, and $\arg\det(M_u M_d)$ is the anomalous contribution from the quark masses. While θ and $\arg\det(M_u M_d)$ are transformed into each other via a chiral transformation, the combination $\bar{\theta}$ stays invariant. Experiments put stringent bounds on $\bar{\theta} \lesssim 10^{-11}$, see Refs. 3 and 4, which is much smaller than the Jarlskog invariant, $J = (2.96^{+0.20}_{-0.16}) \times 10^{-5}$, see Ref. 3. Therefore, the essence of the strong CP problem is the question why the two contributions to $\bar{\theta}$ sum up to such a small number.

There are three main ideas put forward to explain the smallness of $\bar{\theta}$. The first and simplest solution is that one of the quarks is massless.[2] In this case the strong CP phase $\bar{\theta}$ is unphysical. However, recent data strongly suggests that all quarks are massive.[3]

The second popular solution is the so-called axion[5] where $\bar{\theta}$ is promoted to a dynamical degree of freedom which is set to small values by a potential. This solution is very elegant but albeit there have been extensive searches for axions there have been no convincing experimental hints for their existence so far.[3]

The third approach solves the strong CP problem by breaking parity (or CP) spontaneously. Then on the fundamental level the term $\alpha_s/(8\pi)\tilde{G}_{\mu\nu}G^{\mu\nu}$, which violates parity as well as CP, is forbidden by either parity and/or CP, which are assumed to be fundamental symmetries. For a short overview and more references, see Ref. 6, where we introduce the class of models discussed here.

As we discuss in the next section, where we outline our strategy, our class of models is based on a sum rule for the phases in the CKM matrix[7] suggesting a simple structure for quark mass matrices with either real or purely imaginary elements.[8] For an alternative class of textures models see, for instance, Refs. 9–13. As we will see our structure is realised in a simple manner in flavour models based on discrete symmetries where the CP symmetry is spontaneously broken using a method dubbed discrete vacuum alignment.[14] This method was used as well in various flavour models[15,16] which nevertheless usually put a stronger focus on the lepton sector.

2. The Strategy

If CP is a fundamental symmetry of the Lagrangian, the strong CP phase $\bar{\theta}$ vanishes on the fundamental level. However, in order to explain CP violation in weak interactions, CP has to be broken spontaneously. And this has to be done in a controlled way to keep the strong CP phase $\bar{\theta}$ at least tiny enough to be in agreement with experimental data.

In our class of models[6] we have quark mass matrices with $\arg \det(M_u M_d) = 0$ but still the value for the CKM phase is realistic. Furthermore, we disfavour unnatural cancellations between the phases in the up-type and the down-type quark sector. Hence, $\det M_u$ and $\det M_d$ should be real (and positive) by itself already.

One possible choice is, for instance, that M_u is completely real and has a negligible 1-3 mixing (1-3 element), and that

$$M_d = \begin{pmatrix} 0 & * & 0 \\ * & i* & * \\ 0 & 0 & * \end{pmatrix}, \tag{2}$$

where "$*$" are arbitrary but real entries. The only nontrivial complex phase appears in the purely imaginary 2-2 element of M_d. Then the determinants of both mass matrices are real.

This structure of the mass matrices can be realised from the spontaneous breaking of CP and we indeed have a solution for the strong CP problem as we will show in the following. And furthermore this very simple structure can also correctly reproduce the right quark unitarity triangle, as it was demonstrated in Ref. 7, since it satisfies the phase sum rule

$$\alpha \approx \delta_{12}^d - \delta_{12}^u \approx 90^\circ\,, \tag{3}$$

where α is the angle of the CKM unitarity triangle measured to be close[3] to 90° and $\delta_{12}^{d/u}$ are the phases of the complex 1-2 mixing angles diagonalising the quark mass matrices (for the conventions used, see Ref. 7). Now any model, which generates such a structure could do the trick, and in the following we will discuss one possible example.

Suppose we have a (discrete, non-Abelian) family symmetry $G_{\rm F}$ with triplet representations (we use as an example A_4, but S_4, T', $\Delta(27)$, etc. would work equally well). See Ref. 17 for a recent review on family symmetries. In our toy model we assume the right-handed down-type quarks to transform as triplets under $G_{\rm F}$ while all other quarks are singlets. Then the rows of M_d are proportional to the vacuum expectation values (vevs) of family symmetry breaking Higgs fields, so-called flavon fields, which are triplets under $G_{\rm F}$ as well. M_u is generated by vevs of singlet flavon fields.

We introduce four flavon triplets with the following alignments in flavour space

$$\langle\phi_1\rangle \sim \begin{pmatrix}1\\0\\0\end{pmatrix}, \quad \langle\phi_2\rangle \sim \begin{pmatrix}0\\1\\0\end{pmatrix}, \quad \langle\phi_3\rangle \sim \begin{pmatrix}0\\0\\1\end{pmatrix}, \quad \langle\tilde{\phi}_2\rangle \sim i\begin{pmatrix}0\\1\\0\end{pmatrix}, \tag{4}$$

where we have explicitly shown the phases and which can be achieved by standard vacuum alignment techniques. Note that only $\tilde{\phi}_2$ has a complex (imaginary) vev.

To fix the phases of these vevs we use the method described in Ref. 14, which we want to sketch here for a singlet flavon field ξ. Suppose ξ is charged under a discrete Z_n symmetry and apart from that neutral then we can write down a superpotential for ξ,

$$\mathcal{W} = P\left(\frac{\xi^n}{\Lambda^{n-2}} \mp M^2\right), \tag{5}$$

where P is a total singlet, and M and Λ mass parameters. We have dropped couplings for brevity and since we assume fundamental CP symmetry, the couplings and the mass parameters are real.[a] For the scalar potential for ξ we find

$$V = |F_P|^2 = \left|\frac{\xi^n}{\Lambda^{n-2}} \mp M^2\right|^2 \tag{6}$$

and since $|F_P| \overset{!}{=} 0$ the vev of ξ has to satisfy

$$\langle\xi^n\rangle = \pm\Lambda^{n-2}M^2\,. \tag{7}$$

[a]Note that we use the generalised CP transformation, which is trivial with respect to A_4. It agrees with the ordinary CP transformation for real representations of A_4. See Refs. 18 and 19 for a recent discussion of generalised CP in the context of non-Abelian discrete symmetries.

This means

$$\arg(\langle\xi\rangle) = \begin{cases} \dfrac{2\pi}{n}q\,, \quad q = 1,\ldots,n & \text{for “$-$” in Eq. (6)}\,, \\ \dfrac{2\pi}{n}q + \dfrac{\pi}{n}\,, \quad q = 1,\ldots,n & \text{for “$+$” in Eq. (6)}\,. \end{cases} \tag{8}$$

Here the phases of the vevs do not depend on potential parameters, a situation which has been dubbed "calculable phases" in the literature.[20] In Ref. 19 this was understood as the result of an accidental CP symmetry of the potential.

Due to the stringent constraints on $\bar{\theta}$, special care needs to be taken with possible corrections to this parameter. The most important corrections are:

(1) Higher-dimensional operators in the superpotential that could spoil the structure of the mass matrices and hence generate a non-vanishing arg $\det(M_u M_d)$.
(2) Corrections which are induced from the soft SUSY breaking terms.

Here we are only going to touch the first point. For the second point we refer to the discussion in Ref. 6.

3. The Model

In this section we briefly sketch the toy model presented in Ref. 6 which serves as a proof that the strategy outlined before can be realised in an explicit model controlling higher-dimensional operators in the superpotential.

As gauge symmetry we stick to the Standard Model gauge group and impose CP to be a fundamental symmetry. We choose here as non-Abelian discrete family symmetry A_4 which is frequently used in flavour model building, since it allows to readily realise the observed large lepton mixing (which we will not consider here) and since it is the smallest discrete group with triplet representations. To avoid unwanted operators and to implement the discrete vacuum alignment mechanism we have additionally the shaping symmetry $Z_4^5 \times Z_2 \times U(1)_R$. The family symmetry is broken by the ϕ_i, $i = 1, 2, 3$, and $\tilde{\phi}_2$ which are triplets under A_4, cf. Eq. (4). Additionally there are five singlet flavons ξ_i, $i = u, c, t, d, s$, which all receive real vevs.

To arrange for the flavon vev configuration to be dynamically realised along the lines outlined in Sec. 2, additional symmetries and fields have to be introduced. This discussion is somewhat lengthy and technical such that we will skip the detailed discussion of this nevertheless important ingredient. The interested reader can find the full superpotential to align the flavon vevs in Ref. 6.

Instead we want to discuss in somewhat more detail the couplings of the flavons to the matter sector and the corrections from higher-dimensional effective operators. After symmetry breaking, the mass matrices will be generated by the superpotential (remember that the right-handed down-type quarks form A_4 triplets while all other matter fields are A_4 singlets)

$$\mathcal{W}_d = Q_1 \bar{d} H_d \frac{\phi_2 \xi_d}{\Lambda^2} + Q_2 \bar{d} H_d \frac{\phi_1 \xi_d + \tilde{\phi}_2 \xi_s + \phi_3 \xi_t}{\Lambda^2} + Q_3 \bar{d} H_d \frac{\phi_3}{\Lambda} , \tag{9}$$

$$\mathcal{W}_u = Q_1 \bar{u}_1 H_u \frac{\xi_u^2}{\Lambda^2} + Q_1 \bar{u}_2 H_u \frac{\xi_u \xi_c}{\Lambda^2} + Q_2 \bar{u}_2 H_u \frac{\xi_c}{\Lambda} + (Q_2 \bar{u}_3 + Q_3 \bar{u}_2) H_u \frac{\xi_t}{\Lambda} + Q_3 \bar{u}_3 H_u , \tag{10}$$

which results from integrating out the heavy messenger fields and where we dropped couplings for the sake of brevity. Trivial A_4 contractions are not explicitly shown[b] and Λ denotes a generic messenger scale which is larger than the family symmetry breaking scale M_F.

Replacing Higgs and flavon fields with their respective vevs we find the following quark mass matrices

$$M_d = \begin{pmatrix} 0 & b_d & 0 \\ b_d' & i c_d & d_d \\ 0 & 0 & e_d \end{pmatrix} \quad \text{and} \quad M_u = \begin{pmatrix} a_u & b_u & 0 \\ 0 & c_u & d_u \\ 0 & d_u' & e_u \end{pmatrix} , \tag{11}$$

where we use the left–right convention $-\mathcal{L} = \overline{u_L^i} (M_u)_{ij} u_R^j + \overline{d_L^i} (M_d)_{ij} d_R^j + \text{h.c.}$ Note that due to the fundamental CP symmetry and its peculiar breaking pattern, all entries are real apart from the 2-2 element of M_d. As discussed before in the strategy section, it predicts the right quark unitarity triangle[7] in terms of a phase sum rule

$$\alpha \approx \delta_{12}^d - \delta_{12}^u \approx 90^\circ , \tag{12}$$

where the angle α of the CKM unitarity triangle is close to 90°.[3]

In this toy model, we concentrate on the explanation of CP violation in strong and weak interactions. Therefore, we are content with the prediction of the smallness of the strong CP phase and the correct CP phase in the CKM matrix. We are able to fit all masses and mixing angles, cf. Ref. 7. A more realistic model should obviously aim at predicting the masses and mixing angles as well, which happens quite naturally in a GUT context, for instance. In fact, a similar texture has been obtained in a GUT based model in Ref. 16, which could solve the strong CP problem as well.

We sketch now the UV completion of our toy model which justifies completely the effective operators we have given before. We will furthermore discuss all higher-dimensional operators which give corrections to the mass matrices and to the flavon alignment. They will not alter the structure of the mass matrices and hence our conclusions remain unchanged.

[b]The only nontrivial contraction is between $\bar{d}$ and the ϕ_i, which form a singlet contracted by the SO(3)-type inner product "$\cdot$".

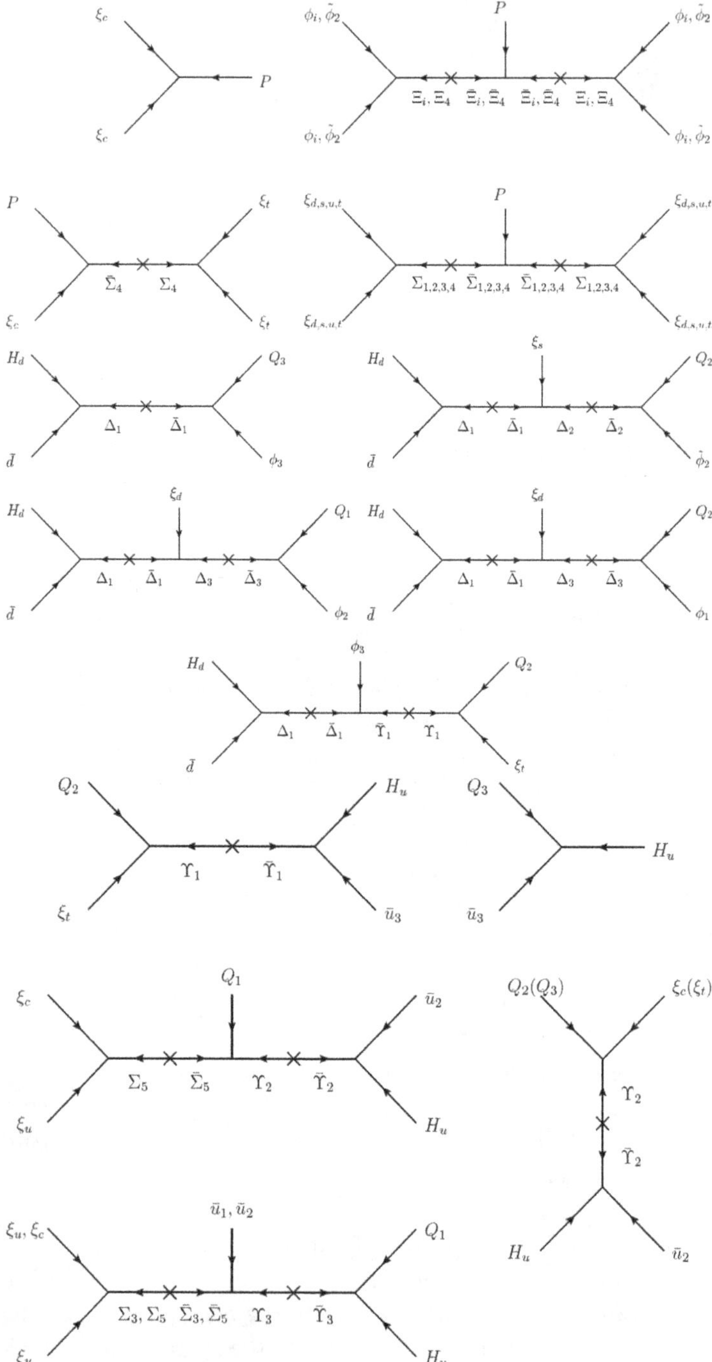

Fig. 1. The supergraphs before integrating out the messengers in our model. For the flavon sector only the diagrams are shown which fix the phases of the flavon vevs. For more details, see Ref. 6.

We will not go through the details of the full renormalisable superpotential here, which can be found in Ref. 6. The relevant point is that the symmetries and the chosen field content allow only for certain higher-dimensional operators depicted by their respective supergraphs in Fig. 1. After the heavy messenger fields are integrated out we end up first of all with the leading operators which we needed to get the right flavon alignment and the right quark mass matrices.

Beyond those operators we did not find any higher-dimensional operators produced at tree-level that would contribute to the down-type quark sector. In contrast, for the up-type quarks there are some additional operators allowed which give (real) corrections to the entries of the up-type quark mass matrix:

$$\mathcal{W}_{\mathrm{u}}^{\mathrm{corr}} = Q_1 \bar{u}_1 H_u \left(\frac{\xi_c^2 \xi_u^2 + \phi_1^2 \phi_2^2}{\Lambda^4} + \frac{\xi_c \xi_t^2 \xi_u^2}{\Lambda^5} + \frac{\xi_t^4 \xi_u^2}{\Lambda^6} \right)$$
$$+ Q_1 \bar{u}_2 H_u \frac{\xi_t^2 \xi_u}{\Lambda^3} + Q_2 \bar{u}_2 H_u \frac{\xi_t^2}{\Lambda^2} \,. \tag{13}$$

These corrections are subleading real corrections to real entries of the Yukawa matrix and hence do not alter the fact that $\bar{\theta} = 0$.

Comparatively complicated are the additional effective operators for the flavon alignment

$$\begin{aligned}
\mathcal{W}_{\mathrm{flavon}}^{\mathrm{corr}} = {} & \frac{P}{\Lambda^4}(\phi_1^2\phi_2^2\xi_u^2 + \xi_u^4\xi_c^2) + \frac{P}{\Lambda^5}\xi_c\xi_t^2\xi_u^4 \\
& + \frac{P}{\Lambda^6}(\xi_t^4\xi_u^4 + \xi_c^2\xi_u^2\phi_1^2\phi_2^2 + \phi_1^4\phi_2^4 + \xi_u^4(\phi_1^4 + \phi_2^4)) \\
& + \frac{P}{\Lambda^7}\xi_c\xi_t^2\xi_u^2\phi_1^2\phi_2^2 + \frac{P}{\Lambda^8}\xi_u^2(\xi_c^2\xi_u^2(\phi_1^4 + \phi_2^4) + \phi_1^2\phi_2^2(\xi_t^4 + \phi_1^4 + \phi_2^4)) \\
& + \frac{P}{\Lambda^9}\xi_c\xi_t^2\xi_u^4(\phi_1^4 + \phi_2^4) + \frac{P}{\Lambda^{10}}(\phi_1^4 + \phi_2^4)(\xi_t^4\xi_u^4 + \xi_c^2\xi_u^2\phi_1^2\phi_2^2 + \phi_1^4\phi_2^4) \\
& + \frac{P}{\Lambda^{11}}\xi_c\xi_t^2\xi_u^2\phi_1^2\phi_2^2(\phi_1^4 + \phi_2^4) + \frac{P}{\Lambda^{12}}\xi_t^4\xi_u^2\phi_1^2\phi_2^2(\phi_1^4 + \phi_2^4) \,.
\end{aligned} \tag{14}$$

Nevertheless, a close inspection reveals that our alignment including the phases of the flavon vevs is not altered by these additional operators which can be supported by symmetry arguments.[6]

Finally, let us briefly comment on the effects anomalies might have on our results. The gauge symmetries remain anomaly free (after adding the leptons), because we do not add new chiral fermions, which are charged under the Standard Model gauge group. In addition, as we do not introduce nontrivial singlet representations of A_4, the A_4 group is anomaly free, but some of the auxiliary Z_n symmetries appear to be anomalous.[c] However, since we do not specify here a complete model (including leptons, a SUSY breaking sector, etc.), we cannot make definite statements about

[c]For a general discussion of anomalies of discrete symmetry groups the reader is referred to Refs. 21 and 22.

anomalies but we assume that the effects of anomalies are either cancelled in the complete theory or sufficiently small.

4. Relation to Other Models

In this section we want to discuss briefly how our class of models is related to other models explaining the smallness of the strong CP phase by a spontaneous breaking of CP. We will especially focus on the Nelson–Barr models of spontaneous CP violation[23,24] being the first and most studied models. Although there are certain similarities, our model, for instance, does not fulfil the Barr criteria.[24]

We do not want to repeat the whole discussion. Instead, we just give the mass matrices in both setups. In the Nelson–Barr setup the mass matrix for the down-type quarks including heavy vector-like quarks (what we call messenger fields) would read

$$M_D \sim \begin{pmatrix} Y v_d & 0 \\ \langle\phi\rangle & M_\Upsilon \end{pmatrix}, \tag{15}$$

where they assume $Y v_d$ and M_Υ to be real by CP symmetry and only the vev of some symmetry breaking fields which governs the couplings of the light to the heavy fields induces CP violation. In such a setup one could get weak CP violation while $\bar\theta \sim \arg\det\ M_D$ still vanishes.

In our toy model we can explicitly write down the corresponding mass matrix

$$M_D \sim \begin{pmatrix} 0 & 0 & 0 & \langle\phi_2\rangle^T & 0 & 0 \\ 0 & 0 & \langle\tilde\phi_2\rangle^T & \langle\phi_1\rangle^T & \langle\xi_t\rangle & \langle\xi_c\rangle \\ 0 & \langle\phi_3\rangle^T & 0 & 0 & 0 & \langle\xi_t\rangle \\ \langle H_d\rangle & M_{\Delta_1} & 0 & 0 & 0 & 0 \\ 0 & \langle\xi_s\rangle & M_{\Delta_2} & 0 & 0 & 0 \\ 0 & \langle\xi_d\rangle & 0 & M_{\Delta_3} & 0 & 0 \\ 0 & \langle\phi_3\rangle^T & 0 & 0 & M_{\Upsilon_1} & \langle\xi_t\rangle \\ 0 & 0 & 0 & 0 & 0 & M_{\Upsilon_2} \end{pmatrix}, \tag{16}$$

which has the determinant

$$\det M_D \sim \langle H_d\rangle^3 M_{\Delta_2}^3 M_{\Delta_3}^3 M_{\Upsilon_1} M_{\Upsilon_2} \langle\xi_d^2\rangle\langle\phi_1\rangle\langle\phi_2\rangle\langle\phi_3\rangle, \tag{17}$$

which is real because $\langle\tilde\phi_2\rangle$ does not appear. This is only due to our alignment.

5. Summary and Conclusions

Here we have discussed a recently proposed novel approach to solve the strong CP problem in the context of spontaneous CP violation without the need for an axion. We assume CP to be a fundamental symmetry of nature and use discrete,

Abelian and non-Abelian (family) symmetries to break it in such a way that the anomalous contribution to the CP violating QCD parameter $\bar{\theta}$ from the quark mass matrices vanishes at tree-level. Simultaneously the CKM phase is predicted to have its observed large value in a simple and transparent way.

An essential ingredient of this approach is that the phases of the symmetry breaking vevs are fixed to certain discrete values with either being real or purely imaginary in the simplest possible setup which is governed in our example by the discrete vacuum alignment method.[14] Nevertheless, other models reproducing the texture from Eq. (2) could do the same trick.

Our toy model is supersymmetric, which helps to fix the flavon vev phases and forbids via the non-renormalisation theorem the appearance of new, unwanted operators in the superpotential from loop corrections, which could spoil our solution for the strong CP problem. Furthermore, the model is based on the family symmetry A_4 with an $U(1)_R$ symmetry and the shaping symmetry $Z_2 \times Z_4^5$ forbidding unwanted operators and providing a mechanism to fix the phases of the flavon vevs via the discrete vacuum alignment method. We discussed an UV completion of the model in that sense that we give a list of heavy messenger fields which generate the desired effective operators after being integrated out. This enables us to show explicitly that our solution for the strong CP problem is not affected by higher order corrections (ignoring non-perturbative and SUSY breaking effects).

Finally, we discussed the relation between our novel class of models to the well-known Nelson–Barr models.[23,24] In the Nelson–Barr models direct couplings between the light sector and the heavy sector are partially forbidden in such a way that the total mass matrix exhibits a special block structure. This is different in our class of models, where all light fields can couple to all heavy messenger fields in principle. The determinant of the total mass matrix in their case is real due to the mentioned block structure, while in our case it is real due to our vacuum alignment (including phases).

The class of models presented here casts new light on an old problem, the strong CP problem. There have been several previous attempts to solve it in terms of spontaneous CP violation in combination with flavour symmetries but our strategy differs significantly from these previous approaches. Most notably, we simultaneously have large CP violation in the CKM matrix with a right-angled unitarity triangle in a simple way, without any contribution to $\bar{\theta}$ from the quark mass matrices. Furthermore, the techniques to handle the symmetry breaking of discrete non-Abelian family symmetries, like in our example model A_4, was first developed in the context of the large leptonic mixing angles and finds here an unexpected new application. Also the method to fix the flavon vev phases was developed to give a dynamical explanation for the phase sum rule but was then in succeeding papers used in the lepton sector as well.

References

1. A. A. Belavin, A. M. Polyakov, A. S. Schwartz and Y. S. Tyupkin, *Phys. Lett. B* **59**, 85 (1975); R. Jackiw and C. Rebbi, *Phys. Rev. Lett.* **37**, 172 (1976); C. G. Callan, Jr., R. F. Dashen and D. J. Gross, *Phys. Lett. B* **63**, 334 (1976).
2. G. 't Hooft, *Phys. Rev. Lett.* **37**, 8 (1976).
3. Particle Data Group Collab. (J. Beringer *et al.*), *Phys. Rev. D* **86**, 010001 (2012).
4. M. Burghoff *et al.*, arXiv:1110.1505.
5. R. D. Peccei and H. R. Quinn, *Phys. Rev. Lett.* **38**, 1440 (1977); *Phys. Rev. D* **16**, 1791 (1977); S. Weinberg, *Phys. Rev. Lett.* **40**, 223 (1978); F. Wilczek, *ibid.* **40**, 279 (1978).
6. S. Antusch, M. Holthausen, M. A. Schmidt and M. Spinrath, *Nucl. Phys. B* **877**, 752 (2013).
7. S. Antusch, S. F. King, M. Malinsky and M. Spinrath, *Phys. Rev. D* **81**, 033008 (2010).
8. See also I. Masina, C. A. Savoy, *Nucl. Phys. B* **755**, 1 (2006); *Phys. Lett. B* **642**, 472 (2006).
9. S. M. Barr, *Phys. Rev. D* **56**, 1475 (1997).
10. S. M. Barr, *Phys. Rev. D* **56**, 5761 (1997).
11. A. Masiero and T. Yanagida, arXiv:hep-ph/9812225.
12. S. L. Glashow, arXiv:hep-ph/0110178.
13. D. Chang and W.-Y. Keung, *Phys. Rev. D* **70**, 051901 (2004).
14. S. Antusch, S. F. King, C. Luhn and M. Spinrath, *Nucl. Phys. B* **850**, 477 (2011).
15. A. Meroni, S. T. Petcov and M. Spinrath, *Phys. Rev. D* **86**, 113003 (2012); S. Antusch, S. F. King and M. Spinrath, *ibid.* **87**, 096018 (2013); S. F. King, *JHEP* **1307**, 137 (2013); C. Luhn, *Nucl. Phys. B* **875**, 80 (2013); S. Antusch, C. Gross, V. Maurer and C. Sluka, *ibid.* **879**, 19 (2014); G. J. Ding, S. F. King and A. J. Stuart, *JHEP* **1312**, 006 (2013); S. F. King, *ibid.* **1401**, 119 (2014); I. Girardi, A. Meroni, S. T. Petcov and M. Spinrath, *ibid.* **1402**, 050 (2014); F. Björkeroth, F. J. de Anda, I. D. M. Varzielas and S. F. King, arXiv:1503.03306.
16. S. Antusch, C. Gross, V. Maurer and C. Sluka, *Nucl. Phys. B* **877**, 772 (2013).
17. S. F. King and C. Luhn, *Rep. Prog. Phys.* **76**, 056201 (2013).
18. F. Feruglio, C. Hagedorn and R. Ziegler, arXiv:1211.5560.
19. M. Holthausen, M. Lindner and M. A. Schmidt, *JHEP* **1304**, 122 (2013).
20. G. C. Branco, J. M. Gerard and W. Grimus, *Phys. Lett. B* **136**, 383 (1984).
21. T. Araki, *Prog. Theor. Phys.* **117**, 1119 (2007); T. Araki, T. Kobayashi, J. Kubo, S. Ramos-Sanchez, M. Ratz and P. K. S. Vaudrevange, *Nucl. Phys. B* **805**, 124 (2008); C. Luhn and P. Ramond, *JHEP* **0807**, 085 (2008).
22. H. Ishimori, T. Kobayashi, H. Ohki, H. Okada, Y. Shimizu and M. Tanimoto, Lect. Notes Phys., Vol. 858 (Springer, 2012).
23. A. E. Nelson, *Phys. Lett. B* **136**, 387 (1984); *ibid.* **143**, 165 (1984).
24. S. M. Barr, *Phys. Rev. Lett.* **53**, 329 (1984); *Phys. Rev. D* **30**, 1805 (1984).

Neutrino Interaction with Background Matter in a Noninertial Frame

Maxim Dvornikov

Institute of Physics, University of São Paulo,
CP 66318, CEP 05315-970 São Paulo, SP, Brazil
Pushkov Institute of Terrestrial Magnetism, Ionosphere and Radiowave Propagation (IZMIRAN), 142190 Troitsk, Moscow, Russia
Physics Faculty, National Research Tomsk State University,
36 Lenin Ave., 634050 Tomsk, Russia
maxim.dvornikov@usp.br

We study Dirac neutrinos propagating in rotating background matter. First we derive the Dirac equation for a single massive neutrino in the noninertial frame, where matter is at rest. This equation is written in the effective curved space-time corresponding to the corotating frame. We find the exact solution of the Dirac equation. The neutrino energy levels for ultrarelativistic particles are obtained. Then we discuss several neutrino mass eigenstates, with a nonzero mixing between them, interacting with rotating background matter. We derive the effective Schrödinger equation governing neutrino flavor oscillations in rotating matter. The new resonance condition for neutrino oscillations is obtained. We also examine the correction to the resonance condition caused by the matter rotation.

Keywords: Massive and mixed neutrinos; Dirac equation in curved space-time; exact solution; neutrino oscillations in matter; noninertial effects, rotation.

1. Introduction

Nowadays it is commonly believed that neutrinos are massive particles and there is a nonzero mixing between different neutrino generations. These neutrino properties result in transitions between neutrino flavors, or neutrino oscillations. It is also known that various external fields, like electroweak interaction of neutrinos with background fermions, neutrino electromagnetic interaction, and neutrino interaction with a strong gravitational field, can also influence the process of neutrino oscillations.

As shown in Ref. 1, noninertial effects in accelerated and rotating frames can affect neutrino propagation and oscillations. The consideration of the reference frame rotation is particularly important for astrophysical neutrinos emitted by a rapidly rotating compact star, e.g., a pulsar. For example, the possibility of the pulsar spin down by the neutrino emission and interaction with rotating matter was recently discussed in Ref. 2. It should be noted that besides elementary particle physics, various processes in noninertial frames are actively studied in condensed matter physics. For instance, the enhancement of the spin current in a semiconductor moving with an acceleration was recently predicted in Ref. 3.

In the present work we summarize the results of Ref. 4, where the neutrino interaction with background matter in a rotating frame was studied. We assume that neutrinos can interact with background fermions by means of electroweak forces. We also take that neutrino mass eigenstates are Dirac particles. In our treatment we account for the noninertial effects since we find the exact solution of a Dirac equation for a neutrino moving in the curved space-time with a metric corresponding a rotating frame.

This work is organized in the following way. In Sec. 2, we start with the brief description of the neutrino interaction with background matter in Minkowski space-time. We consider matter moving with a constant velocity and discuss both neutrino flavor and mass eigenstates. Then, in Sec. 3, we study background matter moving with an acceleration. The Dirac equation, including noninertial effects, for a neutrino interacting with such matter is written down in a frame where matter is at rest. In Subsec. 3.1, we solve the Dirac equation and find the neutrino energy spectrum for ultrarelativistic neutrinos moving in matter rotating with a constant angular velocity. Then, in Sec. 4, we apply our results for the description of neutrino oscillations in rotating background matter. The effective Schrödinger equation governing neutrino oscillations is derived and the new resonance condition is obtained. We consider how the matter rotation can affect the resonance in neutrino oscillation in a realistic astrophysical situation. Finally, in Sec. 5, we summarize our results.

2. Neutrino Interaction with Background Matter

In this section we describe the interaction of different neutrino flavors with background matter in a flat space-time. We discuss a general case of matter moving with a constant mean velocity and having a mean polarization. Then we consider the matter interaction of neutrino mass eigenstates, which are supposed to be Dirac particles.

The interaction of the neutrino flavor eigenstates ν_α, $\alpha = e, \mu, \tau$, with background matter in the flat space-time is described by the following effective Lagrangian:[5]

$$\mathcal{L}_{\text{eff}} = -\sum_\alpha \bar{\nu}_\alpha \gamma_\mu^{\text{L}} \nu_\alpha \cdot f_\alpha^\mu, \tag{1}$$

where $\gamma_\mu^{\text{L}} = \gamma_\mu(1-\gamma^5)/2$, $\gamma^\mu = (\gamma^0, \boldsymbol{\gamma})$ are the Dirac matrices, and $\gamma^5 = \mathrm{i}\gamma^0\gamma^1\gamma^2\gamma^3$.

The interaction Lagrangian in Eq. (1) is derived in the mean field approximation using the effective external currents f_α^μ depending on the characteristics of background matter as[6]

$$f_{\alpha}^{\mu} = \sqrt{2} G_{\mathrm{F}} \sum_{f} \left(q_{\alpha,f}^{(1)} j_{f}^{\mu} + q_{\alpha,f}^{(2)} \lambda_{f}^{\mu} \right), \tag{2}$$

where G_{F} is the Fermi constant and the sum is taken over all background fermions f. Here

$$j_{f}^{\mu} = n_{f} u_{f}^{\mu}, \tag{3}$$

is the hydrodynamic current and

$$\lambda_{f}^{\mu} = n_{f} \left((\boldsymbol{\zeta}_{f} \mathbf{u}_{f}), \boldsymbol{\zeta}_{f} + \frac{\mathbf{u}_{f} (\boldsymbol{\zeta}_{f} \mathbf{u}_{f})}{1 + u_{f}^{0}} \right), \tag{4}$$

is the four vector of the matter polarization. In Eqs. (3) and (4), n_f is the invariant number density (the density in the rest frame of fermions), $\boldsymbol{\zeta}_f$ is the invariant polarization (the polarization in the rest frame of fermions), and $u_{f}^{\mu} = \left(u_{f}^{0}, \mathbf{u}_{f} \right)$ is the four velocity. To derive Eqs. (2)–(4) it is crucial that background fermions have constant velocity. Only in this situation one can make a boost to the rest frame of the fermions where n_f and $\boldsymbol{\zeta}_f$ are defined. The explicit form of the coefficients $q_{\alpha,f}^{(1,2)}$ in Eq. (2) can be found in Ref. 6.

Nowadays it is experimentally confirmed that the flavor neutrino eigenstates are the superposition of the neutrino mass eigenstates, ψ_i, $i = 1, 2, \ldots$,

$$\nu_{\alpha} = \sum_{i} U_{\alpha i} \psi_{i}, \tag{5}$$

where $(U_{\alpha i})$ is the unitary mixing matrix. The transformation in Eq. (5) diagonalizes the neutrino mass matrix. Only using the neutrino mass eigenstates we can reveal the nature of neutrinos, i.e. say whether they are Dirac or Majorana particles. Despite the great experimental efforts to shed light upon the nature of neutrinos, this issue still remains open. Here we shall suppose that ψ_i correspond to Dirac fields.

The effective Lagrangian for the interaction of ψ_i with background matter can be obtained using Eqs. (1) and (5),

$$\mathcal{L}_{\mathrm{eff}} = - \sum_{ij} \bar{\psi}_{i} \gamma_{\mu}^{\mathrm{L}} \psi_{j} \cdot g_{ij}^{\mu}, \tag{6}$$

where

$$g_{ij}^{\mu} = \sum_{\alpha} U_{\alpha i}^{*} U_{\alpha j} f_{\alpha}^{\mu}, \tag{7}$$

is the nondiagonal effective potential in the mass eigenstates basis.

Using Eq. (6) one obtains that the corresponding Dirac equations for the neutrino mass eigenstates are coupled,

$$\left[\mathrm{i} \gamma^{\mu} \partial_{\mu} - m_{i} - \gamma_{\mu}^{\mathrm{L}} g_{ii}^{\mu} \right] \psi_{i} = \sum_{j \neq i} \gamma_{\mu}^{\mathrm{L}} g_{ij}^{\mu} \psi_{j}, \tag{8}$$

where m_i is the mass of ψ_i. One can proceed in the analytical analysis of Eq. (8) if we exactly account for only the diagonal effective potentials g_{ii}^{μ}. To take into account the r.h.s. of Eq. (8), depending on the nondiagonal elements of the matrix (g_{ij}^{μ}), with $i \neq j$, one should apply a perturbative method (see Sec. 4 below).

3. Massive Neutrinos in Noninertial Frames

In this section we generalize the Dirac equation for a neutrino interacting with a background matter to the situation when the velocity of the matter motion is not constant. In particular, we study the case of the matter rotation with a constant angular velocity. Then we obtain the solution of the Dirac equation and find the energy spectrum.

If we discuss a neutrino mass eigenstate propagating in a nonuniformly moving matter, the expressions for f_{α}^{μ} in Eqs. (2)–(4) become invalid since they are derived under the assumption of the unbroken Lorentz invariance. The most straightforward way to describe the neutrino evolution in matter moving with an acceleration is to rewrite the Dirac equation for a neutrino in the noninertial frame where matter is at rest. In this case one can unambiguously define the components of f_{α}^{μ}. Assuming that background fermions are unpolarized, we find that in this reference frame

$$f_{\alpha}^{0} = \sqrt{2} G_{\mathrm{F}} \sum_{f} q_{\alpha,f}^{(1)} n_f \neq 0, \tag{9}$$

with the rest of the effective potentials being equal to zero.

It is known that the motion of a test particle in a noninertial frame is equivalent to the interaction of this particle with a gravitational field. The Dirac equation for a massive neutrino moving in a curved space-time and interacting with background matter can be obtained by the generalization of Eq. (8) (see also Ref. 7),

$$[\mathrm{i}\gamma^{\mu}(x)\nabla_{\mu} - m]\,\psi = \frac{1}{2}\gamma_{\mu}(x) g^{\mu}\left[1 - \gamma^{5}(x)\right]\psi, \tag{10}$$

where $\gamma_{\mu}(x)$ are the coordinate dependent Dirac matrices, $\nabla_{\mu} = \partial_{\mu} + \Gamma_{\mu}$ is the covariant derivative, Γ_{μ} is the spin connection, $\gamma^{5}(x) = -\frac{\mathrm{i}}{4!}E^{\mu\nu\alpha\beta}\gamma_{\mu}(x)\gamma_{\nu}(x)\gamma_{\alpha}(x)\gamma_{\beta}(x)$, $E^{\mu\nu\alpha\beta} = \frac{1}{\sqrt{-g}}\varepsilon^{\mu\nu\alpha\beta}$ is the covariant antisymmetric tensor in curved space-time, and $g = \det(g_{\mu\nu})$ is the determinant of the metric tensor $g_{\mu\nu}$. Note that in Eq. (10) we account for only the diagonal neutrino interaction with matter. That is why we omit the index i in order not to encumber the notation: $m \equiv m_i$, etc. It should be noted that analogous Dirac equation was discussed in Ref. 8.

We shall be interested in the neutrino motion in matter rotating with the constant angular velocity ω. Choosing the corotating frame we get that only $g^{0} \equiv g_{ii}^{0}$ is nonvanishing, cf. Eqs. (7) and (9).

3.1. *Neutrino motion in a rotating frame*

The interval in the rotating frame is[9]

$$ds^2 = g_{\mu\nu}dx^\mu dx^\nu = (1-\omega^2r^2)dt^2 - dr^2 - 2\omega r^2 dt d\phi - r^2 d\phi^2 - dz^2, \tag{11}$$

where we use the cylindrical coordinates $x^\mu = (t, r, \phi, z)$. One can check that the metric tensor in Eq. (11) can be diagonalized, $\eta_{ab} = e_a{}^\mu e_b{}^\nu g_{\mu\nu}$, if we use the following vierbein vectors:

$$\begin{aligned}
e_0{}^\mu &= \left(\frac{1}{\sqrt{1-\omega^2r^2}}, 0, 0, 0\right), \\
e_1{}^\mu &= (0, 1, 0, 0), \\
e_2{}^\mu &= \left(\frac{\omega r}{\sqrt{1-\omega^2r^2}}, 0, \frac{\sqrt{1-\omega^2r^2}}{r}, 0\right), \\
e_3{}^\mu &= (0, 0, 0, 1).
\end{aligned} \tag{12}$$

Here $\eta_{ab} = \mathrm{diag}(1, -1, -1, -1)$ is the metric in a locally Minkowskian frame.

Let us introduce the Dirac matrices in a locally Minkowskian frame by $\gamma^{\bar{a}} = e^a{}_\mu \gamma^\mu(x)$, where $e^a{}_\mu$ is the inverse vierbein: $e^a{}_\mu e_b{}^\mu = \delta^a_b$. Starting from now, we shall mark an index with a bar to demonstrate that a gamma matrix is defined in a locally Minkowskian frame. As shown in Ref. 4, $\gamma^5(x) = i\gamma^{\bar{0}}\gamma^{\bar{1}}\gamma^{\bar{2}}\gamma^{\bar{3}} = \gamma^{\bar{5}}$ does not depend on coordinates.

After the straightforward calculation of the spin connection on the basis of Eq. (12), the Dirac Eq. (10) can be rewritten as

$$\begin{aligned}
[\mathcal{D} - m]\psi &= \frac{1}{2}\sqrt{1-\omega^2r^2}\gamma^{\bar{0}}g^0(1-\gamma^{\bar{5}})\psi, \\
\mathcal{D} &= i\frac{\gamma^{\bar{0}} + \omega r\gamma^{\bar{2}}}{\sqrt{1-\omega^2r^2}}\partial_0 + i\gamma^{\bar{1}}\left(\partial_r + \frac{1}{2r}\right) + i\gamma^{\bar{2}}\frac{\sqrt{1-\omega^2r^2}}{r}\partial_\phi + i\gamma^{\bar{3}}\partial_z \\
&\quad - \frac{\omega}{2(1-\omega^2r^2)}\gamma^{\bar{3}}\gamma^{\bar{5}}.
\end{aligned} \tag{13}$$

The analogous Dirac equation was recently derived in Ref. 10. Since Eq. (13) does not explicitly contain t, ϕ, and z, its solution can be expressed as

$$\psi = \exp\left(-iEt + iJ_z\phi + ip_z z\right)\psi_r, \tag{14}$$

where $\psi_r = \psi_r(r)$ is the spinor depending on the radial coordinate, $J_z = \frac{1}{2} - l$ (see, e.g., Ref. 11), and $l = 0, \pm 1, \pm 2, \ldots$.

In Eq. (13) one can neglect terms $\sim (\omega r)^2$. Indeed, if we study a neutrino in a rotating pulsar, then $r \lesssim 10\,\mathrm{km}$ and $\omega \lesssim 10^3\,\mathrm{s}^{-1}$. Thus $(\omega r)^2 \lesssim 1.1 \times 10^{-3}$ is a small parameter. Therefore Eq. (13) can be transformed to

$$\begin{aligned}
\Bigg[i\gamma^{\bar{1}}\left(\partial_r + \frac{1}{2r}\right) &- \gamma^{\bar{2}}\left(\frac{J_z}{r} - \omega r E\right) + \gamma^{\bar{0}}\left(E - \frac{g^0}{2}\right) - \gamma^{\bar{3}}p_z \\
&+ \frac{g^0}{2}\gamma^{\bar{0}}\gamma^{\bar{5}} - \frac{\omega}{2}\gamma^{\bar{3}}\gamma^{\bar{5}} - m\Bigg]\psi_r = 0,
\end{aligned} \tag{15}$$

where we keep only the terms linear in ω. It should be noted that the term $\sim \omega\gamma^{\bar{3}}\gamma^{\bar{5}}$ in Eq. (15) is equivalent to the neutrino interaction with matter moving with an effective velocity.

The solution of Eq. (15) can be presented in the form,[4] $\psi_r^{\mathrm{L}} = (0,\eta)^{\mathrm{T}}$ and $\psi_r^{\mathrm{R}} = (\xi,0)^{\mathrm{T}}$, where

$$\eta = \begin{pmatrix} -\mathrm{i}C_1 I_{N,s} \\ C_2 I_{N-1,s} \end{pmatrix}, \quad \xi = \begin{pmatrix} C_3 I_{N,s} \\ -\mathrm{i}C_4 I_{N-1,s} \end{pmatrix}. \tag{16}$$

Here $N = 0,1,2,\ldots$, $s = N - l$, and $I_{N,s} = I_{N,s}(\rho)$ is the Laguerre function. The explicit form of the Laguerre function can be found, e.g., in Ref. 4. To derive Eq. (16) we use the Dirac matrices in the chiral representation.[12]

In the important case when $\omega \ll g^0$, the coefficients C_i, $i = 1,\ldots,4$, in Eq. (16) are expressed in the following way:[4]

$$\begin{aligned} C_1^2 \approx & \frac{E_A\omega}{2\pi}\frac{E_A - p_z - g^0}{E_A - g^0}, \quad C_3^2 \approx \frac{\omega}{2\pi}\left(E_S + p_z\right), \\ C_2^2 \approx & \frac{E_A\omega}{2\pi}\frac{E_A + p_z - g^0}{E_A - g^0}, \quad C_4^2 \approx \frac{\omega}{2\pi}\left(E_S - p_z\right). \end{aligned} \tag{17}$$

It should be noted that the solutions presented in Eqs. (16) and (17) satisfy the normalization condition,

$$\int \psi_{N,s,p_z}^{\dagger}(x)\psi_{N',s',p_z'}(x)\sqrt{-g}\mathrm{d}^3x = \delta_{NN'}\delta_{ss'}\delta\left(p_z - p_z'\right). \tag{18}$$

Here ψ and ψ_r are related by Eq. (14).

The energy levels in Eq. (17) are

$$\begin{aligned} \left[E_A - 2N\omega - g^0\right]^2 =& (2N\omega)^2 + 4N\omega g^0 + \left(p_z - \frac{\omega}{2}\right)^2, \\ \left[E_S - 2N\omega\right]^2 =& (2N\omega)^2 + \left(p_z + \frac{\omega}{2}\right)^2, \end{aligned} \tag{19}$$

where E_A and E_S are the energies of active and sterile neutrinos respectively. Comparing the expression for $E_S \approx 2N\omega + \sqrt{(2N\omega)^2 + p_z^2}$ with the energy of a neutrino in an inertial nonrotating frame $\sqrt{\mathbf{p}_\perp^2 + p_z^2}$, where $\mathbf{p}_\perp$ is the momentum in the equatorial plane, we can identify $2N\omega$ inside the square root as $|\mathbf{p}_\perp|$. It should be also noted that the term $2N\omega$, which additively enters to both E_A and E_S, is due to the noninertial effects for a Dirac fermion in a rotating frame.[13]

We can also get the corrections to the energy levels due to the nonzero mass, $E_{A,S} \to E_{A,S} + E_{A,S}^{(1)}$. On the basis of Eq. (16) and (17) one finds the expression for $E_{A,S}^{(1)}$ in the limit $\omega \ll g^0$,

$$E_A^{(1)} = \frac{m^2}{2\left(E_A - 2N\omega - g^0\right)}, \quad E_S^{(1)} = \frac{m^2}{2\left(E_S - 2N\omega\right)}. \tag{20}$$

If we discuss neutrinos moving along the rotation axis, then $2N\omega \ll |p_z|$. Using Eq. (19) we get the energy levels of active neutrinos in this case

$$E_A = |p_z| + g^0 \left(1 + \frac{2N\omega}{|p_z|}\right) + 2N\omega + \frac{2(N\omega)^2}{|p_z|} + \frac{m^2}{2|p_z|}, \tag{21}$$

where we also keep the mass correction in Eq. (20). One can see in Eq. (21) that $|p_z| + g^0 + \frac{m^2}{2|p_z|}$ corresponds to the energy of a left-handed neutrino interacting with background matter in a flat space-time. The rest of the terms in Eq. (21) are the corrections due to the matter rotation.

4. Flavor Oscillations of Dirac Neutrinos in Rotating matter

In this section we study the evolution of the system of massive mixed neutrinos in rotating matter. We formulate the initial condition for this system and derive the effective Schrödinger equation which governs neutrino flavor oscillations. Then we find the correction to the resonance condition owing to the matter rotation and estimate its value for a millisecond pulsar.

We can generalize the results of Sec. 3 to include different neutrino eigenstates. The interaction of neutrino mass eigenstates with background matter is nondiagonal, cf. Eq. (6). Therefore the generalization of Eq. (13) for several mass eigenstates ψ_i reads

$$[\mathcal{D} - m_i]\,\psi_i = \frac{1}{2}\gamma^{\bar{0}} g_i^0 (1 - \gamma^{\bar{5}})\psi_i + \frac{1}{2}\gamma^{\bar{0}} \sum_{j \neq i} g_{ij}^0 (1 - \gamma^{\bar{5}})\psi_j, \tag{22}$$

where $g_i^0 \equiv g_{ii}^0$ and g_{ij}^0 are the time components of the matrix (g_{ij}^μ) given in Eq. (7), m_i is the mass of ψ_i, and $\mathcal{D}$ can be found in Eq. (13). As in Sec. 3, we omitted the term $(\omega r)^2 \ll 1$ in Eq. (22). Note that Eq. (22) is a generalization of Eq. (8) for a system of the neutrino mass eigenstates moving in a rotating frame.

We shall study the evolution of active ultrarelativistic neutrinos and neglect neutrino-antineutrino transitions. In this case we can restrict ourselves to the analysis of two component spinors. The general solution of Eq. (22) has the form,

$$\eta_i(x) = \sum_{N,s} \int \frac{\mathrm{d}p_z}{\sqrt{2\pi}} a_{N,s,p_z}^{(i)} e^{\mathrm{i}p_z z + \mathrm{i}J_z \phi} u_{N,s,p_z}(r) e^{-\mathrm{i}E_i t}, \tag{23}$$

where u_{N,s,p_z} are the basis spinors and $a_{N,s,p_z}^{(i)} = a_{N,s,p_z}^{(i)}(t)$ are the c-number functions. The energy levels E_i are given in Eq. (21) with $m \to m_i$. Here we omit the subscript A in order not to encumber the notation. Our goal is to find the coefficient $a_{N,s,p_z}^{(i)} = a_{N,s,p_z}^{(i)}(t)$. We neglect the small ratio ω/g_i^0 in Eq. (23).

Considering the system of two neutrino mass eigenstates, $i = 1, 2$, parametrized with one mixing angle θ, and choosing the appropriate initial condition,[4] on the basis of Eq. (22) we get the effective Schrödinger equation for $\tilde{\Psi}^{\mathrm{T}} = (a_1, a_2)$,

$$\mathrm{i}\frac{\mathrm{d}\tilde{\Psi}}{\mathrm{d}t} = \begin{pmatrix} 0 & g_{12}^0 \exp\left[\mathrm{i}\left(E_1 - E_2\right)t\right] \\ g_{12}^0 \exp\left[\mathrm{i}\left(E_2 - E_1\right)t\right] & 0 \end{pmatrix} \tilde{\Psi}. \tag{24}$$

Here we omitted all the indexes of a_i besides $i = 1, 2$. It is convenient to introduce the modified effective wave function $\Psi = \mathcal{U}_3\tilde{\Psi}$, where $\mathcal{U}_3 = \mathrm{diag}\left(e^{i\Omega t/2}, e^{-i\Omega t/2}\right)$, $\Omega = E_1 - E_2$. Using Eq. (24), we get for Ψ

$$\mathrm{i}\frac{\mathrm{d}\Psi}{\mathrm{d}t} = \begin{pmatrix} \Omega/2 & g_{12}^0 \\ g_{12}^0 & -\Omega/2 \end{pmatrix} \Psi. \tag{25}$$

Note that Eq. (25) has the form of the effective Schrödinger equation one typically deals with in the study of neutrino flavor oscillations in background matter.

If the transition probability for $\nu_\alpha \leftrightarrow \nu_\beta$ is close to one, i.e. $P_{\nu_\beta \to \nu_\alpha} = |\langle \nu_\alpha(t)|\nu_\beta(0)\rangle|^2 \approx 1$, flavor oscillations of neutrinos are said to be at resonance. Using Eqs. (5), (9), (21), and (25), the resonance condition can be written as,

$$\left(f_\alpha^0 - f_\beta^0\right)\left(1 + \frac{2N\omega}{|p_z|}\right) + \frac{\Delta m^2}{2|p_z|}\cos 2\theta = 0, \tag{26}$$

where $\Delta m^2 = m_1^2 - m_2^2$ is the mass squared difference.

Let us consider electroneutral background matter composed of electrons, protons, and neutrons. If we study the $\nu_e \to \nu_\alpha$ oscillation channel, where $\alpha = \mu, \tau$, we get that $f_{\nu_\alpha}^0 = -\frac{1}{\sqrt{2}}G_\mathrm{F}n_n$ and $f_{\nu_\beta}^0 \equiv f_{\nu_e}^0 = \sqrt{2}G_\mathrm{F}\left(n_e - \frac{1}{2}n_n\right)$, where n_e and n_n are the densities of electrons and neutrons. Using Eq. (26), we obtain that

$$\sqrt{2}G_\mathrm{F}n_e\left(1 + \frac{2N\omega}{|p_z|}\right) = \frac{\Delta m^2}{2|p_z|}\cos 2\theta. \tag{27}$$

At the absence of rotation, $\omega = 0$, Eq. (27) is equivalent to the Mikheyev–Smirnov–Wolfenstein resonance condition in background matter.[14]

Let us evaluate the contribution of the matter rotation to the resonance condition in Eq. (27) for a neutrino emitted inside a rotating pulsar. We make a natural assumption that for a corotating observer neutrinos are emitted in a spherically symmetric way from a neutrinosphere. That is we should take that $l \approx 0$ and $N \approx s$. Then the trajectory of a neutrino is deflected because of the noninertial effects and the interaction with background matter. The radius $\mathcal{R}$ of the trajectory can be found from

$$\mathcal{R}^2 = 2|p_z|\omega \int_0^\infty r^2 |u_{N,s,p_z}(r)|^2 r\mathrm{d}r \approx \frac{2N}{|p_z|\omega}, \tag{28}$$

where we take into account that $N \gg 1$.

We shall assume that $\mathcal{R} \sim R_0$, where $R_0 = 10\,\mathrm{km}$ is the pulsar radius. In this case neutrinos escape a pulsar. Taking that $\omega = 10^3\,\mathrm{s}^{-1}$ and using Eq. (28), we get that the correction to the resonance condition in Eq. (27) is $\frac{2N\omega}{|p_z|} \approx (R_0\omega)^2 \approx 10^{-3}$. The obtained correction to the effective number density is small but nonzero. This result corrects our previous statement[2] that a matter rotation does not contribute neutrino flavor oscillations.

5. Conclusion

In conclusion we notice that we have studied the evolution of massive mixed neutrinos in nonuniformly moving background matter. The interaction of neutrinos with background fermions is described in frames of the Fermi theory (see Sec. 2). A particular case of the matter rotating with a constant angular velocity has been studied in Subsec. 3.1. We have derived the Dirac equation for a weakly interacting neutrino in a rotating frame and found its solution in case of ultrarelativistic neutrinos, cf. Eqs. (16) and (17). The energy spectrum obtained in Eqs. (19) and (20) includes the correction owing to the nonzero neutrino mass.

We have used the Dirac equation in a noninertial frame, cf. Eq. (10), as a main tool for the study of the neutrino motion in matter moving with an acceleration. To develop the quantum mechanical description of such a neutrino we have chosen a noninertial frame where matter is at rest. In this frame the effective potential of the neutrino-matter interaction is well defined. However, the wave equation for a neutrino turns out to be more complicated since one has to deal with noninertial effects.

In Sec. 4 we have generalized our results to include various neutrino generations as well as mixing between them. We have derived the effective Schrödinger equation which governs neutrino flavor oscillations. We have obtained the correction to the resonance condition in background matter owing to the matter rotation. Studying neutrino oscillations in a millisecond pulsar, we have obtained that the effective number density changes by 0.1 % owing to the matter rotation.

Despite the obtained correction is small, we may suggest that our results can have some implication to the explanation of great peculiar velocities of pulsars. It was suggested in Ref. 15 that an asymmetry in neutrino oscillations in a magnetized pulsar can explain a great peculiar velocity of the compact star. An evidence for the alignment of the angular and the linear velocity vectors of pulsars was reported in Ref. 16. Therefore we may suggest that neutrino flavor oscillations in a rapidly rotating pulsar can contribute to its peculiar velocity. It should be noted that neutrino spin-flavor oscillations, including noninertial effects, in a rapidly rotating magnetized star were studied in Ref. 17 in the context of the explanation of high peculiar velocities of pulsars.

Finally, we mention that the Dirac equation for a fermion, electroweakly interacting with the rotating background matter, was recently solved.[18] The vierbein vectors, different from these in Eq. (12), were used in Ref. 18. Comparing the energy levels obtained in Ref. 18 with the results of the general analysis,[13] one concludes that the vierbein used in Ref. 18 is more appropriate for the description of ultrarelativistic particles like neutrinos.

Acknowledgments

I am thankful to S. P. Gavrilov for helpful comments, to FAPESP (Brazil) for the Grant No. 2011/50309-2, to the Tomsk State University Competitiveness Improvement Program and to RFBR (research project No. 15-02-00293) for partial support.

References

1. G. Lambiase, Neutrino oscillations in non-inertial frames and the violation of the equivalence principle. Neutrino mixing induced by the equivalence principle violation, *Eur. Phys. J. C* **19**, 553 (2001).
2. M. Dvornikov and C. O. Dib, Spin-down of neutron stars by neutrino emission, *Phys. Rev. D* **82**, 043006 (2010) [arXiv:0907.1445].
3. B. Basu and D. Chowdhury, Inertial effect on spin orbit coupling and spin transport, *Ann. Phys. (N.Y.)* **335**, 47 (2013) [arXiv:1302.1063].
4. M. Dvornikov, Neutrino interaction with matter in a noninertial frame, *J. High Energy Phys.* **10** (2014) 053 [arXiv:1408.2735].
5. C. Giunti and C. W. Kim, *Fundamentals of Neutrino Physics and Astrophysics* (Oxford University Press, Oxford, 2007), pp. 137–179.
6. M. Dvornikov and A. Studenikin, Neutrino spin evolution in presence of general external fields, *J. High Energy Phys.* **09** (2002) 016 [hep-ph/0202113].
7. A. A. Grib, S. G. Mamaev and V. M. Mostepanenko, *Quantum Effects in Intense External Fields: Methods and Results not Related to the Perturbation Theory* (Atomizdat, Moscow, 1980), pp. 13–15.
8. D. Píriz, M. Roy and J. Wudka, Neutrino oscillations in strong gravitational fields, *Phys. Rev. D* **54**, 1587 (1996) [hep-ph/9604403].
9. L. D. Landau and E. M. Lifshitz, *The Classical Theory of Fields* (Butterworth Heinemann, Amsterdam, 1994), 4th ed., pp. 329–330.
10. K. Bakke, Rotating effects on the Dirac oscillator in the cosmic string spacetime, *Gen. Relativ. Grav.* **45**, 1845 (2013) [arXiv:1307.2847].
11. P. Schluter, K.-H. Wietschorke and W. Greiner, The Dirac equation in orthogonal coordinate systems: I. The local representation, *J. Phys. A: Math. Gen.* **16**, 1999 (1983).
12. C. Itzykson and J.-B. Zuber, *Quantum Field Theory* (McGraw-Hill, New York, 1980), pp. 691–696.
13. F. W. Hehl and W.-T. Ni, Inertial effects of a Dirac particle, *Phys. Rev. D* **42**, 2045 (1990).
14. M. Blennow and A. Yu. Smirnov, Neutrino propagation in matter, *Adv. High Energy Phys.* **2013**, 972485 (2013) [arXiv:1306.2903].
15. A. Kusenko and G. Segrè, Velocities of pulsars and neutrino oscillations, *Phys. Rev. Lett.* **77**, 4872 (1996) [hep-ph/9606428].
16. S. Johnston, G. Hobbs, S. Vigeland, M. Kramer, J. M. Weisberg and A. G. Lyne, Evidence for alignment of the rotation and velocity vectors in pulsars, *Mon. Not. Roy. Astron. Soc.* **364**, 1397 (2005) [astro-ph/0510260].
17. G. Lambiase, Pulsar kicks induced by spin flavor oscillations of neutrinos in gravitational fields, *Mon. Not. Roy. Astron. Soc.* **362**, 867 (2005) [astro-ph/0411242].
18. M. Dvornikov, Galvano-rotational effect in a pulsar induced by the electroweak interaction [arXiv:1503.00608].

Seesaw Models with Minimal Flavor Violation

Xiao-Gang He

INPAC, SKLPPC, and Department of Physics,
Shanghai Jiao Tong University, Shanghai 200240, China
National Center for Theoretical Sciences and
Physics Department of National Tsing Hua University, Hsinchu 300, Taiwan
CTS, CASTS, and Physics Department,
National Taiwan University, Taipei 106, Taiwan
hexg@phys.ntu.edu.tw

In this talk, I discuss implementation of minimal flavor violation (MFV) in seesaw models based on work appeared in arXiv:1401.2615,[1] arXiv:1404.4436[2] and arXiv:1411.6612.[3] Phenomenological implications on flavor-changing interactions related to leptons are studied by considering some effective dimension-six operators. We also comment on how one of the new effective operators can induce flavor-changing dilepton decays of the Higgs boson.

Keywords: Minimal flavor violation; seesaw models; CP violation.

1. The Minimal Flavor Violation

The minimal flavor violation (MFV) framework offers a predictive and systematic way to explore new physics which does not conserve quark and lepton flavor and CP symmetries. MFV postulates that the sources of all FCNC and CP violation reside in standard model (SM) renormalizable Yukawa couplings defined at tree level.[4] The implementation of the MFV principle for quarks is straightforward, but for leptons there are ambiguities.[5] At present whether neutrinos are Dirac fermions or Majorana fermions are not known. The mass-generation mechanisms and Yukawa couplings for the neutrinos in the two cases differ significantly. Since the MFV hypothesis is closely associated with Yukawa couplings, one expects that the resulting phenomenologies in the two scenarios are also different.

For Dirac neutrino case, one extends the SM by including three right-handed neutrinos, $\nu_{k,R}$, transforming as $(1,1,0)$ under the SM gauge group $\mathcal{G}_{\rm SM} = {\rm SU}(3)_C \times {\rm SU}(2)_L \times {\rm U}(1)_Y$. The Lagrangian for lepton masses is given by

$$\mathcal{L}_{\rm m} = -(Y_\nu)_{kl} \bar{L}_{k,L} \nu_{l,R} \tilde{H} - (Y_e)_{kl} \bar{L}_{k,L} E_{l,R} H \,, \tag{1}$$

where summation over $k, l = 1, 2, 3$ is implicit, $Y_{\nu,e}$ are Yukawa coupling matrices,

$L_{k,L}$ represents left-handed lepton doublets, $\nu_{k,R}$ and $E_{k,R}$ denote right-handed neutrinos and charged leptons, H is the Higgs boson doublet, and $\tilde{H} = i\tau_2 H^*$ involving the second Pauli matrix τ_2. Under the SM gauge group, $L_{k,L}$, $E_{k,R}$, and H transform as $(1,2,-1/2)$, $(1,1,-1)$, and $(1,2,1/2)$, respectively.

The MFV hypothesis[1–3,5] implies that $\mathcal{L}_{\rm m}$ is formally invariant under the global group $\mathrm{U}(3)_L \times \mathrm{U}(3)_\nu \times \mathrm{U}(3)_E = G_\ell \times \mathrm{U}(1)_L \times \mathrm{U}(1)_\nu \times \mathrm{U}(1)_E$ with $G_\ell = \mathrm{SU}(3)_L \times \mathrm{SU}(3)_\nu \times \mathrm{SU}(3)_E$. This entails that $L_{k,L}$, $\nu_{k,R}$, and $E_{k,R}$ belong to the fundamental representations of the $\mathrm{SU}(3)_{L,\nu,E}$, respectively,

$$L_L \to V_L L_L\,, \quad \nu_R \to V_\nu \nu_R\,, \quad E_R \to V_E E_R\,, \quad V_{L,\nu,E} \in \mathrm{SU}(3)_{L,\nu,E}\,, \tag{2}$$

and under G_ℓ the Yukawa couplings transform in the spurion sense according to

$$Y_\nu \to V_L Y_\nu V_\nu^\dagger \sim (3,\bar{3},1)\,, \quad Y_e \to V_L Y_e V_E^\dagger \sim (3,1,\bar{3})\,. \tag{3}$$

Taking advantage of the requirement that the final effective Lagrangian be invariant under G_ℓ, without loss of generality one can always work in the basis where Y_e is diagonal, $Y_e = (\sqrt{2}/v)\mathrm{diag}(m_e, m_\mu, m_\tau)$, with $v \simeq 246$ GeV being the Higgs' vacuum expectation value (VEV), and $\nu_{k,L}$, $\nu_{k,R}$, $E_{k,L}$, and $E_{k,R}$ refer to the mass eigenstates. Consequently, one can express $L_{k,L}$ and Y_ν in terms of the Pontecorvo–Maki–Nakagawa–Sakata neutrino mixing matrix $U_{\rm PMNS}$ as

$$L_{k,L} = \begin{pmatrix} (U_{\rm PMNS})_{kl}\nu_{l,L} \\ E_{k,L} \end{pmatrix}, \quad Y_\nu = \frac{\sqrt{2}}{v} U_{\rm PMNS}\hat{m}_\nu\,, \quad \hat{m}_\nu = \mathrm{diag}(m_1, m_2, m_3)\,, \tag{4}$$

where $m_{1,2,3}$ are neutrino masses and in the standard parametrization[23]

$$U_{\rm PMNS} = \begin{pmatrix} c_{12}c_{13} & s_{12}c_{13} & s_{13}e^{-i\delta} \\ -s_{12}c_{23} - c_{12}s_{23}s_{13}e^{i\delta} & c_{12}c_{23} - s_{12}s_{23}s_{13}e^{i\delta} & s_{23}c_{13} \\ s_{12}s_{23} - c_{12}c_{23}s_{13}e^{i\delta} & -c_{12}s_{23} - s_{12}c_{23}s_{13}e^{i\delta} & c_{23}c_{13} \end{pmatrix}, \tag{5}$$

with δ being the Dirac CP-violation phase, $c_{kl} = \cos\theta_{kl}$, and $s_{kl} = \sin\theta_{kl}$.

Based on the transformation properties of the fields and Yukawa spurions, one then uses an arbitrary number of Yukawa coupling matrices to put together G_ℓ-invariant objects which can induce the desired FCNC and CP-violating interactions. Thus, for operators involving two lepton fields, the pertinent building blocks are

$$\begin{gathered} \bar{L}_L\gamma_\alpha\Delta_\ell L_L\,, \quad \bar{\nu}_R\gamma_\alpha\Delta_{\nu 8}\nu_R\,, \quad \bar{E}_R\gamma_\alpha\Delta_{e8}E_R\,, \\ \bar{\nu}_R(1,\sigma_{\alpha\beta})\Delta_\nu L_L\,, \quad \bar{E}_R(1,\sigma_{\alpha\beta})\Delta_e L_L\,. \end{gathered} \tag{6}$$

For these to be G_ℓ invariant, the Δ's should transform according to

$$\begin{gathered} \Delta_\ell \sim (1\oplus 8,1,1)\,, \quad \Delta_{\nu 8} \sim (1,1\oplus 8,1)\,, \quad \Delta_{e8} \sim (1,1,1\oplus 8)\,, \\ \Delta_\nu \sim (\bar{3},3,1)\,, \quad \Delta_e \sim (\bar{3},1,3)\,. \end{gathered} \tag{7}$$

Since $\bar{L}_L\gamma_\alpha\Delta_\ell L_L$, $\bar{\nu}_R\gamma_\alpha\Delta_{\nu 8}\nu_R$, and $\bar{E}_R\gamma_\alpha\Delta_{e8}E_R$ must be Hermitian, $\Delta_{\ell,\nu 8,e8}$ must be Hermitian as well. To be acceptable terms in the Lagrangian, the above objects should be combined with appropriate numbers of other SM fields into singlets under the SM gauge group, with all the Lorentz indices contracted.

The MFV principle dictates that these Δ's are built up from the Yukawa coupling matrices $Y_{\nu,e}$ and $Y_{\nu,e}^\dagger$. Let us first discuss a nontrivial Δ which transforms as $(1 \oplus 8, 1, 1)$ under G_ℓ and consists of terms in powers of

$$\mathsf{A} = Y_\nu Y_\nu^\dagger = \frac{2}{v^2} U_{\mathrm{PMNS}} \hat{m}_\nu^2 U_{\mathrm{PMNS}}^\dagger \,, \quad \mathsf{B} = Y_e Y_e^\dagger = \frac{2}{v^2} \mathrm{diag}(m_e^2, m_\mu^2, m_\tau^2) \,, \tag{8}$$

both of which also transform as $(1 \oplus 8, 1, 1)$. Formally, Δ is a sum of infinitely many terms, $\Delta = \sum \xi_{jkl\cdots} \mathsf{A}^j \mathsf{B}^k \mathsf{A}^l \cdots$ with coefficients $\xi_{jkl\cdots}$ expected to be at most of $\mathcal{O}(1)$. These coefficients must be real because otherwise they would introduce new sources of CP-violation beyond those in the Yukawa couplings. With the Cayley–Hamilton identity $X^3 = X^2 \operatorname{Tr} X + \frac{1}{2} X[\operatorname{Tr} X^2 - (\operatorname{Tr} X)^2] + I \operatorname{Det} X$ for an invertible 3×3 matrix X, one can resum the series into a limited number of terms[6,7]

$$\begin{aligned} \Delta = {} & \xi_1 I + \xi_2 \mathsf{A} + \xi_3 \mathsf{B} + \xi_4 \mathsf{A}^2 + \xi_5 \mathsf{B}^2 + \xi_6 \mathsf{AB} + \xi_7 \mathsf{BA} + \xi_8 \mathsf{ABA} \\ & + \xi_9 \mathsf{BA}^2 + \xi_{10} \mathsf{BAB} \xi_{11} \mathsf{AB}^2 + \xi_{12} \mathsf{ABA}^2 + \xi_{13} \mathsf{A}^2 \mathsf{B}^2 \\ & + \xi_{14} \mathsf{B}^2 \mathsf{A}^2 + \xi_{15} \mathsf{B}^2 \mathsf{AB} + \xi_{16} \mathsf{AB}^2 \mathsf{A}^2 + \xi_{17} \mathsf{B}^2 \mathsf{A}^2 \mathsf{B} \,, \end{aligned} \tag{9}$$

where I stands for the 3×3 unit matrix. Though one starts with all $\xi_{jkl\cdots}$ being real, the resummation process generally renders the coefficients ξ_r in Eq. (9) complex due to imaginary parts created among the traces of the matrix products $\mathsf{A}^j \mathsf{B}^k \mathsf{A}^l \cdots$ with $j + k + l + \cdots \geq 6$ after the application of the Cayley–Hamilton identity. The imaginary contributions turn out to be reducible to factors proportional to a Jarlskog invariant quantity, $\operatorname{Im} \operatorname{Tr}(\mathsf{A}^2 \mathsf{BAB}^2) = (i/2) \operatorname{Det}[\mathsf{A}, \mathsf{B}]$, which is much smaller than unity.[6,7]

With Eq. (9), one can devise the objects in Eq. (7). Thus, the first of the Hermitian combinations can be $\Delta_\ell = \Delta + \Delta^\dagger$. To obtain nontrivial $\Delta_{\nu,e}$, one can take $\Delta_\nu = Y_\nu^\dagger \Delta$ and $\Delta_e = Y_e^\dagger \Delta$. The construction of $\Delta_{\nu 8,e8}$ can be carried out in a similar way, except A and B are replaced by $\tilde{A} = Y_\nu^\dagger Y_\nu$ and $\tilde{B} = Y_e^\dagger Y_e$. Since $\tilde{A}$ and $\tilde{B}$ are diagonal, so are any powers of them. Therefore, $\Delta_{\nu 8,e8}$ do not produce any FCNC and CP-violation effects.

The above discussion can be easily applied to the quark sector with the renormalizable Yukawa Lagrangian given by,[4]

$$\mathcal{L}_{\mathrm{m}} = -(Y_u)_{kl} \bar{Q}_{k,L} U_{l,R} \tilde{H} - (Y_d)_{kl} \bar{Q}_{k,L} D_{l,R} H \,, \tag{10}$$

where $Y_{u,d}$ are Yukawa coupling matrices, $Q_{k,L}$ represents left-handed quark doublets, and $U_{k,R}$ ($D_{k,R}$) denote right-handed up-type (down-type) quarks. These fields transform as $(3, 2, 1/6)$, $(3, 1, 2/3)$, and $(3, 1, -1/3)$, respectively, under the SM gauge group $\mathcal{G}_{\mathrm{SM}}$. In the basis where Y_d is diagonalized,

$$Y_d = \frac{\sqrt{2}}{v} \mathrm{diag}(m_d, m_s, m_b) \,, \quad Y_u = \frac{\sqrt{2}}{v} V_{\mathrm{CKM}}^\dagger \hat{m}_u \,, \quad \hat{m}_u = \mathrm{diag}(m_u, m_c, m_t) \,, \tag{11}$$

where $V_{\rm CKM}$ is the Cabibbo–Kobayashi–Maskawa matrix which has the same standard parametrization as in Eq. (5). For MFV interactions, employing $Y_{u,d}$ along with $\mathsf{A} = Y_u Y_u^\dagger$ and $\mathsf{B} = Y_d Y_d^\dagger$ as building blocks, one can construct objects such as Δ_q, Δ_u, and Δ_d, which are the quark counterparts of Δ_ℓ, Δ_ν and Δ_e, respectively.[1]

2. Seesaw Models with MFV

If neutrinos are Majorana particles, the Yukawa couplings that take part in generating their masses differ from those in the Dirac neutrino case and depend on the model details. We work with seesaw type I, II and III models.[8–20] A crucial step in the implementation of MFV in a given model is to identify the quantities A and B in terms of the relevant Yukawa couplings.[3,21]

MFV in type-I seesaw

In the type-I seesaw model the SM is expanded to allow the right-handed neutrinos, $\nu_{k,R}$, have to possess have Majorana masses.[8–14] The renormalizable Lagrangian for the lepton masses is

$$\mathcal{L}_{\rm m}^{\rm I} = -(Y_\nu)_{kl}\bar{L}_{k,L}\nu_{l,R}\tilde{H} - (Y_e)_{kl}\bar{L}_{k,L}E_{l,R}H - \frac{1}{2}(M_\nu)_{kl}\overline{\nu_{k,R}^{\rm c}}\nu_{l,R}\,, \tag{12}$$

where $M_\nu = {\rm diag}(M_1, M_2, M_3)$ contains the right-handed neutrinos' Majorana masses and the superscript c refers to charge conjugation. The presence of M_ν breaks the global $\mathrm{U}(3)_\nu$ completely if $M_{1,2,3}$ are unequal and partially into $\mathrm{O}(3)_\nu$ if $M_{1,2,3}$ are equal.[5] We will work with the case with degenerate right-handed mass. The mass matrix for neutrinos is given by

$$\mathsf{M} = \begin{pmatrix} 0 & M_{\rm D} \\ M_{\rm D}^{\rm T} & M_\nu \end{pmatrix} \tag{13}$$

in the $\left(U_{\rm PMNS}^*(\nu_L)^{\rm c}, \nu_R\right)^{\rm T}$ basis, where $M_{\rm D} = vY_\nu/\sqrt{2}$. If the eigenvalues of M_ν are much greater than the elements of $M_{\rm D}$, the seesaw mechanism becomes operational,[8–14] and the seesaw mass matrix is

$$m_\nu = -\frac{v^2}{2}Y_\nu M_\nu^{-1} Y_\nu^{\rm T} = U_{\rm PMNS}\hat{m}_\nu U_{\rm PMNS}^{\rm T}\,, \tag{14}$$

where now $U_{\rm PMNS}$ contains the diagonal matrix $P = {\rm diag}(e^{i\alpha_1/2}, e^{i\alpha_2/2}, 1)$ multiplied from the right, $\alpha_{1,2}$ being the CP-violating Majorana phases. It follows that Y_ν in Eq. (4) is no longer valid, and one can instead pick Y_ν to be[22]

$$Y_\nu = \frac{i\sqrt{2}}{v}U_{\rm PMNS}\hat{m}_\nu^{1/2} O M_\nu^{1/2}\,, \tag{15}$$

where O is a matrix satisfying $OO^{\rm T} = I$. Since O can be complex, it is a potentially important new source of CP violation besides $U_{\rm PMNS}$. From Eq. (15)

$$\mathsf{A} = Y_\nu Y_\nu^\dagger = \frac{2}{v^2}U_{\rm PMNS}\hat{m}_\nu^{1/2} O M_\nu O^\dagger \hat{m}_\nu^{1/2} U_{\rm PMNS}^\dagger\,, \tag{16}$$

whereas $\mathsf{B} = Y_e Y_e^\dagger$ in Eq. (8) is unchanged. These matrices are to be incorporated into the MFV objects $\Delta_{\ell,\nu,e}$. We note that, since $\tilde{A} = Y_\nu^\dagger Y_\nu = (2/v^2) M_\nu^{1/2} O^\dagger \hat{m}_\nu O M_\nu^{1/2}$, here $\Delta_{\nu 8}$ is no longer diagonal, but it pertains only to ν_R interactions.

MFV in type-II seesaw model

The type-II seesaw model extends the SM with the addition of only one scalar $\mathrm{SU}(2)_L$ triplet given by[15–19]

$$T = \begin{pmatrix} T^+/\sqrt{2} & T^{++} \\ T^0 & -T^+/\sqrt{2} \end{pmatrix} \tag{17}$$

which transforms as $(1,3,1)$ under the SM gauge group $\mathcal{G}_{\mathrm{SM}}$. Accordingly, the Lagrangian describing the Yukawa couplings of leptons is

$$\mathcal{L}_{\mathrm{m}}^{\mathrm{II}} = -(Y_e)_{kl} \bar{L}_{k,L} E_{l,R} H - \frac{1}{2} (Y_T)_{kl} \bar{L}_{k,L} \tilde{T} L_{l,L}^{\mathrm{c}} \,, \tag{18}$$

with $\tilde{T} = i\tau_2 T^*$. It respects lepton-number conservation if T is assigned a lepton number of -2. After the VEV of the neutral component of T becomes nonzero, $\langle T^0 \rangle = v_T/\sqrt{2}$, one obtains in $\mathcal{L}_{\mathrm{m}}^{\mathrm{II}}$ the neutrino mass matrix

$$m_\nu = \frac{1}{\sqrt{2}} v_T Y_T = U_{\mathrm{PMNS}} \hat{m}_\nu U_{\mathrm{PMNS}}^{\mathrm{T}} \tag{19}$$

in the basis where the charged lepton's Yukawa coupling matrix, Y_e, has been diagonalized. If the nonzero elements of Y_T are of $\mathcal{O}(1)$, the tiny size of neutrino masses then comes from the suppression of v_T due to certain choices of the parameters in the scalar potential.

To implement the MFV hypothesis in this seesaw scheme, one observes that the Lagrangian in Eq. (18) possesses formal invariance under the global group $\mathrm{U}(3)_L \times \mathrm{U}(3)_E = G'_\ell \times \mathrm{U}(1)_L \times \mathrm{U}(1)_E$, with $G'_\ell = \mathrm{SU}(3)_L \times \mathrm{SU}(3)_E$, if L_L and E_R belong to the fundamental representation of $\mathrm{SU}(3)_{L,E}$, respectively,

$$L_L \to V_L L_L \,, \quad E_R \to V_E E_R \,, \quad V_{L,E} \in \mathrm{SU}(3)_{L,E} \,, \tag{20}$$

and the Yukawa couplings are spurions transforming according to

$$Y_e \to V_L Y_e V_E^\dagger \,, \quad Y_T \to V_L Y_T V_L^{\mathrm{T}} \,. \tag{21}$$

Here the building block Δ still has the expression in Eq. (9), with $\mathsf{B} = Y_e Y_e^\dagger$ being the same as in Eq. (8), but unlike before

$$\mathsf{A} = Y_T Y_T^\dagger = \frac{2}{v_T^2} U_{\mathrm{PMNS}} \hat{m}_\nu^2 U_{\mathrm{PMNS}}^\dagger \,, \quad Y_T = \frac{\sqrt{2}}{v_T} \,, \quad U_{\mathrm{PMNS}} \hat{m}_\nu U_{\mathrm{PMNS}}^{\mathrm{T}} \,. \tag{22}$$

It is interesting to notice that A in Eq. (22) is the same as its Dirac-neutrino counterpart in Eq. (8), up to an overall factor. Due to this difference, whereas the elements of the latter are tiny, those in Eq. (22) can be of $\mathcal{O}(1)$ if v_T is similar in order of magnitude to the neutrino masses.

MFV in type-III seesaw model

In this case the new particles consist only of three fermionic $\mathrm{SU}(2)_L$ triplets[20]

$$\Sigma_k = \begin{pmatrix} \Sigma_k^0/\sqrt{2} & \Sigma_k^+ \\ \Sigma_k^- & -\Sigma_k^0/\sqrt{2} \end{pmatrix}, \quad k = 1, 2, 3, \tag{23}$$

which transform as $(1, 3, 0)$ under the SM gauge group $\mathcal{G}_{\mathrm{SM}}$. The Lagrangian responsible for the lepton masses is then

$$\mathcal{L}_{\mathrm{m}}^{\mathrm{III}} = -(Y_e)_{kl}\bar{L}_{k,L}E_{l,R}H - \sqrt{2}(Y_\Sigma)_{kl}\bar{L}_{k,L}\Sigma_l\tilde{H} - \frac{1}{2}(M_\Sigma)_{kl}\,\mathrm{Tr}\big(\overline{\Sigma_k^{\mathrm{c}}}\Sigma_l\big), \tag{24}$$

where Σ_k^{c} is the charge conjugate of Σ_k. For convenience, we define the right-handed fields $\mathcal{E}_{k,R} = \Sigma_k^-$ and $\mathcal{N}_{k,R} = \Sigma_k^0$ and left-handed fields $\mathcal{E}_{k,L} = \left(\Sigma_k^+\right)^{\mathrm{c}}$ and $\mathcal{N}_{k,L} = \left(\Sigma_k^0\right)^{\mathrm{c}}$. In terms of matrices containing them and SM leptons, one can express the mass terms in $\mathcal{L}_{\mathrm{m}}^{\mathrm{III}}$ after electroweak symmetry breaking as

$$\begin{gathered} (\bar{E}_L \;\; \bar{\mathcal{E}}_L)\begin{pmatrix} M_\ell & \sqrt{2}M_{\mathrm{D}} \\ 0 & M_\Sigma \end{pmatrix}\begin{pmatrix} E_R \\ \mathcal{E}_R \end{pmatrix}, \\ \frac{1}{2}(\bar{\nu}_L' \;\; \bar{\mathcal{N}}_L)\begin{pmatrix} 0 & M_{\mathrm{D}} \\ M_{\mathrm{D}}^{\mathrm{T}} & M_\Sigma \end{pmatrix}\begin{pmatrix} (\nu_L')^{\mathrm{c}} \\ \mathcal{N}_R \end{pmatrix}, \end{gathered} \tag{25}$$

where $M_\ell = vY_e/\sqrt{2}$ and $M_{\mathrm{D}} = vY_\Sigma/\sqrt{2}$ are 3×3 matrices and $\nu_L' = U_{\mathrm{PMNS}}\nu_L$. For $M_\Sigma \gg M_{\mathrm{D}}$ in their nonzero elements, a seesaw mechanism like that in type. Hence it is tempting simply to write Y_Σ in a similar way to Y_ν in type I,

$$Y_\Sigma = \frac{i\sqrt{2}}{v}U_{\mathrm{PMNS}}\hat{m}_\nu^{1/2}OM_\Sigma^{1/2}. \tag{26}$$

One needs to justify this approximation because the light charged leptons, E_k, mix with the heavy ones, $\mathcal{E}_k$, as can be deduced from Eq. (25). This mixing alters U_{PMNS} in Eq. (26) to $(U_{EE})_L^\dagger U_{\mathrm{PMNS}}$ as well as Y_e to $(U_{EE})_L^\dagger Y_e(U_{EE})_R$. At leading order, $(U_{EE})_L = I - M_D M_\Sigma^{-2}M_D^\dagger$ and $(U_{EE})_R = I$ for $M_D \ll M_\Sigma$. Thus, the deviations of $(U_{EE})_{L,R}$ from the unit matrix are negligible, and the approximation of Y_Σ in Eq. (26) is justified. Therefore, in type III one can still work with

$$\mathsf{A} = Y_\Sigma Y_\Sigma^\dagger = \frac{2}{v^2}U_{\mathrm{PMNS}}\hat{m}_\nu^{1/2}OM_\Sigma O^\dagger\hat{m}_\nu^{1/2}U_{\mathrm{PMNS}}^\dagger \tag{27}$$

and B in Eq. (8). These are the same form as in type I.

To proceed, we also need to specify the A matrix, which is model dependent. For the type-I or -III seesaw scenario, A in Eq. (16) [or (27)] can have many different realizations, depending on M_ν and O. We consider first the least complicated possibility that the right-handed neutrinos $\nu_{k,R}$ are degenerate, with $M_\nu = \mathcal{M}I$, and O is a real orthogonal matrix, in which case $\mathsf{A} = (2\mathcal{M}/v^2)U_{\mathrm{PMNS}}\hat{m}_\nu U_{\mathrm{PMNS}}^\dagger$, with eigenvalues $\hat{\mathsf{A}}_{1,2,3} = 2\mathcal{M}m_{1,2,3}/v^2$.

To ensure that the sum for Δ converges, the eigenvaues of A should have an upper bound. We assume $\max(\hat{\mathsf{A}}_1, \hat{\mathsf{A}}_2, \hat{\mathsf{A}}_3) = 1$. This requires that the right-handed neutrino mass is $\mathcal{M} \simeq 6.0 \times 10^{14}$ GeV. For the type-II scheme, A is given only in Eq. (22), which has eigenvalues $\hat{\mathcal{A}}_{1,2,3} = 2m^2_{1,2,3}/v^2_T$ Utilising these matrix elements, one should adjust v_T such that $\max(\hat{\mathcal{A}}_1, \hat{\mathcal{A}}_2, \hat{\mathcal{A}}_3) = 1$. In our numerical calculations, we use $v_T \sim 0.07$ eV.

3. Some Phenomenological Implications

To explore the phenomenological consequences of MFV, one adopts an effective theory approach,[4,5] assuming that the heavy degrees of freedom in the full theory have been integrated out. We will take the following dipole operators as examples[5]

$$\begin{aligned} O^{(e1)}_{RL} &= g'\bar{E}_R Y_e^\dagger \Delta_{\ell 1} \sigma_{\kappa\omega} H^\dagger L_L B^{\kappa\omega}\,, \\ O^{(e2)}_{RL} &= g\bar{E}_R Y_e^\dagger \Delta_{\ell 2} \sigma_{\kappa\omega} H^\dagger \tau_a L_L W_a^{\kappa\omega}\,, \end{aligned} \tag{28}$$

where W and B stand for the usual $\mathrm{SU}(2)_L \times \mathrm{U}(1)_Y$ gauge fields with coupling constants g and g', respectively, τ_a are Pauli matrices, summation over $a = 1, 2, 3$ is implicit, and $\Delta_{\ell 1,\ell 2}$ have the same form as Δ in Eq. (9), but with generally different ξ_r. One can write the effective Lagrangian for $O^{(e1,e2)}_{RL}$ as

$$\mathcal{L}_{\text{eff}} = \frac{1}{\Lambda^2}\left[O^{(e1)}_{RL} + O^{(e2)}_{RL}\right] + \text{H.c.}\,, \tag{29}$$

where Λ is the scale of MFV and their own coefficients in this Lagrangian have been absorbed by the ξ_r in their respective Δ's. Since ξ and Λ are not known, in our numerical analysis, we will take ξ to be of order one and put constraints on Λ to have some feeling about possible MFV scale. The relevant values for neutrino mixing and masses to be used in the discuss are given in Table 1.

These interactions contribute to $\mu \to e$ the anomalous magnetic moments $(g-2)$ and electric dipole moments (EDMs) of charged leptons, directly. When the photon is connected to charged leptons or quarks, $E_l \to E_k E_j^- E_j^+$ and nuclear conversion can also be induced.

Table 1. Parameters for neutrino mixing and masses.[23,24] Mass squares in unit 10^{-5} eV^2.

Parameter	NH	IH
$\sin^2\theta_{12}$	0.308 ± 0.017	0.308 ± 0.017
$\sin^2\theta_{23}$	$0.437^{+0.033}_{-0.023}$	$0.455^{+0.139}_{-0.031}$
$\sin^2\theta_{13}$	$0.0234^{+0.0020}_{-0.0019}$	$0.0240^{+0.0019}_{-0.0022}$
δ/π	$1.39^{+0.38}_{-0.27}$	$1.31^{+0.29}_{-0.33}$
$\delta m^2 = m_2^2 - m_1^2$	$(7.54^{+0.26}_{-0.22})$	$(7.54^{+0.26}_{-0.22})$
$\Delta m^2 = \lvert m_3^2 - (m_1^2 + m_2^2)/2\rvert$	(2.43 ± 0.06)	(2.38 ± 0.06)

The edm for a charged lepton is given by $d_{E_k} = (\sqrt{2}ev/\Lambda^2)\,\mathrm{Im}(Y_e^\dagger \Delta_\ell)_{kk}$. The compact form of Δ_ℓ is particularly suited for edm analysis. If ξ_r are real, only a few of the terms contribute, for the electron we have[1]

$$d_e = \frac{\sqrt{2}ev}{\Lambda^2}\left[\xi_{12}^\ell\,\mathrm{Im}\big(Y_e^\dagger \mathsf{ABA}^2\big)_{11} + \xi_{16}^\ell\,\mathrm{Im}\big(Y_e^\dagger \mathsf{AB}^2\mathsf{A}^2\big)_{11}\right]. \tag{30}$$

In general ξ can be complex and will induce edm. However, they are higher order in A and B and can be neglected.

If neutrinos are Dirac particles, d_e has the form[1,2]

$$d_e^{\mathrm{D}} = \frac{32em_e}{\Lambda^2 v^8}\big(m_\mu^2 - m_\tau^2\big)\big(m_1^2 - m_2^2\big)\big(m_2^2 - m_3^2\big)\big(m_3^2 - m_1^2\big)\xi_{12}^\ell J_\ell\,, \tag{31}$$

where $J_\ell = \mathrm{Im}(U_{e2}U_{\mu 3}U_{e3}^* U_{\mu 2}^*)$ is a Jarlskog invariant for U_{PMNS}.

On the other hand, in types I and III with degenerate $\nu_{k,R}$ and a real O matrix, in which case A is given by

$$d_e^{\mathrm{I,III}} = \frac{32em_e\mathcal{M}^3}{\Lambda^2 v^8}\big(m_\mu^2 - m_\tau^2\big)\big(m_1 - m_2\big)\big(m_2 - m_3\big)\big(m_3 - m_1\big)\xi_{12}^\ell J_\ell\,. \tag{32}$$

Since $m_k \ll \mathcal{M}$, one can see that d_e^{D} is considerably suppressed relative to d_e^{M}.

In contrast, for type II one derives $d_e^{\mathrm{II}} = (v/v_T)^6 d_e^{\mathrm{D}}$, which is far above d_e^{D} due to $v_T \ll v$. From these formulas, one can readily find those for $d_{\mu,\tau}$ by cyclically changing the mass subscripts.

Numerically, $d_e^{\mathrm{D}} = 1.3\times 10^{-99}\,e$ cm $\mathrm{GeV}^2/\hat{\Lambda}^2$,[1,2] which is too minuscule to yield any useful restraint on $\hat{\Lambda}$ from the newest data $|d_e|_{\mathrm{exp}} < 8.7\times 10^{-29}\,e$ cm from the ACME experiment.[26] Here $\hat{\Lambda}^2$ is defined as Λ^2/ξ_i. In the Majorana neutrino case, the type-I (or -III) prediction in Eq. (32) has been evaluated in Refs. 1 and 2 to yield the limit $\hat{\Lambda} > 0.36(0.12)$ TeV corresponding to $\mathcal{M} = 6.16(6.22)\times 10^{14}$ GeV for the NH (IH) of neutrino masses.

In type II we arrive at, $d_e^{\mathrm{II}}/e\,\mathrm{cm} = 2.7\ (2.6)\times 10^{-31}\ (\mathrm{eV}/v_T)^6\ (\mathrm{GeV}/\hat{\Lambda})^2$, for the NH (IH) case. Then $|d_e|_{\mathrm{exp}} < 8.7\times 10^{-29}\,e$ cm translates into, $\hat{\Lambda} > 0.055\ (0.054)$ GeV $(\mathrm{eV}/v_T)^3$. With $v_T \simeq 0.069$ eV from the requirment that the largest eigenvalue of A in Eq. (22) be unity, it follows that $\hat{\Lambda} > 0.17$ TeV, which is roughly comparable to its counterparts in type I (or III) quoted above.

In the preceding discussion, d_e is caused by the CP-violating Dirac phase δ in U_{PMNS}, and the Majorana phases $\alpha_{1,2}$ therein do not take part. However, if O is complex, the phases in it may give rise to an extra contribution to d_e and the Majorana phases can modify it further. As investigated in detail in Refs. 1 and 2, these new CP-violating contributions to d_e can be more important than those of δ. Such effects do not occur in type II, as d_e^{II} does not have dependence on O or $\alpha_{1,2}$.

The electron edm provides the strongest constraints on MFV scale Λ. Contributions to FCNC from CP conserving interactions occur at low powers of A and B in Δ_ℓ. The leading terms $\Delta = \xi_2 A + \xi_3 A^2$ will dominate. The strongest constraints are from $\mu \to e\gamma$. Using current experimental limits on various FCNC processes, one can obtain constraints for the MFV scale. The results from several other processes

Table 2. Limits for $\hat{\Lambda}$ with $m_1(m_3) = 0$ for NH (IH).

Observable	Experimental upper bound[23]	$\hat{\Lambda}_{\rm min}$/TeV	
		Types I & III	Type II
$\mathcal{B}(\mu \to e\gamma)$	5.7×10^{-13} (Ref. 23)	338 (307)	294 (312)
$\mathcal{B}(\mu\,{\rm Ti} \to e\,{\rm Ti})$	6.1×10^{-13} (Ref. 25)	85 (77)	73 (78)
$\mathcal{B}(\mu{\rm Au} \to e\,{\rm Au})$	7.0×10^{-13} (Ref. 23)	80 (73)	70 (74)
$\mathcal{B}(\mu^- \to e^-e^-e^+)$	1.0×10^{-12} (Ref. 23)	81 (74)	70 (75)
$\mathcal{B}(\tau \to \mu\gamma)$	4.4×10^{-8} (Ref. 23)	22 (24)	23 (23)
$\mathcal{B}(\tau^- \to \mu^-\mu^-\mu^+)$	2.1×10^{-8} (Ref. 23)	5.6 (5.9)	5.9 (5.9)
$\mathcal{B}(\tau^- \to \mu^-e^-e^+)$	1.8×10^{-8} (Ref. 23)	8.7 (9.3)	9.2 (9.3)
$\mathcal{B}(\tau \to e\gamma)$	3.3×10^{-8} (Ref. 23)	15 (13)	13 (13)
$\mathcal{B}(\tau^- \to e^-e^-e^+)$	2.7×10^{-8} (Ref. 23)	4.9 (4.2)	4.3 (4.2)
$\mathcal{B}(\tau^- \to e^-\mu^-\mu^+)$	2.7×10^{-8} (Ref. 23)	3.2 (2.7)	2.8 (2.7)

along with that from $\mu \to e\gamma$ are listed in Table 2, where $\hat{\lambda}^2_{\rm min}$ is the lower limits constrained from data. Potentially severe restrictions will be supplied by future measurements on $\mu \to e$ conversion in nuclei which will begin in a few years and are expected to achieve sensitivity two orders of magnitudes or better eventually.[27,28] In due time, more stringent constraints can be obtained.

Besides the dipole operators, there are other dimension-six operators that can arise in the three simplest seesaw scenarios. We will not display constraints for these operators here, but to point out at dimension six, there is one operator which can induce flavor violating decay of the Higgs boson, namely[3]

$$O^{(e3)}_{RL} = (\mathcal{D}_\mu H)^\dagger \bar{E}_R Y_e^\dagger \Delta_{RL} \mathcal{D}^\mu L_L \,, \tag{33}$$

with $\mathcal{D}_\mu$ being the usual covariant derivative involving the electroweak gauge bosons.

This operator may have potential impact on the properties of the recently discovered Higgs boson. The latest LHC measurements have begun to reveal the Higgs couplings to charged leptons. The ATLAS and CMS Collaborations have observed $h \to \tau^+\tau^-$ and measured its signal strength to be $\sigma/\sigma_{\rm SM} = 1.42^{+0.44}_{-0.38}$ and 0.91 ± 0.27, respectively.[29,30] In contrast, the only experimental information available on $h \to \mu^-\mu^+$ to date are the bounds $\mathcal{B}(h \to \mu^-\mu^+) < 1.5\times10^{-3}$ and 1.6×10^{-3} from ATLAS and CMS, respectively.[31,32] On the other hand, CMS[33] has intriguingly reported the detection of a slight excess of flavor-violating $h \to \mu^\pm\tau^\mp$ events with a significance of 2.5σ. If the finding is interpreted as a statistical fluctuation, it translates into the limit $\mathcal{B}(h \to \mu\tau) = \mathcal{B}(h \to \mu^-\tau^+) + \mathcal{B}(h \to \mu^+\tau^-) < 1.57\%$ at 95% C.L.[33] If the requirements from $h \to \mu^-\mu^+$, $\tau^-\tau^+$ data, which led to the higher $\hat{\Lambda}_{\rm min}$ values in the table, are to be satisfied also by the contributions to $h \to \mu^\pm\tau^\mp$, we find that $\Gamma_{h\to\mu\tau}/\Gamma_h^{\rm SM}$ cannot be more than about 0.11%. This can be translated into constraints on Λ. The results are given in Table 3.

Table 3. Limits from $h \to l_i \bar{l}_j$ on $\hat{\Lambda}$.

Process	$\hat{\Lambda}_{\min}/\mathrm{GeV}$	
	Types I & III	Type II
$h \to \mu^-\mu^+, \tau^-\tau^+$	175 (170)	168 (170)
$h \to \mu^\mp\tau^\pm$	83 (88)	88 (87)

Acknowledgments

This research was supported in part by MOE Academic Excellence Program (Grant No. 102R891505) and ROC NSC and by PRC NSFC (Grant No. 11175115) and Shanghai Science and Technology Commission (Grant No. 11DZ2260700). I would like to thank Lee, Tandean and Zheng for enjoyable collaborations.

References

1. X. G. He, C. J. Lee, S. F. Li and J. Tandean, *Phys. Rev. D* **89**, 091901 (2014).
2. X. G. He, C. J. Lee, S. F. Li and J. Tandean, *J. High Energy Phys.* **1408**, 019 (2014).
3. X. G. He, C. J. Lee, J. Tandean and Y. J. Zheng, arXiv:1411.6612 [hep-ph].
4. G. D'Ambrosio, G. F. Giudice, G. Isidori and A. Strumia, *Nucl. Phys. B* **645**, 155 (2002).
5. V. Cirigliano, B. Grinstein, G. Isidori and M. B. Wise, *Nucl. Phys. B* **728**, 121 (2005).
6. G. Colangelo, E. Nikolidakis and C. Smith, *Eur. Phys. J. C* **59**, 75 (2009).
7. L. Mercolli and C. Smith, *Nucl. Phys. B* **817**, 1 (2009).
8. P. Minkowski, *Phys. Lett. B* **67**, 421 (1977).
9. T. Yanagida, *Proc. Workshop on the Unified Theory and the Baryon Number in the Universe*, eds. O. Sawada and A. Sugamoto (KEK, Tsukuba, 1979), p. 95.
10. T. Yanagida, *Prog. Theor. Phys.* **64**, 1103 (1980).
11. M. Gell-Mann, P. Ramond and R. Slansky, *Supergravity*, eds. P. van Nieuwenhuizen and D. Freedman (North-Holland, Amsterdam, 1979), p. 315.
12. P. Ramond, arXiv:hep-ph/9809459.
13. S. L. Glashow, *Proc. 1979 Cargese Summer Institute on Quarks and Leptons*, eds. M. Levy *et al.* (Plenum Press, New York, 1980), p. 687.
14. R. N. Mohapatra and G. Senjanovic, *Phys. Rev. Lett.* **44**, 912 (1980).
15. J. Schechter and J. W. F. Valle, *Phys. Rev. D* **22**, 2227 (1980).
16. M. Magg and C. Wetterich, *Phys. Lett. B* **94**, 61 (1980).
17. T. P. Cheng and L. F. Li, *Phys. Rev. D* **22**, 2860 (1980).
18. R. N. Mohapatra and G. Senjanovic, *Phys. Rev. D* **23**, 165 (1981).
19. G. Lazarides, Q. Shafi and C. Wetterich, *Nucl. Phys. B* **181**, 287 (1981).
20. R. Foot, H. Lew, X. G. He and G. C. Joshi, *Z. Phys. C* **44**, 441 (1989).
21. M. B. Gavela, T. Hambye, D. Hernandez and P. Hernandez, *J. High Energy Phys.* **0909**, 038 (2009).
22. J. A. Casas and A. Ibarra, *Nucl. Phys. B* **618**, 171 (2001).
23. Particle Data Group (K. A. Olive *et al.*), *Chin. Phys. C* **38**, 090001 (2014).
24. F. Capozzi *et al.*, *Phys. Rev. D* **89**, 093018 (2014).
25. D. K. Papoulias and T. S. Kosmas, *Phys. Lett. B* **728**, 482 (2014), arXiv:1312.2460 [nucl-th].

26. ACME Collab. (J. Baron *et al.*), *Science* **343**, 269 (2014).
27. F. Cei and D. Nicolo, *Adv. High Energy Phys.* **2014**, 282915 (2014).
28. T. Mori and W. Ootani, *Prog. Part. Nucl. Phys.* **79**, 57 (2014).
29. CMS Collab., Report No. CMS-PAS-HIG-14-009, July 2014.
30. ATLAS Collab., Report No. ATLAS-CONF-2014-061, ATLAS-COM-CONF-2014-080, October 2014.
31. ATLAS Collab. (G. Aad *et al.*), *Phys. Lett. B* **738**, 68 (2014).
32. CMS Collab. (V. Khachatryan *et al.*), arXiv:1410.6679 [hep-ex].
33. CMS Collab., Report No. CMS-PAS-HIG-14-005, July 2014.

Generating Majorana Neutrino Masses with Loops

Chao-Qiang Geng

Chongqing University of Posts & Telecommunications, Chongqing, 400065, China
Department of Physics, National Tsing Hua University, Hsinchu, Taiwan
Physics Division, National Center for Theoretical Sciences, Hsinchu, Taiwan
geng@phys.nthu.edu.tw

We give a review on neutrino models in which Majorana neutrino masses are generated radiatively through loop diagrams. In particular, we concentrate on the two-loop models which contain extra doubly charged singlet Φ and triplet Δ scalars beyond the standard model so that the new Yukawa $\Psi \bar{\ell}^c_R \ell_R$ and effective $\Psi^{\pm\pm} W^{\mp} W^{\mp}$ couplings can be induced. In these two-loop models, we find that the neutrino mass spectrum is a normal hierarchy and the rate of the neutrinoless double beta decay ($0\nu\beta\beta$) can be large as it is dominated by the short-distance tree contribution. In addition, by using the neutrino oscillation data and comparing with the global fitting result in the literature, we find a unique neutrino mass matrix and predict the Dirac and two Majorana CP phases to be $3(\pi - \eta)/2$, $\pi + 3\eta/2$ and $(3\pi - \eta)/2$ with $\eta = 0.07\pi$, respectively.

Keywords: Majornara neutrino; neutrino mass; neutrinoless double beta decay; loop diagram.

1. Introduction

Neutrino oscillations observed by the solar, atmospheric, and reactor neutrino experiments[1] have revealed that neutrinos are massive but tiny and mix with each other.[2] The neutrino mixing matrix[3] $V_{\rm PMNS}$ can be parametrized as follows:[2,4]

$$V_{\rm PMNS} = \begin{pmatrix} c_{12}c_{13} & s_{12}c_{13} & s_{13}e^{-i\delta} \\ -s_{12}c_{23} - c_{12}s_{23}s_{13}e^{i\delta} & c_{12}c_{23} - s_{12}s_{23}s_{13}e^{i\delta} & s_{23}c_{13} \\ s_{12}s_{23} - c_{12}c_{23}s_{13}e^{i\delta} & -c_{12}s_{23} - s_{12}c_{23}s_{13}e^{i\delta} & c_{23}c_{13} \end{pmatrix} \times \begin{pmatrix} 1 & 0 & 0 \\ 0 & e^{i\alpha_{21}/2} & 0 \\ 0 & 0 & e^{i\alpha_{31}/2} \end{pmatrix}, \tag{1}$$

where $s_{ij}(c_{ij})$=$\sin\theta_{ij}$ $(\cos\theta_{ij})$ with θ_{ij} being the mixing angles, δ is the Dirac CP violation phase, and α_{21} and α_{31} are the two Majorana CP violation phases. The recent global best-fit ($\pm 1\sigma$) values from the neutrino oscillation data are given by[2]

$$\begin{aligned}
\sin^2\theta_{12} &= 0.308 \pm 0.017\,,\ \sin^2\theta_{23} = 0.437^{+0.033}_{-0.023}\,,\\
\sin^2\theta_{13} &= 0.0234^{+0.0020}_{-0.0019}\ \left(0.0240^{+0.0019}_{-0.0022}\right),\\
\Delta m^2_{21} &= \left(7.54^{+0.26}_{-0.22}\right)\times 10^{-5}\,\mathrm{eV}\,,\\
\Delta m^2_{32} &= (2.43\pm 0.06)\times 10^{-3}\,\mathrm{eV}\ \left((2.38\pm 0.06)\times 10^{-3}\,\mathrm{eV}\right),\\
\delta/\pi &= 1.39^{+0.38}_{-0.27}\ \left(1.31^{+0.29}_{-0.33}\right),
\end{aligned} \tag{2}$$

with a normal (inverted) neutrino mass hierarchy and 2σ range for the Dirac CP violation phase δ.

From the results in Eq. (2), it is clear that at least two neutrinos carry nonzero masses. However, the origin of these small masses is still a mystery. In the standard model, the neutrino masses have to be all zero duo to the absence of the right-handed neutrinos (ν_R) and the chiral nature of the left-handed ones (ν_L). For theories beyond the standard model, the simplest way to obtain nonzero neutrino masses is to include ν_R so that the Yukawa interaction for neutrinos exists, leading to Dirac neutrino masses after the electroweak symmetry breaking, just like the charged fermions. In order to account for the neutrino data in Eq. (2), extreme small Yukawa couplings of $O(10^{-13} - 10^{-12})$ are inevitably required, which are commonly believed to be too small to be natural.

Apart from the mass generation of Dirac neutrinos with ν_R, seesaw mechanisms with type-I,[5] type-II[6] and type-III[7] have been proposed to generate masses for Majorana neutrinos by realizing the Weinberg operator[8] $(\bar{L}^c_L H)(H^T L_L)$ at tree-level, where H and L_L are the doublets of Higgs and left-handed lepton fields, respectively. In these scenarios, either heavy degrees of freedom or tiny coupling constants are needed in order to conceive the small neutrino masses. On the other hand, models with the Majorana neutrino masses generated at one-loop,[9,10] two-loop[11–13] and higher loop[14–16] diagrams have also been proposed without introducing ν_R. Due to the loop suppression factors, the strong bounds on the coupling constants and heavy states are relaxed, resulting in a somewhat natural explanation for the smallness of neutrino masses. However, in most of the above radiative neutrino models, since only $SU(2)_L$ singlet scalars are introduced, new physics effects are limited in the lepton sector, whereas those involving hadrons, such as the neutrinoless double beta decay ($0\nu\beta\beta$) believed as a benchmark of the Majorana nature of neutrinos, do not show up.

In this talk, I will concentrate on a special type of neutrino models,[12,13] in which a doubly charged singlet scalar $\Psi : (1, 4)$ and a triplet $\Delta : (3, 2)$[a] under $\mathrm{SU}(2)_L \times \mathrm{U}(1)_Y$ are introduced to yield the new Yukawa coupling $\Psi\bar{\ell}^c_R\ell_R$ with the right-handed charged lepton ℓ_R as well as the effective gauge coupling $\Psi^{\pm\pm}W^{\mp}W^{\mp}$ due

[a] The convention for the electroweak quantum numbers (I, Y) with $Q = I + Y/2$ is used throughout this paper.

to the mixing between $\Psi^{\pm\pm}$ and $\Delta^{\pm\pm}$, leading to the neutrino masses through two-loop diagrams as shown in Fig. 1(a),[12] where $P^{\pm\pm}_{1,2}$ are the two mass eigenstates of the doubly charged scalars $\Psi^{\pm\pm}$ and $\Delta^{\pm\pm}$. It is interesting to note that $\Psi^{\pm\pm}W^{\mp}W^{\mp}$ can also be induced from non-renomalizable high-order operators.[17–19] One of the most interesting features of these models is that $0\nu\beta\beta$ is dominated by the short-range contribution at tree level due to the effective coupling of $\Psi^{\pm\pm}W^{\mp}W^{\mp}$ (see Fig. 1(b), unlike other radiative Majorana neutrino mass models in which $0\nu\beta\beta$ is suppressed as it arises from the traditional long-range one proportional to neutrino masses from loops.

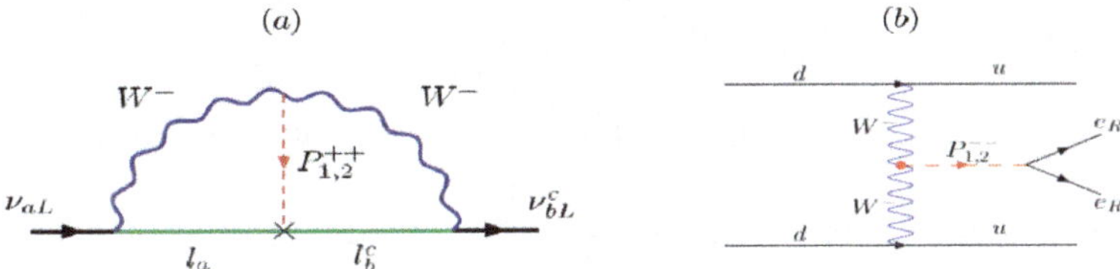

Fig. 1. Diagrams for (a) the neutrino mass generation and (b) the neutrinoless double beta decay.

2. Two-loop Neutrino Models

For all non-Higgs like scalars with non-trivial $SU(2)_L \times U(1)_Y$ quantum numbers,[b] there are only three possible renormalizable Yukawa interactions, given by[20]

$$f_{\ell\ell'}\bar{L}^c_{L_\ell}L_{L_{\ell'}}s\,,\quad y_{\ell\ell'}\bar{\ell}^c_{R_\ell}\ell_{R_{\ell'}}\Psi\,,\quad \text{and}\quad g_{\ell\ell'}\bar{L}^c_{L_\ell}L_{L_{\ell'}}\Delta\,, \tag{3}$$

where ℓ or ℓ' denotes e, μ and τ, c represents the charged conjugation, and $s : (1,2)$ is the singlet scalar field. Note that Δ, s and Ψ are used in the Type-II seesaw mechanism,[6] Zee[9] and Zee–Babu[11] models, respectively.

In Eq. (3), the third Yukawa interaction is the most troublesome one as it generates neutrino masses at tree level with an extreme small value of the vacuum expectation value (VEV) or Yukawa couplings, which is obviously un-natural. Without introducing the triplet Δ, the interactions in Eq. (3) are precisely given by the Zee–Babu model,[11] which has been extensively studied in the literature, in particular its phenomenology of the doubly charged scalar at the LHC. In this talk, we consider the models in which the first Yukawa interaction in Eq. (3) is absent due to the non-existence of s, while the third one is forbidden in a natural way by introducing a new doublet with an Z_2 symmetry[21] or replacing Δ by a higher multiplet, such as $\xi : (5,2)$ without the discrete symmetry.[20] In this type of the models, the lepton number is broken as a result of a nonzero VEV of Δ, v_Δ, which is is constrained to be $\lesssim 5\,\text{GeV}$ due to the global fitting result of $\rho_0 = 1.0000 \pm 0.0009$.[2] We take

[b]We assume that scalars carry no color.

the masses of the mass eigenstates $P_{1,2}$ to be $M_{1,2}$ ($M_i > M_W$) and θ their mixing angle.

In the Zee–Babu model, neutrino masses are generated from the two-loop diagrams due to the couplings of $s^{\pm}s^{\pm}\Psi^{\mp\mp}$. whereas in our models without $s^{\pm}$, they can be induced from similar two-loop diagrams with $\Psi^{\pm\pm}$ coupling to $W^{\mp}W^{\mp}$ though the mixing of $\Delta^{\pm\pm}$ as shown in Fig. 1(a). In this mechanism, neutrino masses are calculable, given by[20]

$$(M_\nu)_{\ell\ell'} \simeq \frac{g^4}{\sqrt{2}(4\pi)^4} y_{\ell\ell'} m_\ell m_{\ell'} v_\Delta \sin 2\theta \times \left[\frac{1}{M_1^2}\log^2\left(\frac{M_W^2}{M_1^2}\right) - \frac{1}{M_2^2}\log^2\left(\frac{M_W^2}{M_2^2}\right)\right], \quad (4)$$

where $m_{\ell,\ell'}$ correspond to charged lepton masses; $P_{1,2}$ are the mass eigenstates of doubly charged scalars with θ representing their mixing angle and $M_i > M_W$ assumed. Note that the neutrino masses are suppressed by the two-loop factor, $SU(2)_L$ gauge coupling, charged lepton masses, mixing angle θ, and v_Δ, respectively, without fine-tuning Yukawa couplings $y_{\ell\ell'}$.

It is crucial that the above types of the two-loop neutrino mass generation, in which $\bar{\ell}_R^c \ell_R \Psi$ is the only source of the lepton flavor voilation, can lead to an interesting structure for the neutrino mass matrix. The relative sizes among the matrix elements are determined by the combination factors of $y_{\ell\ell'} m_\ell m_{\ell'}$. Assuming that each value of $y_{\ell\ell'}$ is at the same order, there exist interesting hierarchies for the mass matrix elements, given by

$$(M_\nu)_{ee} \ll (M_\nu)_{e\mu} \ll (M_\nu)_{e\tau} \ll (M_\nu)_{\mu\mu} \ll (M_\nu)_{\mu\tau} \ll (M_\nu)_{\tau\tau} . \quad (5)$$

In particular, $(M_\nu)_{ee}$ is much less than $(M_\nu)_{\tau\tau}$ due to $m_e^2/m_\tau^2 \sim 10^{-7}$. In Ref. 23, it has been shown that only the normal hierarchy for the neutrino mass spectrum can have the matrix textures in which $(M_\nu)_{ee}$ together with another matrix element is zero. Clearly, as the mass hierarchies in Eq. (5) naturally realize $(M_\nu)_{ee} \simeq (M_\nu)_{e\mu} \simeq 0$, our two-loop models generically predict the normal neutrino mass hierarchy.

Recently, we have shown[13] that for given values of mass square splittings and mixing angle, such as those central values from the neutrino oscillation data, there are only two solutions for the three CP phases of δ, α_{21} and α_{31}, along with the lightest neutrino mass m_0, to satisfy the mass hierarchies in Eq. (5). In particular, by using the central values of the global fitting result for the normal hierarchy mass spectrum in Eq. (2), we find a unique solution, given by

$$m_0 \simeq 5.1 \times 10^{-3}\,\text{eV}\,, \ \alpha_{21} = \pi + \frac{3}{2}\eta\,, \ \alpha_{31} = \frac{3}{2}\pi - \frac{1}{2}\eta\,, \quad (6)$$

which leads to

$$M_\nu \simeq \begin{pmatrix} 0 & 0 & 1.0e^{-i\eta} \\ 0 & 2.4\,e^{i(\frac{\pi}{2}+\eta)} & 2.3\,e^{i\frac{\pi}{2}} \\ 1.0\,e^{-i\eta} & 2.3\,e^{i\frac{\pi}{2}} & 2.8\,e^{i(\frac{\pi}{2}+\frac{2\eta}{3})} \end{pmatrix} \times 10^{-11}\,\text{GeV}\,, \quad (7)$$

where $\eta = 0.07\pi$. Note that the empty values for $(M_\nu)_{ee}$ and $(M_\nu)_{e\mu}$ in Eq. (7) can be placed by some small nonzero values when any of the parameters in Eq. (6) is under slightly shifting. It is interesting to point out that the Dirac CP violation phase $\delta \simeq 3(\pi - \eta)/2$ along with the two Majorana ones in Eq. (6) is a nature consequence of our data fitting result instead of an input parameter.

As the coupling matrix elements $y_{\ell\ell'}$ are the only sources of the lepton flavor voilation, the processes of $\ell \to \ell'\ell''\ell'''$ ($\ell \to \ell'\gamma$) with the tree-level (one-loop) contributions involving $\Psi^{\pm\pm}$ could give significant constraints on $y_{\ell\ell'}$. However, those on y_{ee} and $y_{e\mu}$ can be ignored since they do not affect the tiny matrix elements $(M_\nu)_{ee}$ and $(M_\nu)_{e\mu}$ when we discuss the neutrino mass spectrum. Among the current experimental bounds, $\mathrm{Br}(\mu^+ \to e^+\gamma) < 5.7 \times 10^{-13}$ in Ref. 24 is the most stringent one to limit $y_{\ell\ell'}$. In particular, we can obtain[16]

$$|y_{e\tau}|^2 \frac{\cos\theta^2}{M_1^2} < \left(\frac{0.168}{\mathrm{TeV}}\right)^2 , \tag{8}$$

where we have assumed that $M_2 \gg M_1 \tan\theta$. To account for the current experimental data on the neutrino masses as obtained in Eq. (7), the matrix element $(M_\nu)_{e\tau}$ should be around 1.04×10^{-11} GeV. As a result, we can use this value to check whether the mechanism of the neutrino mass generation can work and constrain the doubly charged scalars.

It is interesting to note that the neutrinoless double beta decay in our models can have a significant different feature from other models with radiative neutrino mass generations. In our models, the short-range contributions to the decay dominate over the traditional long-range ones,[12] with the decay amplitudes proportional to $(M_\nu)_{ee}$. It is clear that the long-range parts can be safely neglected due to the small electron mass in $(M_\nu)_{ee}$, whereas the short-range ones are proportional only to the Yukawa coupling y_{ee}. By calculating $0\nu\beta\beta$, the upper limit on $|y_{ee}|$ could be derived, despite of the fact that it is ignored when discussing the neutrino mass matrix. The half life for $0\nu\beta\beta$ is given by[27,28]

$$T_{1/2}^{0\nu\beta\beta} = (G_{01}|\epsilon_3^{LLL}|^2|\mathcal{M}_3|^2)^{-1} , \tag{9}$$

which leads to[17]

$$\epsilon_3^{LLL} = m_p\,(y_{ee}^* \sin 2\theta)\frac{v_\Delta}{\sqrt{2}}\frac{M_1^2 - M_2^2}{M_1^2 M_2^2} , \tag{10}$$

where G_{01} and $|\mathcal{M}_3|$ are the phase space factor and the matrix element for the hadronic sector, respectively, and ϵ_3^{LLL} is the coefficient, which is effectively related to the dimension-9 operator $(\bar{u}_L\gamma_\mu d_L)(\bar{u}_L\gamma^\mu d_L)(\bar{e}_R e_R^c)$, defined in Refs. 27 and 28, and m_p is the proton mass. Note that the coefficient in Eq. (10) has no explicit dependence on the electron mass. By using the $0\nu\beta\beta$ experimental limit[29] and taking $M_1 = 1$ TeV, $M_2 = 1.5$ TeV and $|\sin 2\theta| = 0.04$, we get

$$|y_{ee}| < 3.4 \times 10^{-2} . \tag{11}$$

3. Conclusions

We have studied the Majorana neutrino masses generated radiatively by two-loop diagrams due to the Yukawa $y_{\ell\ell'}\Psi\bar{\ell}^c_R\ell'_R$ and effective $\Psi^{\pm\pm}W^{\mp}W^{\mp}$ couplings. We have shown that the lepton violating processes, in particular, $\mu^+ \to e^+\gamma$ can give stringent constraints on the new Yukawa coupling $y_{\ell\ell'}$. We have illustrated that the normal neutrino mass hierarchy is a generic feature in these two-loop neutrino mass generation models. Moreover, by using the central values of the neutrino oscillation data and comparing with the global fitting result in the literature, we have obtained the unique neutrino mass matrix in Eq. (7) and predicted the Dirac and two Majorana CP phases to be $3(\pi - \eta)/2$, $\pi + 3\eta/2$ and $(3\pi - \eta)/2$ with $\eta = 0.07\pi$, respectively. Finally, we emphasize that the neutrinoless double beta decays can be very large as they are dominated by the short-distance contributions at tree-level, which can be tested in the future experiments and used to constrain the Yukawa coupling of y_{ee}.

Acknowledgments

I am grateful to the organizers: H. Fritzsch and Z.-Z. Xing for the invitation to this conference, and to Institute of Advanced Studies at Nanyang Technological University for the hospitality. This work was supported in part by National Center for Theoretical Sciences, National Science Council (Grant No. NSC-101-2112-M-007-006-MY3) and National Tsing Hua University (Grant No. 104N2724E1).

References

1. P. Anselmann *et al.* [GALLEX Collaboration], Phys. Lett. B**285**, 390 (1992); Y. Fukuda *et al.* [Super-Kamiokande Collaboration], Phys. Rev. Lett. **81**, 1562 (1998); Q. R. Ahmad *et al.* [SNO Collaboration], Phys. Rev. Lett. **89**, 011301 (2002); M. H. Ahn *et al.* [K2K Collaboration], Phys. Rev. D**74**, 072003 (2006).
2. K. A. Olive *et al.* [Particle Data Group], Chin. Phys. C**38**, 090001 (2014).
3. Z. Maki, M. Nakagawa and S. Sakata, Prog. Theor. Phys. **28**, 870 (1962); B. Pontecorvo, Sov. Phys. JETP **26**, 984 (1968).
4. L. L. Chau and W. Y. Keung, Phys. Rev. Lett. **53**, 1802 (1984).
5. H. Fritzsch, M. Gell-Mann and P. Minkowski, Phys. Lett. B**59**, 256 (1975); P. Minkowski, Phys. Lett. B**67**, 421 (1977); M. Gell-Mann, P. Ramond, and R Slansky, Proceedings of the Supergravity Stony Brook Workshop, New York, 1979; T. Yanagida, Proceedings of the Workshop on the Baryon Number of the Universe and Unified Theories, Tsukuba, Japan, 1979; R. N. Mohapatra and G. Senjanovic, Phys. Rev. Lett. **44**, 912 (1980).
6. M. Magg and C. Wetterich, Phys. Lett. B**94**, 61 (1980); J. Schechter and J. W. F. Valle, Phys. Rev. D**22**, 2227 (1980); **25**, 774 (1982). T. P. Cheng and L. F. Li, Phys. Rev. D**22**, 2860 (1980).
7. R. Foot, H. Lew, X. G. He and G. C. Joshi, Z. Phys. C**44**, 441 (1989).
8. S. Weinberg, Phys. Rev. D**22**, 1694 (1980).
9. A. Zee, Phys. Lett. B**93**, 389 (1980).

10. E. Ma, Phys. Rev. D**73**, 077301 (2006).
11. A. Zee, Nucl. Phys. B**264**, 99 (1986); K. S. Babu, Phys. Lett. B**203**, 132 (1988).
12. C. S. Chen, C. Q. Geng and J. N. Ng, Phys. Rev. D**75**, 053004 (2007); C. S. Chen, C. Q. Geng, J. N. Ng and J. M. S. Wu, JHEP **0708**, 022 (2007).
13. C. Q. Geng and L. H. Tsai, arXiv:1503.06987 [hep-ph].
14. L. M. Krauss, S. Nasri and M. Trodden, Phys. Rev. D**67**, 085002 (2003); M. Aoki, S. Kanemura and O. Seto, Phys. Rev. Lett. **102**, 051805 (2009).
15. M. Gustafsson, J. M. No and M. A. Rivera, Phys. Rev. Lett. **110**, no. 21, 211802 (2013) [Erratum-ibid. **112**, no. 25, 259902 (2014)].
16. C. Q. Geng, D. Huang and L. H. Tsai, Phys. Rev. D**90**, 113005 (2014).
17. M. Gustafsson, J. M. No and M. A. Rivera, Phys. Rev. D**90**, 013012 (2014).
18. S. F. King, A. Merle and L. Panizzi, JHEP **1411**, 124 (2014).
19. D. A. Sierra, A. Degee, L. Dorame and M. Hirsch, arXiv:1411.7038 [hep-ph].
20. C. S. Chen, C. Q. Geng, D. Huang and L. H. Tsai, Phys. Rev. D**87**, 077702 (2013).
21. C. S. Chen and C. Q. Geng, Phys. Rev. D**82**, 105004 (2010).
22. V. D. Barger, J. L. Hewett and R. J. N. Phillips, Phys. Rev. D**41**, 3421 (1990).
23. Z. z. Xing, Phys. Lett. B**530**, 159 (2002); B**539**, 85 (2002); P. H. Frampton, S. L. Glashow and D. Marfatia, Phys. Lett. B**536**, 79 (2002); B. R. Desai, D. P. Roy and A. R. Vaucher, Mod. Phys. Lett. A**18**, 1355 (2003); W. l. Guo and Z. z. Xing, Phys. Rev. D**67**, 053002 (2003); M. Honda, S. Kaneko and M. Tanimoto, JHEP **0309**, 028 (2003).
24. J. Adam *et al.* [MEG Collaboration], Phys. Rev. Lett. **110**, 201801 (2013).
25. M. Nebot, J. F. Oliver, D. Palao and A. Santamaria, Phys. Rev. D**77**, 093013 (2008).
26. W. H. Bertl *et al.* [SINDRUM II Collaboration], Eur. Phys. J. C**47**, 337 (2006).
27. H. Pas, M. Hirsch, H. V. Klapdor-Kleingrothaus and S. G. Kovalenko, Phys. Lett. B**498**, 35 (2001).
28. F. F. Deppisch, M. Hirsch and H. Pas, J. Phys. G**39**, 124007 (2012).
29. A. Gando *et al.* [KamLAND-Zen Collaboration], Phys. Rev. C**85**, 045504 (2012); Phys. Rev. Lett. **110**, 062502 (2013).

Three-Neutrino Oscillation Parameters: Status and Prospects

F. Capozzi,[a,b] G. L. Fogli,[a,b] E. Lisi,[b,*] A. Marrone,[a,b] D. Montanino[c,d] and A. Palazzo[e]

[a] *Dipartimento Interateneo di Fisica, Via Amendola 173, 70126 Bari, Italy*
[b] *Istituto Nazionale di Fisica Nucleare, Sezione di Bari, Via Orabona 4, 70126 Bari, Italy*
[c] *Dipartimento di Matematica e Fisica, Via Arnesano, 73100 Lecce, Italy*
[d] *Istituto Nazionale di Fisica Nucleare, Sez. di Lecce, Via Arnesano, 73100 Lecce, Italy*
[e] *Max-Planck-Institut für Physik, Föhringer Ring 6, 80805 München, Germany*

Neutrino oscillation searches using a variety of sources (solar, atmospheric, accelerator and reactor neutrinos) have established a standard three-neutrino (3ν) mass-mixing framework and five of its parameters: the two squared mass gaps (δm^2, Δm^2) and the three mixing angles (θ_{12}, θ_{13}, θ_{23}). At present, a single class of experiments dominates each of these parameters, while only combined analyses of various (eventually all) data sets are needed to constrain the still unknown mass hierarchy [sign(Δm^2)], θ_{23} octant and CP-violating phase δ. We review the status of the known and unknown parameters (as emerging from a global analysis of the oscillation data), investigate the correlations and stability of the such parameters within different combinations of data sets, and discuss the near-term prospects in this field.

1. Introduction

Since the discovery of atmospheric ν oscillations in 1998, a new paradigm — the 3ν mass-mixing framework — has emerged in particle physics.[1] Indeed, the vast majority of ν oscillation data can be explained by assuming that the three known flavor states $\nu_\alpha = (\nu_e, \nu_\mu, \nu_\tau)$ are mixed with three massive states $\nu_i = (\nu_1, \nu_3, \nu_3)$ via three mixing angles $(\theta_{12}, \theta_{13}, \theta_{23})$ and a possible CP-violating phase δ. Oscillations are driven by two independent differences between the squared masses m_i^2, which can be defined as $\delta m^2 = m_2^2 - m_1^2 > 0$ and $\Delta m^2 = m_3^2 - (m_1^2 + m_2^2)/2$, where $\Delta m^2 > 0$ and < 0 correspond to normal (NH) and inverted (IH) hierarchy, respectively.[2]

At present, five of the above 3ν oscillation parameters have been measured, with an accuracy largely dominated by a specific class of experiments, namely:

*Speaker. Email: eligio.lisi@ba.infn.it

θ_{12} by solar data,[3] θ_{13} by short-baseline (SBL) reactor data,[4] θ_{23} by atmospheric data, mainly from Super-Kamiokande (SK),[3] δm^2 by long-baseline reactor data from KamLAND (KL),[3] and Δm^2 by long-baseline (LBL) accelerator data, mainly from MINOS[1] and T2K.[5] However, the mass hierarchy, the θ_{23} octant, and the CP-violating phase δ are still unknown and will be addresses by future experiments.[6]

In this context, global neutrino data analyses may be useful to assess the overall consistency and accuracy of the known parameters, as well as to squeeze possible hints about the unknown ones. In the following, we report and discuss the results of a recent global analysis which include data available at the time of this Conference.[2] The reader is referred to Ref. 2 for further details and references not reported herein.

It should be noted that, in the 3ν framework, there are other unknowns not accessible to oscillation experiments, namely: the absolute neutrino mass scale[1] (possibly from cosmology[7]), the Dirac or Majorana nature of the neutrino fields[8,9] and, in the latter case, the associated Majorana phases.[8] Current constraints and prospects on these unknowns, which are crucial for theoretical model building,[6,10,11] will also be briefly commented below. Finally, it should be mentioned that some controversial results (not discussed herein) might indicate possible extensions of the above 3ν framework in terms of one or more additional mass states ν_j ($j \geq 4$), mostly sterile and with mass gaps at the (sub)eV scale. The reader is referred to Refs. 1 and 12 for up-to-date discussions of the sterile neutrino phenomenology.

2. Global Analysis: Methodology

In this Section we briefly discuss the various data sets and their combination in global fits.

LBL Acc. + Solar + KL data. The oscillation phenomenology of LBL accelerator experiments is dominated by the oscillation parameters (Δm^2, θ_{23}) in the $\nu_\mu \to \nu_\mu$ disappearance channel, supplemented by θ_{13} in the $\nu_\mu \to \nu_e$ appearance channel. However, the current accuracy of MINOS and T2K data requires that the oscillation probability is precisely calculated in terms of all the input parameters, including matter effects and subdominant terms driven by (δm^2, θ_{12}, δ). Since (δm^2, θ_{12}) are essentially fixed by the Solar and KL experiments, it makes sense to combine these data with LBL accelerator data from the very beginning. We remark that "Solar + KL" data provide a preference for $\sin^2\theta_{13} \sim 0.02$ in our analysis, which plays a role in the combination "LBL Acc. + Solar + KL," as discussed below.

Adding SBL reactor data. After the recent T2K observation of electron flavor appearance, the combination of LBL Acc. + Solar + KL data can provide a highly significant measurement of θ_{13} which, however, depends on the unknown CP violating phase δ and θ_{23} octant. SBL reactor experiments (Daya Bay, RENO, Double Chooz) provide (δ, θ_{23})-independent and accurate measurements of θ_{13}, which play a crucial role in the "LBL Acc. + Solar + KL + SBL Reac." combination.

Adding atmospheric neutrino data. Atmospheric data involve a very rich oscillation phenomenology in both appearance and disappearance modes involving ν_μ and ν_e. In principle, the high-statistics Super-Kamiokande experiment (phases I–IV) is thus sensitive to subleading effects related to the mass hierarchy, the θ_{23} octant and the CP phase δ. However, within the current experimental and theoretical systematic uncertainties, it remains difficult to disentangle and probe such small effects at a level exceeding $\sim 1\sigma$–2σ. Moreover, different and independent analyses of SK data, at comparable levels of refinement, do not necessarily provide similar hints about subleading effects. Therefore, we prefer to add these data only in the final "LBL Acc. + Solar + KL + SBL Reac. + SK Atm." combination, in order to separately gauge their effects on the various 3ν parameters.

Conventions for allowed regions. The data are compared to theoretical expectations via a refined χ^2 function which accounts for all known sources of correlated and uncorrelated uncertainties. In each of the above combined data analyses, the six oscillation parameters (Δm^2, δm^2, θ_{12}, θ_{13}, θ_{23}) are unconstrained in any given hierarchy (normal or inverted). Parameter ranges at N standard deviations are defined as $N\sigma = \sqrt(\chi^2 - \chi^2_{\rm min})$. This definition holds also in two-dimensional plots, where it is understood that the previous $N\sigma$ ranges are reproduced by projecting 2D contours over one parameter axis. All undisplayed parameters are marginalized away. Finally, the relative preference of the data for either NH or IH is measured by the quantity $\Delta\chi^2_{\rm I-N} = \chi^2_{\rm min}({\rm IH}) - \chi^2_{\rm min}({\rm NH})$, with the caveat that it cannot immediately be translated into "$N\sigma$" by taking the square root of its absolute value, because it refers to two discrete hypotheses.

3. Results on Single Oscillation Parameters

In this Section we graphically report the results of our global analysis of increasingly richer data sets, grouped in accordance to the previous discussion, in terms of single oscillation parameters.

Figures 1, 2 and 3 show the $N\sigma$ curves for the data sets defined in the previous section. In each figure, the solid (dashed) curves refer to NH (IH); the two curves basically coincide for the δm^2 and θ_{12} parameters, since they are determined by Solar+KL data which are largely insensitive to the hierarchy. For each parameter in Figs. 1–3, the more linear and symmetrical are the curves, the more gaussian is the associated probability distribution.

Figure 1 refers to the combination LBL Acc. + Solar + KL which, by itself, sets highly significant lower and upper bounds on all the oscillation parameters but δ. In this figure, the relatively strong appearance signal in T2K[5] dominates the lower bound on θ_{13}, and also drives the slight but intriguing preference for $\delta \simeq 1.5\pi$: indeed, for $\sin\delta \sim -1$, the CP-odd term in the $\nu_\mu \to \nu_e$ appearance probability is maximized. It should be noted that current MINOS appearance data generally prefer $\sin\delta > 0$;[2] however, the stronger T2K appearance signal largely dominates in the global fit. On the other hand, MINOS disappearance data drive the slight

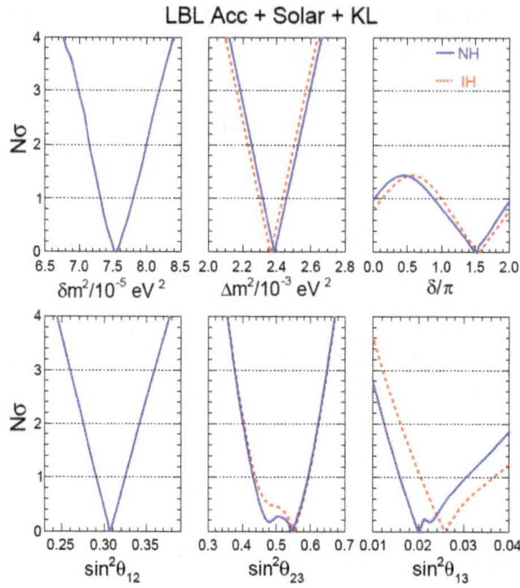

Fig. 1. Combined 3ν analysis of LBL Acc. + Solar + KL data: Bounds on the oscillation parameters in terms of standard deviations $N\sigma$ from the best fit. Solid (dashed) lines refer to NH (IH). The horizontal dotted lines mark the 1σ, 2σ and 3σ levels for each parameter.

preference for nonmaximal θ_{23}, as compared with nearly maximal θ_{23} in T2K.[5] The (even slighter) preference for the second θ_{23} octant is due to the interplay of LBL accelerator and Solar + KL data, as discussed in the next Section.

Figure 2 shows the results obtained by adding the SBL reactor data, which strongly reduce the θ_{13} uncertainty. Further effects of these data include: (i) a slightly more pronounced preference for $\delta \simeq 1.5\pi$ and $\sin\delta < 0$, and (ii) a swap of the preferred θ_{23} octant with the hierarchy ($\theta_{23} < \pi/4$ in NH and $\theta_{23} > \pi/4$ in IH). These features will be interpreted in terms of parameter covariances in the next Section.

Figure 3 shows the results obtained by adding the SK atmospheric data, thus obtaining the most complete data set. The main differences with respect to Fig. 2 include: (i) an even more pronounced preference for $\sin\delta < 0$, with a slightly lower best fit at $\delta \simeq 1.4\pi$; (ii) a slight reduction of the errors on Δm^2 and a relatively larger variation of its best-fit value with the hierarchy; (iii) a preference for θ_{23} in the first octant for both NH and IH, which is a persisting feature of our analyses. The effects (ii) and (iii) show that atmospheric neutrino data have the potential to probe subleading hierarchy effects, although they do not yet emerge in a stable or significant way.

In Figs. 1–3, an intriguing feature is the increasingly pronounced preference for nonzero CP violation with increasingly rich data sets, although the two CP-conserving cases ($\delta = 0,\ \pi$) remain allowed at $< 2\sigma$ in both NH and IH, even when all data are combined (see Fig. 3). It is worth noticing that the two maximally CP-violating cases ($\sin\delta = \pm 1$) have opposite likelihood: while the range around

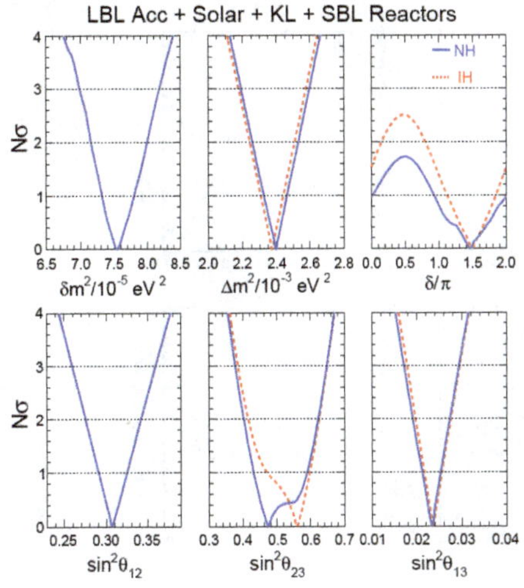

Fig. 2. As in Fig. 1, but adding SBL reactor data in the fit.

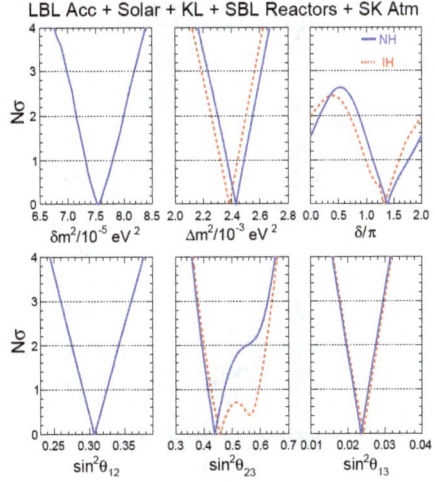

Fig. 3. As in Fig. 2, but adding SK atm. data (global fit to all ν data).

$\delta \sim 1.5\pi$ ($\sin\delta \sim -1$) is consistently preferred, small ranges around $\delta \sim 0.5\pi$ ($\sin\delta \sim +1$) appear to be disfavored (at $> 2\sigma$ in Fig. 3), In the next few years, the appearance channel in LBL accelerator experiments will provide crucial data to investigate these hints about ν CP violation,[2] with relevant implications for models of leptogenesis.[11]

From the comparison of Figs. 1–3 one can also notice a generic preference for nonmaximal mixing ($\theta_{23} \neq 0$), although it appears to be weaker than in our past analyses, essentially because the most recent T2K data prefer nearly maximal mixing, and thus "dilute" the opposite preference coming from MINOS and atmospheric data. Moreover, the indications about the octant appear to be somewhat unstable in different combinations of data. In the present analysis, only atmospheric data consistently prefer the first octant in both hierarchies, but the overall significance remains at the level $\sim 2\sigma$ in NH and is much lower in IH. These fluctuations show how difficult it is to reduce the allowed range of θ_{23}. In this context, the disappearance channel in LBL accelerator experiments will provide crucial data to address the issue of nonmaximal θ_{23} in the next few years.[5,6]

Finally, we comment on the size of $\Delta\chi^2_{\mathrm{I-N}}$ which, by construction, is not apparent in Figs. 1–3. We find $\Delta\chi^2_{\mathrm{I-N}} = -1.3, -1.4, +0.3$, for the data sets in Figs. 1, 2, and 3, respectively. Unfortunately, such values are both small and with unstable sign, and do not provide us with any relevant indication about the hierarchy.

4. Selected Parameter Covariances

In this Section we show the allowed regions for selected couples of oscillation parameters, and discuss some interesting correlations.

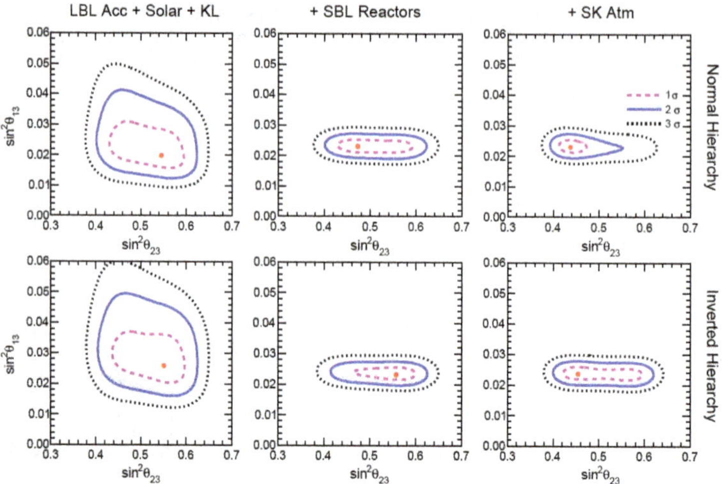

Fig. 4. Results of the analysis in the plane charted by $(\sin^2\theta_{23}, \sin^2\theta_{13})$, all other parameters being marginalized away. From left to right, the regions allowed at 1, 2 and 3σ refer to increasingly rich datasets: LBL accelerator + solar + KamLAND data (left panels), plus SBL reactor data (middle panels), plus SK atmospheric data (right panels). Best fits are marked by dots. The three upper (lower) panels refer to normal (inverted) hierarchy.

Figure 4 shows the allowed regions in the plane $(\sin^2\theta_{23}, \sin^2\theta_{13})$. From left to right, the panels refer to increasingly rich data sets, while upper and lower panels refer to NH and IH, respectively. In the left panels, a slight negative correlation emerges from LBL appearance data, since the dominant oscillation amplitude contains a factor $\sin^2\theta_{23}\sin^2\theta_{13}$. The contours extend towards relatively large values of θ_{13}, especially in IH, in order to accommodate the relatively strong T2K appearance signal.[5] However, solar + KL data provide independent (although weaker) constraints on θ_{13} and, in particular, prefer $\sin^2\theta_{13} \sim 0.02$ in our analysis. This value is on the "low side" of the allowed regions and is thus responsible for the relatively high value of θ_{23} at best fit, namely, for the second-octant preference in both NH and IH. However, when current SBL reactor data are included in the middle panels, a slightly higher value of θ_{13} ($\sin^2\theta_{13} \simeq 0.023$) is preferred with very small uncertainties: this value is high enough to shift the best-fit of θ_{23} from the second to the first octant in NH, but not in IH. Finally, the inclusion of SK atmospheric data (right panels) provides in our analysis an overall preference for the first octant, which is however quite weak in IH. Unfortunately, as previously mentioned, the current hints about the θ_{23} octant do not appear to be particularly stable or convergent.

Figure 5 shows the allowed regions in the plane $(\sin^2\theta_{13}, \delta/\pi)$, which is at the focus of current research in neutrino physics. In the left panels there is a remarkable preference for $\delta \sim 1.5\pi$, where a compromise is reached between the relatively high θ_{13} values preferred by the T2K appearance signal, and the relatively low value

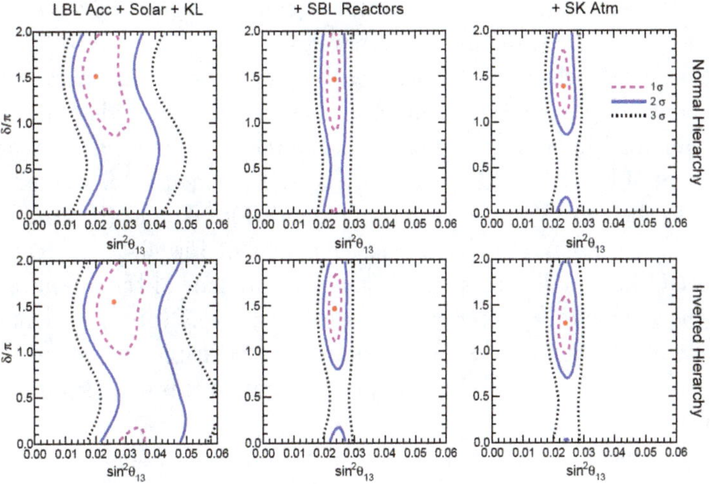

Fig. 5. As in Fig. 4, but in the plane ($\sin^2\theta_{13}$, δ/π).

preferred by solar + KL data. In the middle panel, SBL reactor data strengthen this trend by reducing the covariance between θ_{13} and δ. It is quite clear that we can still learn much from the combination of accelerator and reactor data in the next few years. Finally, the inclusion of SK atmospheric data in the right panels also adds some statistical significance to this trend, with a slight lowering of the best-fit value of δ.

5. Implications on Absolute Mass Observables

In general, absolute neutrino masses can be probed via three main methods. The first, classical one is provided by β decay, sensitive to the so-called "effective electron neutrino mass" m_β,[1]

$$m_\beta = \left[\sum_i |U_{ei}|^2 m_i^2\right]^{\frac{1}{2}} = \left[c_{13}^2 c_{12}^2 m_1^2 + c_{13}^2 s_{12}^2 m_2^2 + s_{13}^2 m_3^2\right]^{\frac{1}{2}} . \tag{1}$$

The second observable — if neutrinos are Majorana spinors — is the effective "Majorana neutrino mass" $m_{\beta\beta}$ in $0\nu\beta\beta$ decay,[8,9]

$$m_{\beta\beta} = \left|\sum_i U_{ei}^2 m_i\right| = \left|c_{13}^2 c_{12}^2 m_1 + c_{13}^2 s_{12}^2 m_2 e^{i\phi_2} + s_{13}^2 m_3 e^{i\phi_3}\right| , \tag{2}$$

where $\phi_{2,3}$ are additional unknown parameters (Majorana phases).[1] Note that nuclear uncertainties might complicate the interpretation of possible future $0\nu\beta\beta$ signals.[8] The third observable is the sum of neutrino masses in standard cosmology:[7]

$$\Sigma = m_1 + m_2 + m_3 . \tag{3}$$

The oscillation constraints reported in the previous Section induce strong correlations among the above three main observables.

Figure 6 shows such correlations in terms of 2σ constraints (bands) in the planes charted by any couple of the absolute mass observables. Note that the bands in the (m_β, Σ) plane of Fig. 6 are quite narrow, due to the high accuracy reached in the determination of all the oscillation parameters. In principle, precise measurements of (m_β, Σ) in the sub-eV range (where the bands for NH and IH branch out) could determine the hierarchy. In the two lower panels of Fig. 6, there remains a large vertical spread in the allowed slanted bands, as a result of the unknown Majorana phases in $m_{\beta\beta}$, which may interfere either constructively (upper part of each band) or destructively (lower part of each band). In principle, precise data in either the $(m_{\beta\beta}, m_\beta)$ plane or the $(m_{\beta\beta}, \Sigma)$ plane might thus provide constraints on the Majorana phases.

At present, there are only safe upper bounds on these absolute mass parameters, at the eV level for m_β,[1] and in the sub-eV range for $m_{\beta\beta}$[8,9] and Σ.[7] A great experimental activity is in progress towards mass sensitivity goals of $O(\sqrt{\Delta m^2})$, at least via $0\nu\beta\beta$ and cosmological probes. Sensitivities of $O(\sqrt{\delta m^2})$ in $0\nu\beta\beta$ decay appear to be extremely challenging at present.

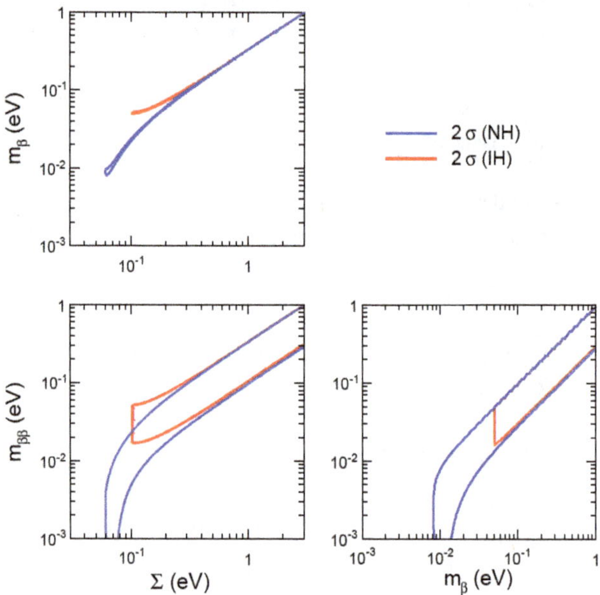

Fig. 6. Constraints induced by oscillation data (at 2σ level) in the planes charted by any two among the absolute mass observables m_β (effective electron neutrino mass), $m_{\beta\beta}$ (effective Majorana mass), and Σ (sum of neutrino masses). Blue (red) bands refer to normal (inverted) hierarchy.

In the most optimistic scenario, the absolute neutrino masses might be all around 0.1–0.2 eV, and thus observable in the next few years through measurements of at least two among the three (m_β, $m_{\beta\beta}$, Σ) parameters. Then, the concordance of two or three of these observables with the oscillation bands in Fig. 6 would provide a fundamental cross-check of the standard framework with three massive and mixed neutrinos. If concordance is not achieved (e.g., if strong cosmological limits on Σ are not compatible with possible signals of $m_{\beta\beta} > 0$ within the bands of Fig. 6, or vice versa), the situation would become even more interesting from a phenomenological viewpoint. In this case, data might suggest modifications of the standard framework either in cosmology (e.g., adopting suitable variants of the concordance cosmological model) or in neutrino physics (e.g., exploring nonstandard mechanisms for $0\nu\beta\beta$ decay — a topic witnessing renewed interest). Conversely, the lack of a signal in any of the observables (m_β, $m_{\beta\beta}$, Σ) in the next few years would make the perspectives for the neutrino mass quest extremely challenging.

6. Summary and Prospects

In the light of recent results coming from reactor and accelerator experiments, and of their interplay with solar and atmospheric data, we have updated the estimated $N\sigma$ ranges of the known 3ν parameters (Δm^2, δm^2, θ_{12}, θ_{13}, θ_{23}), and we have revisited the status of the current unknowns [sign(Δm^2), sign($\theta_{23} - \pi/4$), δ]. The results of the global analysis of all data[2] are shown in Fig. 3 in terms of single parameters. One can appreciate the high accuracy reached in the determination of the known oscillation parameters.

We have also discussed in some detail the status of the unknown parameters. Concerning the hierarchy [sign(Δm^2)], we find no significant difference between normal and inverted mass ordering. However, assuming normal hierarchy, we find possible hints about the other two unknowns, namely: a slight preference for the first θ_{23} octant, and possible indications for nonzero CP violation (with $\sin\delta < 0$), although at a level below $\sim 2\sigma$ in both cases. The second hint appears also in inverted hierarchy, but with even lower statistical significance.

In order to understand how the various constraints and hints emerge from the analysis, and to appreciate their (in)stability, we have considered increasingly rich data sets, starting from the combination of LBL accelerator plus solar plus KamLAND data, then adding SBL reactor data, and finally including atmospheric data. We have discussed the fit results both on single parameters and on selected couples of correlated parameters. It turns out that the hints about the θ_{23} octant appear somewhat unstable at present, while those about δ (despite being statistically weaker) seem to arise from an intriguing convergence of several pieces of data.

Finally, we have discussed the implication of such results for the three observables sensitive to absolute neutrino masses via single- and double-beta decay and cosmology. In general, global analyses of oscillation and non oscillation data appear to provide valuable tools to gauge the overall consistency of the data in a given framework (assumed to be standard 3ν mixing herein). Further experimental data might either confirm the 3ν framework and fix its remaining unknowns (possible CP violation, θ_{23} octant, absolute masses and their ordering, Dirac versus Majorana nature, and Majorana phases in the latter case), or find interesting discrepancies which would require new physics beyond the three known neutrino states and their standard interactions.

In the near or medium term, there are interesting plans to address the hierarchy issue via medium-baseline reactor experiments[4] capable to observe the interference between δm^2 and $\pm\Delta m^2$. Double beta decay[9] searches will also contribute to test the (more favorable) inverted hierarchy range for $m_{\beta\beta}$. Cosmology has, in principle, good chances to test the absolute neutrino mass scale in the near future, if systematics are kept under control.[7] Current long baseline accelerator experiments will probably improve the current indications on the θ_{23} octant and on the favored δ range, but with a significance exceeding $\sim 2\sigma$ only in the most favorable cases.[5] In the far future, more powerful accelerator searches are being planned to get indications at higher confidence levels, especially for CP violation and mass hierarchy;[6] in this context, large-volume atmospheric neutrino detectors may also provide important probes of matter effects, mass hierarchy and θ_{23}.[13] Of course, such expectations and the current planning of near- and far-future projects might be significantly altered by unexpected discoveries, e.g., of new neutrino states or new interactions, which might emerge at any time in this surprising and vibrant field of research.

Acknowledgments

E.L. is grateful to the organizers for their kind invitation and hospitality in Singapore. This work is supported in part through the Theoretical Astroparticle Physics Research Programs by the Italian Istituto Nucleare di Fisica Nucleare (INFN TAsP Initiative) and Ministero dell'Istruzione, Università e Ricerca (MIUR PRIN Project).

References

1. K. A. Olive *et al.* (Particle Data Group), *Chin. Phys. C* **38**, 090001 (2014). See the review therein: "Neutrino mass, mixing and oscillations," by K. Nakamura and S. T. Petcov.
2. F. Capozzi, G. L. Fogli, E. Lisi, A. Marrone, D. Montanino and A. Palazzo, *Phys. Rev. D* **89**, 093018 (2014).
3. A. Suzuki, talk at this Conference.
4. Y. F. Wang, S. H. Seo and S. B. Kim, talks at this Conference.
5. C. Bronner, talk at this Conference.
6. A. Y. Smirnov, talk at this Conference.
7. Y. Y. Y. Wong, talk at this Conference.
8. W. Rodejohann, Z. Z. Xing, and H. Paes, talks at this Conference.

9. G. Gratta and Y. Efremenko, talks at this Conference.
10. H. Fritzsch, talk at this Conference.
11. S. Petcov, M. C. Chen, X. G. He, R. Volkas, J. Valle, S. Antusch, P. Minkowski (and several others), talks at this Conference.
12. C. Rubbia, C. Giunti, J. Rosner, A. Ereditato, M. Danilov, W. Wang, A. De Roeck, talks at this Conference.
13. E. Resconi and N. Sinha, talks at this Conference.

On the Majorana Neutrinos and Neutrinoless Double Beta Decays

Zhi-Zhong Xing
Institute of High Energy Physics, Chinese Academy of Sciences, Beijing 100049, China
Center for High Energy Physics, Peking University, Beijing 100080, China
xingzz@ihep.ac.cn

Ye-Ling Zhou
Institute of High Energy Physics, Chinese Academy of Sciences, Beijing 100049, China
zhouyeling@ihep.ac.cn

The neutrinoless double-beta ($0\nu\beta\beta$) decay is a lepton-number-violating process which is experimentally unique for identifying the Majorana nature of massive neutrinos. We give a brief overview of some theoretical aspects of this process. In particular, a novel "coupling-rod" diagram is introduced to describe the effective Majorana mass $\langle m \rangle_{ee}$ in the complex plane. Possible contributions of new physics to $\langle m \rangle_{ee}$ are also discussed.

Keywords: $0\nu\beta\beta$; Majorana neutrinos; lepton number violation; coupling-rod diagram.

1. Introduction

Given a continuous electron energy spectrum of the beta decay observed by James Chadwick,[1] Wolfgang Pauli conjectured the existence of an electron antineutrino in the final state of this process in 1930.[2] In other words, the beta decay is a three-body decay mode $(A, Z) \to (A, Z+1) + e^- + \overline{\nu}_e$. In 1933, Enrico Fermi proposed the four-fermion interaction to describe the beta decay,[3] which made it possible to calculate the reaction rates of nucleons and electrons (or positrons) interacting with neutrinos (or antineutrinos).

The double-beta ($\beta\beta$) decay is a second-order weak-interaction process that transforms a nuclide of atomic number Z into its isobar with atomic number $Z+2$: $(Z, A) \to (Z+2, A) + 2e^- + 2\overline{\nu}_e$, which was first discussed by Maria Goeppert-Mayer in 1935.[4] The masses of the parent and daughter nuclei must satisfy $m(Z, A) > m(Z+2, A)$ and $m(Z, A) < m(Z+1, A)$, such that the ordinary beta decay from

(Z, A) to $(Z+1, A)$ is forbidden. The first observed $\beta\beta$ decay was $^{82}\mathrm{Se} \to {}^{82}\mathrm{Kr} + 2e^- + 2\overline{\nu}_e$ with a half-life around 10^{20} yr.[5]

In 1937, Ettore Majorana pointed out that a massive neutral fermion could be its own antiparticle.[6] Two years later, Wendell Furry proposed the neutrinoless double-beta ($0\nu\beta\beta$) decay $(Z, A) \to (Z, A+2) + 2e^-$ mediated by the Majorana neutrinos.[7] This is a lepton-number-violating (LNV) process with $\Delta L = 2$, and its effective mass term $\langle m \rangle_{ee}$ is related to the neutrino masses and Majorana CP-violating phases. Today the $0\nu\beta\beta$ decay is recognized as a unique process at low energies to identify the Majorana nature of massive neutrinos, and hence searching for its signal becomes one of the most important tasks in the non-oscillation aspects of experimental neutrino physics.

This short paper gives a brief overview of some theoretical aspects of the $0\nu\beta\beta$ decay. In Section 2, we focus on $\langle m \rangle_{ee}$ and its geometrical description. Section 3 is devoted to some further discussions and comments on the $0\nu\beta\beta$ decay, including the possible influence of new physics on $\langle m \rangle_{ee}$.

2. The Effective Majorana Neutrino Mass $\langle m \rangle_{ee}$

In the lepton sector the 3×3 flavor mixing matrix can be parametrized in terms of three rotation angles and three CP-violating phases:

$$U = \begin{pmatrix} c_{12}c_{13} & s_{12}c_{13} & s_{13}e^{-i\delta} \\ -s_{12}c_{23} - c_{12}s_{13}s_{23}e^{i\delta} & c_{12}c_{23} - s_{12}s_{13}s_{23}e^{i\delta} & c_{13}s_{23} \\ s_{12}s_{23} - c_{12}s_{13}c_{23}e^{i\delta} & -c_{12}s_{23} - s_{12}s_{13}c_{23}e^{i\delta} & c_{13}c_{23} \end{pmatrix} P_\nu \,, \tag{1}$$

where $c_{ij} \equiv \cos\theta_{ij}$, $s_{ij} \equiv \sin\theta_{ij}$ $(ij = 12, 13, 23)$, and $P_\nu = \mathrm{Diag}\left\{e^{i\rho/2}, 1, e^{i(\delta+\sigma/2)}\right\}$ with ρ and σ being the so-called Majorana phases. The effective Majorana mass term of the $0\nu\beta\beta$ decay is given by

$$\begin{aligned} \langle m \rangle_{ee} &\equiv m_1 |U_{e1}|^2 e^{i\rho} + m_2 |U_{e2}|^2 + m_3 |U_{e3}|^2 e^{i\sigma} \\ &= m_1 c_{12}^2 c_{13}^2 e^{i\rho} + m_2 s_{12}^2 c_{13}^2 + m_3 s_{13}^2 e^{i\sigma} \,, \end{aligned} \tag{2}$$

which depends on both the neutrino masses and the Majorana CP-violating phases.

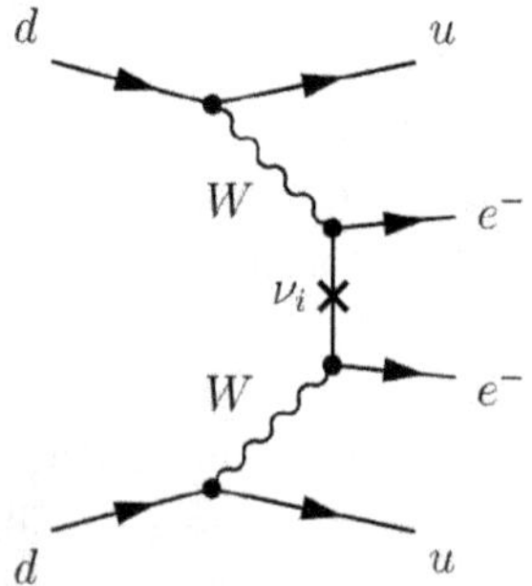

Fig. 1. The Feynman diagram of the $0\nu\beta\beta$ decay in the standard three-flavor scheme, where the cross stands for the insertion of a Majorana mass.

A Feynman diagram of the $0\nu\beta\beta$ decay mediated by the Majorana neutrinos is illustrated in Fig. 1. The half-life of the decaying nuclide can be written as

$$T^{0\nu}_{1/2} = \left(G^{0\nu}\right)^{-1} \left|M^{0\nu}\right|^{-2} \left|\langle m\rangle_{ee}\right|^{-2} , \tag{3}$$

where $G^{0\nu}$ is a phase-space factor depending on the transition Q value and on the atomic number Z, $M^{0\nu}$ is the nuclear matrix element (NME), and $\langle m\rangle_{ee}$ is the effective Majorana mass term defined in Eq. (2). So far $G^{0\nu}$ has been calculated precisely. If the NME is well understood, one can directly transfer the experimental limit on $T^{0\nu}_{1/2}$ to the constraint on $|\langle m\rangle_{ee}|$. Unfortunately, the calculation of the NME is based on some models which describe many-body interactions of nucleons in nuclei. Since different models focus on different aspects of nuclear physics, large uncertainties (at least a factor of 2 or 3) are unavoidable.[8]

The $0\nu\beta\beta$ decay is much weaker than the corresponding $\beta\beta$ decay due to the suppression arising from the tiny neutrino masses. To distinguish the former from the latter, one may analyze the energy spectrum of two electrons. Different from the continuous electron energy spectrum of the $\beta\beta$ decay, the electron energy spectrum of the $0\nu\beta\beta$ decay is discrete. The two electrons emitted from the $0\nu\beta\beta$ decay gain their kinetic energies equal to the Q value of this process. Since neutrinos are massive, these energies are slightly larger than the maximum of the kinetic energies of two electrons released from the $\beta\beta$ decay. In an ideal situation one may find a $0\nu\beta\beta$ line located on the right-hand side of the endpoint of the electron energy spectrum of the $\beta\beta$ decay. In practice it is very challenging to identify a $0\nu\beta\beta$ signal, because the energy resolution of the detector is usually not good enough to pin down the $0\nu\beta\beta$ signal peak overwhelmed by the $\beta\beta$ background.

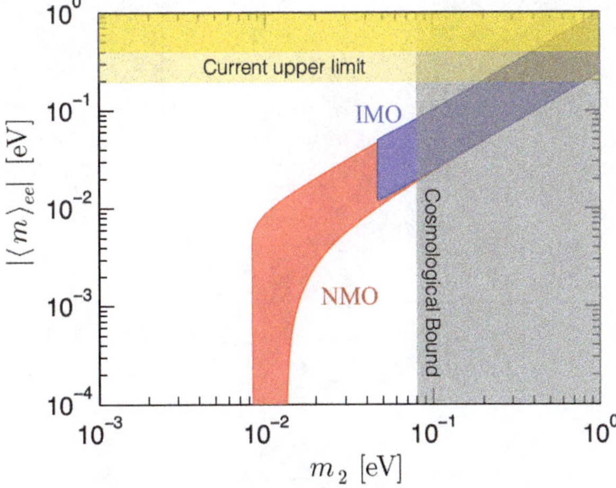

Fig. 2. The dependence of $|\langle m\rangle_{ee}|$ on m_2, where the 3σ ranges of Δm^2_{21}, Δm^2_{31}, θ_{12} and θ_{13} have been input,[12] and the Majorana phases ρ and σ are allowed to vary between 0 and 2π.

So far the $0\nu\beta\beta$ decay has not been established. The ongoing experiments KamLAND-Zen and EXO-200 use ^{136}Xe to search for the $0\nu\beta\beta$ decay, while another experiment GERDA uses ^{76}Ge. They have been operating for several years and set the lower limits on the half-lives of the relevant nuclei roughly around 10^{25} yr.[9–11] This corresponds to the upper limit on $|\langle m\rangle_{ee}|$ in the range 0.2–0.4 eV after the uncertainties of the NMEs are taken into account. We plot $|\langle m\rangle_{ee}|$ changing with m_2 in Fig. 2, where the upper limits on $|\langle m\rangle_{ee}|$ from the $0\nu\beta\beta$-decay experiments and the cosmological constraint $m_1 + m_2 + m_3 < 0.23$ eV from the Planck data[13] at the 95% confidence level are also shown. The next-generation experiments aim at reaching the sensitivity of $|\langle m\rangle_{ee}| \lesssim 0.04$ eV.[8]

Let us introduce the coupling-rod diagram of $\langle m\rangle_{ee}$ to understand its salient features.[14] Since the possibility of $m_1 = 0$ or $m_3 = 0$ is still allowed by current experimental data, we take the nonzero $m_2 U_{e2}^2$ term as the base vector to geometrically describe $\langle m\rangle_{ee}$ in the complex plane. Taking account of the phase convention of P_ν in Eq. (1), we define

$$\overrightarrow{OA} \equiv m_2 |U_{e2}|^2 \; , \quad \overrightarrow{AB} \equiv m_1 |U_{e1}|^2 e^{i\rho} \; , \quad \overrightarrow{CO} \equiv m_3 |U_{e3}|^2 e^{i\sigma} \; . \tag{4}$$

As illustrated in Fig. 3, the vector $\overrightarrow{CB} = \overrightarrow{OA} + \overrightarrow{AB} + \overrightarrow{CO}$ stands for $\langle m\rangle_{ee}$. It connects the two circles formed by the rotations of $\overrightarrow{AB}$ and $\overrightarrow{OC}$ and looks like the "coupling rod" of a locomotive.

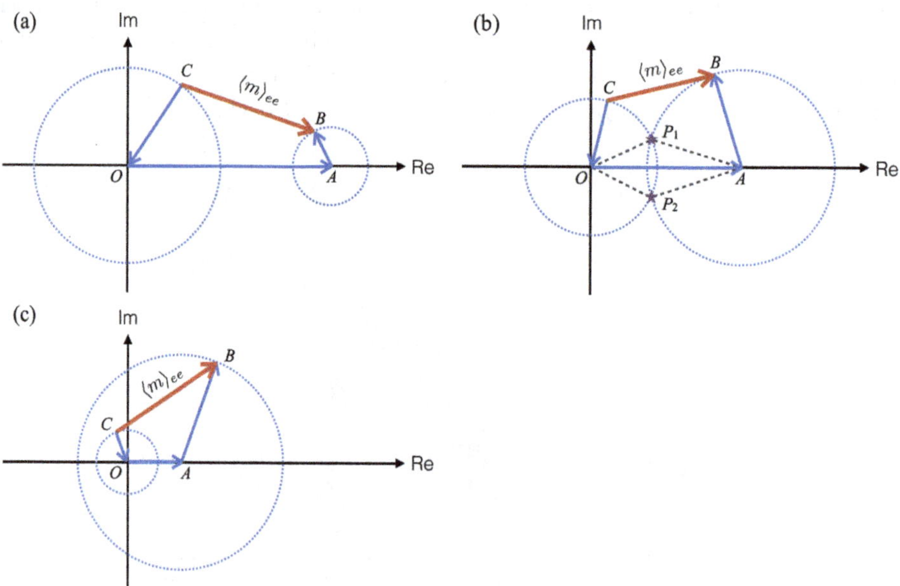

Fig. 3. The coupling-rod diagram of $\langle m\rangle_{ee} \equiv \overrightarrow{CB}$ in the complex plane, where $\overrightarrow{OA} \equiv m_2|U_{e2}|^2$, $\overrightarrow{AB} \equiv m_1|U_{e1}|^2 e^{i\rho}$ and $\overrightarrow{CO} \equiv m_3|U_{e3}|^2 e^{i\sigma}$. For the normal neutrino mass hierarchy, the three configurations of $\langle m\rangle_{ee}$ are all possible; and for the inverted hierarchy, only Fig. 3(c) is allowed.[14]

Depending on the length of $\overrightarrow{OA}$ and the radii of $\bigodot O$ and $\bigodot A$, there are three possibilities for the relative positions of these two circles:[14]

- $AB + OC < OA$ as shown in Fig. 3(a). In this case, $\bigodot O$ and $\bigodot A$ are external to each other, and thus $|\langle m\rangle_{ee}| = BC > 0$ holds. The allowed range of $|\langle m\rangle_{ee}|$ is constrained as $OA - AB - OC \leqslant |\langle m\rangle_{ee}| \leqslant OA + AB + OC$.
- $|AB - OC| \leqslant OA \leqslant AB + OC$ as shown in Fig. 3(b). Namely, $\bigodot O$ and $\bigodot A$ intersect or touch on the horizontal axis. The points of intersection imply $|\langle m\rangle_{ee}| = BC = 0$. In this case, the two Majorana phases are constrained.
- $AB - OC > OA$ as shown in Fig. 3(c). $\bigodot A$ contains $\bigodot O$ and does not touch the latter. The allowed range of $|\langle m\rangle_{ee}|$ turns out to be $AB - OA - OC \leqslant |\langle m\rangle_{ee}| \leqslant AB + OA + OC$.

The above discussions are not apparently subject to the neutrino mass ordering. Given the experimental data,[12] we arrive at the following conclusions: (1) for the normal neutrino mass hierarchy, the possibilities illustrated in Fig. 3(a), 3(b) and 3(c) are all allowed, and they correspond to the values of m_1 which are small ($m_1 \ll m_2 \ll m_3$), medium and large ($m_1 \lesssim m_2 \lesssim m_3$), respectively; (2) for the inverted mass hierarchy, only the possibility shown in Fig. 3(c) is allowed.

We may set the lightest neutrino mass to be zero in the spirit of Occam's razor. In this case m_2 can be fixed and $\langle m\rangle_{ee}$ only involves a single phase parameter. Once $|\langle m\rangle_{ee}|$ is determined from a measurement of the $0\nu\beta\beta$ decay, one will be able to determine the Majorana phase. There are two special cases:

- $m_1 = 0$, which leads to $AB = 0$. In this case $\bigodot A$ shrinks to a point, and thus the quadrilateral in Fig. 3 is simplified to $\triangle OAC$. As a result, $|\langle m\rangle_{ee}|$ only depends on a single CP-violating phase σ:

$$|\langle m\rangle_{ee}| = \sqrt{\Delta m^2_{21} s^4_{12} c^4_{13} + \Delta m^2_{31} s^4_{13} + 2\sqrt{\Delta m^2_{21} \Delta m^2_{31}}\ c^2_{12} s^2_{12} s^2_{13} \cos\sigma}\ . \quad (5)$$

- $m_3 = 0$, which leads to $OC = 0$. In this case the quadrilateral in Fig. 3 is simplified to $\triangle OAB$, and the magnitude of $\langle m\rangle_{ee}$ turns out to be

$$|\langle m\rangle_{ee}| = c^2_{13}\sqrt{-\Delta m^2_{32} s^4_{12} - \Delta m^2_{31} c^4_{12} + 2\sqrt{\Delta m^2_{31} \Delta m^2_{32}}\ c^2_{12} s^2_{12} \cos\rho}\ , \quad (6)$$

in which $\Delta m^2_{32} = \Delta m^2_{31} - \Delta m^2_{21}$.

The corresponding coupling-rod diagrams are simplified to the triangles, as shown in Fig. 4. In either case, the range of $|\langle m\rangle_{ee}|$ can easily be determined by allowing the Majorana phase to vary from 0 to 2π. Taking the 3σ ranges of three flavor mixing angles and two neutrino mass-squared differences[12] as the inputs, we arrive at 0.68 meV $\leqslant |\langle m\rangle_{ee}| \leqslant$ 4.7 meV in the $m_1 = 0$ case, and 12.4 meV $\leqslant |\langle m\rangle_{ee}| \leqslant$ 50.1 meV in the $m_3 = 0$ case.

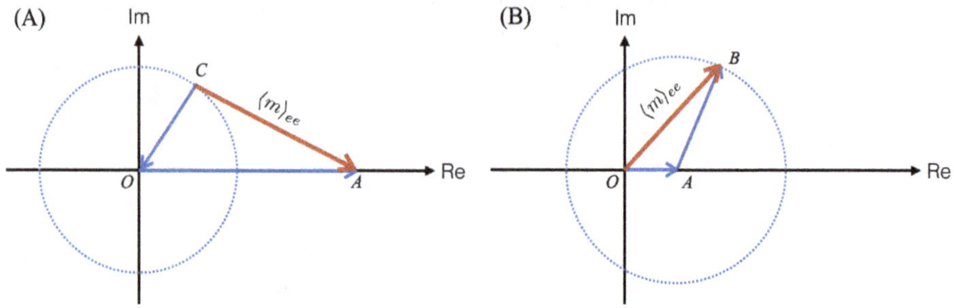

Fig. 4. The triangle of $\langle m\rangle_{ee}$ with the lightest neutrino mass being zero. Fig. 4(A) and 4(B) correspond to the normal and inverted neutrino mass hierarchies, respectively.

3. Further Discussions and Comments

The $0\nu\beta\beta$-decay experiments play a key role in identifying the Majorana nature of massive neutrinos. In the last section we have considered $\langle m\rangle_{ee}$ in the standard three-flavor scheme. However, new physics might contribute to the $0\nu\beta\beta$ decay. So it is natural to ask the following questions:

- Can we claim that massive neutrinos are the Majorana particles if a signal of the $0\nu\beta\beta$ decay is finally established?
- If a signal of the $0\nu\beta\beta$ decay is never observed, can we say that massive neutrinos are the Dirac particles?

The answers to these two questions are YES and NO, respectively.

The answer to the first question is affirmative according to the Schechter–Valle theorem:[15] an observation of the $0\nu\beta\beta$ decay implies the existence of a Majorana mass term for the neutrinos, no matter which mechanism results in the $0\nu\beta\beta$ decay itself. A simple explanation is given below. At the quark level the $0\nu\beta\beta$ decay occurs via $dd \to uue^-e^-$. Connecting u and d with e and ν_e via the standard weak charged-current interactions, one obtains the $\overline{\nu}_e \to \nu_e$ transition as illustrated in Fig. 5, which is essentially a Majorana mass term. The latter is a four-loop effect:

$$\frac{G_{\rm F}^2}{(16\pi^2)^4} \cdot {\rm MeV}^5 \sim \mathcal{O}(10^{-24}\ {\rm eV}) \, , \tag{7}$$

where $1/(16\pi^2)$ is the loop factor and MeV stands for the typical mass scale of the electron and quarks involved in the $0\nu\beta\beta$ process.[16]

The answer to the second question is negative, because the rate of the $0\nu\beta\beta$ decay is likely to be suppressed for some reasons. In the three-flavor scheme with a normal neutrino mass hierarchy, $|\langle m\rangle_{ee}|$ is possible to take a vanishing or vanishingly small value[17] as shown in Fig. 2. In this special region m_1 varies from 10^{-3} eV to 10^{-2} eV, and m_2 is around 0.01 eV to 0.02 eV. If $|\langle m\rangle_{ee}|$ happened to be in this unfortunate region, we would be unable to observe the $0\nu\beta\beta$ decay.

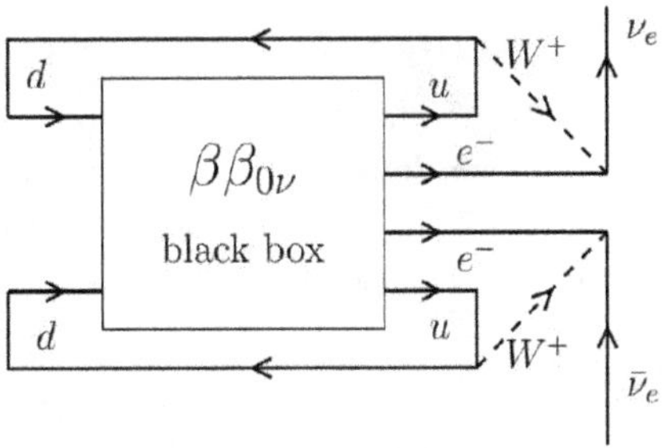

Fig. 5. The diagram for generating a Majorana neutrino mass through the $0\nu\beta\beta$ black box.[15]

There are a number of theoretical models which violate the lepton number. They may contribute to the $0\nu\beta\beta$ decay and modify the effective Majorana mass term $\langle m \rangle_{ee}$.[18] In general, the modification can be written as

$$\langle m \rangle_{ee} \to \langle m \rangle'_{ee} \approx \langle m \rangle_{ee} + m_{\rm NP} \,, \tag{8}$$

where $m_{\rm NP} = |m_{\rm NP}| e^{{\rm i}\phi}$ is a complex mass parameter. A cancellation between $m_{\rm NP}$ and $\langle m \rangle'_{ee}$ is not impossible. There are two categories of new physics models:

- New physics which is directly related to extra species of neutrinos, either light and sterile or heavy and sterile.
- New physics which has little or nothing to do with the neutrino mass issues, such as some supersymmetric models.

A contribution from the light sterile neutrinos is the simplest case, in which $m_{\rm NP} = m_4 R_{e4}^2 + m_5 R_{e5}^2 + \cdots$ with R_{ei} (for $i = 4, 5, \ldots$) standing for the mixing factors between the active and sterile neutrinos.[19] Given the heavy Majorana neutrinos in the type-I seesaw mechanism, they may mediate the $0\nu\beta\beta$ decay and thus contribute to the effective Majorana mass term. In this case $m_{\rm NP}$ depends on the masses and mixing factors of heavy Majorana neutrinos in a more complicated way.[20] Another typical example is the supersymmetric models which can introduce the LNV, either R-parity violating or R-parity conserving. Some of the LNV interactions have little to do with the neutrino mass generation but contribute to the $0\nu\beta\beta$ decay. Their contributions are complicated since they may contribute to both lepton and quark sectors, and modify both the NME and the effective Majorana mass term.[21]

In principle the neutrino–antineutrino oscillations should be the best playground for studying the salient features of Majorana neutrinos,[22] since their probabilities are sensitive to the absolute neutrino mass scale and all the three CP-violating phases.[23] In practice, however, the amplitudes of such LNV processes are suppressed by a factor $m_i/E \lesssim 10^{-6}$. Hence the detection of neutrino–antineutrino oscillations is unfeasible in the foreseeable future unless there will be a significant breakthrough in experimental technology.

Acknowledgments

One of us (Z. Z. Xing) would like to thank H. Fritzsch and K. K. Phua for hospitality at the IAS-NTU, where the International Conference on Massive Neutrinos was held. This work was supported in part by the National Natural Science Foundation of China under Nos. 11375207 and 11135009.

References

1. J. Chadwick, *Verhandle. Deut. Phys.* **16**, 383 (1914).
2. W. Pauli, lecture given in Zürich in 1957, published in *Physik und Erkenntnistheorie* (Friedr, Vieweg, & Sohn, Braunschweig/Wiesbaden, 1984), p. 156.
3. E. Fermi, *La Ricerca Scientifica* **2**, 12 (1933); *Z. Phys.* **88**, 161 (1934).
4. M. Goeppert-Mayer, *Phys. Rev.* **48**, 512 (1935).
5. S. R. Elliott, A. A. Hahn and M. K. Moe, *Phys. Rev. Lett.* **59**, 2020 (1987).
6. E. Majorana, *Nuovo Cimento* **14**, 171 (1937).
7. W. H. Furry, *Phys. Rev.* **15**, 1184 (1939).
8. S. M. Bilenky and C. Giunti, *Int. J. Mod. Phys. A* **30**, 1530001 (2015).
9. A. Gando *et al.* [KamLAND-Zen Collaboration], *Phys. Rev. Lett.* **110**, 062502 (2013).
10. M. Agostini *et al.* [GERDA Collaboration], *Phys. Rev. Lett.* **111**, 122503 (2013).
11. J. B. Albert *et al.* [EXO-200 Collaboration], *Nature* **510**, 229 (2014).
12. F. Capozzi, G.L. Fogli, E. Lisi, A. Marrone, D. Montanino, and A. Palazzo, *Phys. Rev. D* **89**, 093018 (2014).
13. P. A. R. Ade *et al.* [Planck Collaboration], arXiv:1502.01589 [astro-ph.CO].
14. Z. Z. Xing and Y. L. Zhou, *Chin. Phys. C* **39**, 011001 (2015).
15. J. Schechter and J. W. F. Valle, *Phys. Rev. D* **25**, 2951 (1982).
16. M. Duerr, M. Lindner and A. Merle, *JHEP* **1106**, 091 (2011).
17. Z. Z. Xing, *Phys. Rev. D* **68**, 053002 (2003).
18. W. Rodejohann, *Int. J. Mod. Phys. E* **20**, 1833 (2011).
19. Z. Z. Xing, Phys. Rev. D **85**, 013008 (2012); Y. F. Li and S. S. Liu, *Phys. Lett. B* **706**, 406 (2012).
20. Z. Z. Xing, *Phys. Lett. B* **679**, 255 (2009); and references therein.
21. For a review, see: H. V. Klapdor-Kleingrothaus, arXiv:hep-ex/9901021.
22. B. Pontecorvo, *Sov. Phys. JETP* **6**, 429 (1957); *Sov. Phys. JETP* **7**, 172 (1958).
23. Z. Z. Xing, *Phys. Rev. D* **87**, 053019 (2013); Z. Z. Xing and Y. L. Zhou, *Phys. Rev. D* **88**, 033002 (2013).

Dirac or Inverse Seesaw Neutrino Masses from Gauged $B - L$ Symmetry

Ernest Ma
Department of Physics and Astronomy, University of California, Riverside, California 92521, USA
ma@phyun8.ucr.edu

Rahul Srivastava
The Institute of Mathematical Science, Chennai, India
rahuls@imsc.res.in

The gauged $B - L$ symmetry is one of the simplest and well studied extension of standard model. In the conventional case, addition of three singlet right-handed neutrinos each transforming as -1 under the $B - L$ symmetry renders it anomaly free. It is usually assumed that the $B - L$ symmetry is spontaneously broken by a singlet scalar having two units of $B - L$ charge, resulting in a natural implementation of Majorana seesaw mechanism for neutrinos. However, as we discuss in this proceeding, there is another simple anomaly free solution which leads to Dirac or inverse seesaw masses for neutrinos. These new possibilities are explored along with an application to neutrino mixing with S_3 flavour symmetry.

Keywords: Neutrino mass mechanisms; neutrino mixing; right-handed neutrinos; Dirac seesaw; inverse seesaw; $B - L$ symmetry breaking; S(3) flavor symmetry.

1. Introduction

The nature of neutrinos i.e. whether they are Majorana or Dirac particles is one of the most important open questions in neutrino physics. Answering this question is essential to finding the underlying theory of neutrino masses and mixing. This issue can be potentially resolved by neutrinoless double beta decay experiments ($0\nu\beta\beta$). Currently several ongoing experiments are looking for signals of $0\nu\beta\beta$ but no such signal has been observed so far.[1–3] At present there is no compelling evidence from experiments or cosmological observations in favor of either Dirac or Majorana nature of neutrinos. With our current understanding, Dirac neutrinos are as plausible as Majorana ones.

However, theoretically Majorana neutrinos have received considerably more attention than Dirac neutrinos. There are several mechanisms (e.g. seesaw mechanisms) which satisfactorily explain smallness of neutrino masses if neutrinos are Majorana particles. On the other hand, Dirac neutrinos are not so well studied. There are only few models capable of providing a natural explanation for smallness of Dirac neutrino masses. In this proceeding we present one simple model for Dirac neutrinos with naturally small masses based on gauged $B-L$ symmetry. This proceeding is based on our work[4] and interested reader is referred to it for further details.

The plan of this proceeding is as follows. In Section 2 we will briefly review the familiar scenario of gauged $B-L$ symmetry leading to the Majorana neutrinos and seesaw mechanism. The possibility of Dirac neutrinos in such a scenario will also be briefly discussed. In Section 3 we will work out the new anomaly free solutions for gauged $B-L$ symmetry and show how it leads to Dirac neutrinos with naturally small masses. In Section 4 we will expand on our results and using S_3 flavour symmetry will construct realistic neutrino mass matrices consistent with present oscillation data. In Section 5 we will discuss the possibility of inverse seesaw Majorana neutrinos masses arising from the new anomaly free solutions. We will conclude in Section 6.

2. Majorana Neutrinos from $B-L$ Gauge Symmetry

Historically Baryon number (B) and Lepton numbers (L_i) were introduced to explain the stability of proton and absence of lepton flavour changing processes. In Standard Model (SM) the baryon and lepton numbers turn out to be accidentally conserved classical symmetries. The B and L currents are anomalous and only the combination $B-L$ is anomaly free. These accidental symmetries need not be conserved by beyond standard model (BSM) physics. For example if one adds a Majorana mass term for neutrinos then the $B-L$ symmetry gets broken by 2 units. The addition of right-handed neutrinos provides the possibility to promote this global $B-L$ symmetry to a anomaly free gauged $B-L$ symmetry.

Now addition of any new $U(1)_x$ gauge symmetry implies that one needs to cancel anomalies.[5–7] The gauged $B-L$ symmetry can potentially induce both gauge as well as gauge-gravitational anomalies. The triangular gauge anomalies arising from gauged $U(1)_{B-L}$ are:

- $\mathrm{Tr}\left(\mathrm{U(1)_{B-L}}\,[\mathrm{SU(2)_L}]^2\right)$
- $\mathrm{Tr}\left(\mathrm{U(1)_{B-L}}\,[\mathrm{U(1)_Y}]^2\right)$
- $\mathrm{Tr}\left(\mathrm{U(1)_{B-L}}\right)^3$

With the particle content of SM the first two anomalies are automatically canceled. Moreover, if the right-handed neutrinos ν_{iR}; $i=1,2,3$ transform as $\nu_{iR}\sim -1$ under $U(1)_{B-L}$, then $\sum U(1)^3_{B-L}=0$. The gauge-gravitational anomalies also vanish in this case as $-3(-1)=3$.

Lets briefly discuss the possibility of Dirac neutrinos in this scenario before discussing the conventional case of Majorana neutrinos. The addition of right-handed neutrinos with −1 charge under $U(1)_{B-L}$ allows one to have gauge invariant Yukawa coupling for neutrinos. The $SU(3)_C \otimes SU(2)_L \otimes U(1)_Y \otimes U(1)_{B-L}$ invariant Yukawa coupling for neutrinos is then given by

$$-\mathcal{L}_Y^\nu = \sum_{i,j} y_{ij} \bar{L}_{iL} \hat{\Phi}^* \nu_{jR} + \text{h.c.} \tag{1}$$

where $\hat{\Phi}^* = i\tau_2 \Phi^*$ and $\Phi = (\phi^+, \phi^0)^T$ is the SM Higgs doublet. Since the right- as well as left-handed neutrinos transform non-trivially under the gauged $U(1)_{B-L}$ symmetry, this implies that the Majorana mass term for ν_{iR} is forbidden and neutrinos are Dirac particles. In this case the $U(1)_{B-L}$ symmetry remains unbroken.[a] However, in such a scenario the smallness of neutrino masses requires unnaturally small Yukawa couplings and the model does not provide any explanation for their smallness. For more details and variants of this scenario we refer to Refs. 10–12.

A relatively better understanding of smallness of neutrino masses can be obtained if in addition to the right-handed neutrinos one also adds a singlet scalar χ transforming as $\chi \sim 2$ under $U(1)_{B-L}$. In this case the $SU(3)_C \otimes SU(2)_L \otimes U(1)_Y \otimes U(1)_{B-L}$ invariant Yukawa coupling for neutrinos is given by

$$-\mathcal{L}_Y^\nu = \sum_{i,j} y_{ij} \bar{L}_{iL} \hat{\Phi}^* \nu_{jR} + \frac{1}{2} \sum_{i,j} f_{ij} \bar{\nu}_{iR}^c \chi \nu_{jR} + \text{h.c.} \tag{2}$$

The spontaneous symmetry breaking (SSB) of χ then leads to breaking of $B-L$ symmetry. If $\langle \chi \rangle = u$ the right-handed neutrinos get a Majorana mass $M_{ij} = \sqrt{2} f_{ij} u$. If $u \gg v$, i.e. the $B-L$ symmetry breaking scale is far greater than the electroweak scale, then $M_R \gg m_D$ leading to a natural implementation of Type I seesaw mechanism.

Before ending this session we like to remark that the gauged $B-L$ symmetry can be imbedded in other Beyond Standard Model scenarios, e.g. it is an essential ingredient of Left–Right symmetric models. It can also be embedded in GUT groups, e.g. $SO(10)$.

3. Dirac Neutrinos from Gauged $B-L$ Symmetry

In this section we look at the possibility of another simple choice of $B-L$ charges for right-handed neutrinos which leads to anomaly free $U(1)_{B-L}$ gauge symmetry. Unlike the previous case, let the three right-handed neutrinos transform as $\nu_{iR} = (+5, -4, -4)$ under $B-L$ symmetry.[4,13–15] Since $\nu_{iR} \sim (+5, -4, -4)$, therefore

---

[a]The Z' gauge boson associated with $U(1)_{B-L}$ can get mass via Stuckelberg mechanism without breaking the $B-L$ symmetry.[8–10]

$$-(+5)^3 - (-4)^3 - (-4)^3 = +3, \tag{3}$$

$$-(5) - (-4) - (-4) = +3. \tag{4}$$

Thus in this case also the model is free from gauge as well as gauge-gravitational anomalies. Now the SM Higgs doublet $(\phi^+, \phi^0)^T$ does not connect ν_L with ν_R. Therefore the neutrinos do not get mass from the standard electroweak symmetry breaking. To generate neutrino masses let us add three heavy Dirac singlet fermions $N_{L,R}$ transforming as -1 under $B-L$ symmetry. They will not change the anomaly cancellation conditions and the model will remain anomaly free. Also let us add a singlet scalar χ_3 transforming as $+3$ under $B-L$.

Now for ν_{R2} and ν_{R3}, $(\bar{\nu}_L, \bar{N}_L)$ is linked to (ν_R, N_R) through the 2×2 mass matrix as follows

$$M_{\nu,N} = \begin{pmatrix} 0 & m_0 \\ m_3 & M \end{pmatrix} \tag{5}$$

where m_0 comes from $\langle\phi^0\rangle$. Moreover, m_3 comes from $\langle\chi_3\rangle$, due to the Yukawa coupling $\bar{N}_L \nu_R \chi_3$. The invariant mass M is naturally large, so the Dirac seesaw[16] yields a small neutrino mass $m_3 m_0/M$.

In the conventional $U(1)_{B-L}$ model, $\chi_2 \sim +2$ under $B-L$ is chosen to break the gauge symmetry, so that ν_R gets a Majorana mass and lepton number L is broken to $(-1)^L$. Here, $\chi_3 \sim +3$ means that it is impossible to construct an operator of any dimension for a Majorana mass term and L remains a conserved global symmetry, with $\nu_{L,R}$ and $N_{L,R}$ all having $L = 1$.

Since $\nu_{R1} \sim +5$ does not connect with ν_L or N_L directly, there is one massless neutrino in this case. The dimension-five operator $\bar{N}_L \nu_{R3} \chi_3^* \chi_3^*/\Lambda$ is allowed by $U(1)_{B-L}$ and would give it a small Dirac mass. Alternatively, one can add a second scalar $\chi_6 \sim 6$ to the model to account for mass of ν_{R1}.

4. The S_3 Flavour Symmetry

The discussion in previous section was aimed primarily at mass generation for Dirac neutrinos. The $U(1)_{B-L}$ symmetry alone does not provide any explanation for the currently observed PMNS mixing pattern. In this section we will generalize our discussion and will construct phenomenologically viable lepton mass matrices. In order to understand the leptonic family structure consistent with present neutrino oscillation data, we will make use of the non-Abelian discreet symmetry group S_3.

The S_3 group is the smallest non-Abelian discreet symmetry group and is the group of the permutation of three objects. It consists of six elements and is also isomorphic to the symmetry group of the equilateral triangle. It admits three irreducible representations 1, $1'$ and 2 with the tensor product rules,

$$\begin{aligned} &1 \otimes 1' = 1', \quad 1' \otimes 1' = 1, \quad 2 \otimes 1 = 2, \\ &2 \otimes 1' = 2, \quad 2 \otimes 2 = 1 \oplus 1' \oplus 2. \end{aligned} \tag{6}$$

In this proceeding we will use the complex representation of the S_3 group.[17,18] In the complex representation, if

$$\begin{pmatrix}\phi_1\\ \phi_2\end{pmatrix}, \begin{pmatrix}\psi_1\\ \psi_2\end{pmatrix} \in \mathbf{2} \Rightarrow \begin{pmatrix}\phi_2^\dagger\\ \phi_1^\dagger\end{pmatrix}, \begin{pmatrix}\psi_2^\dagger\\ \psi_1^\dagger\end{pmatrix} \in \mathbf{2} \tag{7}$$

then

$$\begin{aligned} \phi_1\psi_2 + \phi_2\psi_1, \quad \phi_2^\dagger\psi_2 + \phi_1^\dagger\psi_1 \quad &\in \mathbf{1} \\ \phi_1\psi_2 - \phi_2\psi_1, \quad \phi_2^\dagger\psi_2 - \phi_1^\dagger\psi_1 \quad &\in \mathbf{1'} \\ \begin{pmatrix}\phi_2\psi_2\\ \phi_1\psi_1\end{pmatrix}, \begin{pmatrix}\phi_1^\dagger\psi_2\\ \phi_2^\dagger\psi_1\end{pmatrix} &\in \mathbf{2}. \end{aligned} \tag{8}$$

With this brief summary of S_3 group and its irreducible representations, we now move on to constructing an S_3 invariant lepton sector. The $B-L$ charge and S_3 assignment of the fields for the lepton sector is as shown in Table 1,

Fields	$B-L$	S_3	Fields	$B-L$	S_3
L^e	-1	$1'$	e_R	-1	$1'$
L^μ	-1	$1'$	μ_R	-1	$1'$
L^τ	-1	1	τ_R	-1	1
N_L^1	-1	$1'$	N_R^1	-1	$1'$
N_L^2	-1	$1'$	N_R^2	-1	$1'$
N_L^3	-1	1	N_R^3	-1	1
Φ	0	1	ν_R^e	5	$1'$
$\begin{pmatrix}\nu_R^\mu\\ \nu_R^\tau\end{pmatrix}$	-4	2	$\begin{pmatrix}\chi_2\\ \chi_3\end{pmatrix}$	3	2

where we denote the left-handed lepton doublets by $L^\alpha = (\nu_L^\alpha, l_L^\alpha)^{\mathrm{T}}$ where $\alpha = e, \mu, \tau$; the right-handed charged leptons are denoted as e_R, μ_R, τ_R and the right-handed neutrinos as $\nu_R^e, \nu_R^\mu, \nu_R^\tau$. Also, let us denote the heavy singlet fermions as $N_{L,R}^i$; $i = 1, 2, 3$. The "Standard Model like" scalar doublet is denoted by $\Phi = (\phi^+, \phi^0)^{\mathrm{T}}$ and the singlet scalars are denoted by $\chi_{2,3}$.

The S_3 and $B-L$ invariant Yukawa interaction $\mathcal{L}_Y$ can then be written as

$$\mathcal{L}_Y = \mathcal{L}_{L^\alpha l_R} + \mathcal{L}_{L^\alpha N_R} + \mathcal{L}_{N_L N_R} + \mathcal{L}_{N_L \nu_R} \tag{9}$$

where

$$
\begin{aligned}
\mathcal{L}_{L^\alpha l_R} &= y'_e\,\bar{L}^e\,\Phi\,e_R + y'_{12}\,\bar{L}^e\,\Phi\,\mu_R + y'_{21}\,\bar{L}^\mu\,\Phi\,e_R + y'_\mu\,\bar{L}^\mu\,\Phi\,\mu_R + y_\tau\,\bar{L}^\tau\,\Phi\,\tau_R,\\
\mathcal{L}_{L^\alpha N_R} &= g'_{11}\,\bar{L}^e\,\hat{\Phi}^*\,N_R^1 + g'_{12}\,\bar{L}^e\,\hat{\Phi}^*\,N_R^2 + g'_{21}\,\bar{L}^\mu\,\hat{\Phi}^*\,N_R^1 + g'_{22}\,\bar{L}^\mu\,\hat{\Phi}^*\,N_R^2\\
&\quad + g_{33}\,\bar{L}^\tau\,\hat{\Phi}^*\,N_R^3,\\
\mathcal{L}_{N_L N_R} &= M'_{11}\,\bar{N}_L^1\,N_R^1 + M'_{12}\,\bar{N}_L^1\,N_R^2 + M'_{21}\,\bar{N}_L^2\,N_R^1 + M'_{22}\,\bar{N}_L^2\,N_R^2 + M_{33}\,\bar{N}_L^3\,N_R^3,\\
\mathcal{L}_{N_L\nu_R} &= \frac{f'_{11}}{\Lambda}\,(\bar{N}_L^1\,\nu_R^e)\otimes\left[\begin{pmatrix}\chi_3^*\\ \chi_2^*\end{pmatrix}\otimes\begin{pmatrix}\chi_3^*\\ \chi_2^*\end{pmatrix}\right]_1 + \frac{f'_{21}}{\Lambda}\,(\bar{N}_L^2\,\nu_R^e)\otimes\left[\begin{pmatrix}\chi_3^*\\ \chi_2^*\end{pmatrix}\otimes\begin{pmatrix}\chi_3^*\\ \chi_2^*\end{pmatrix}\right]_1\\
&\quad + f'_{12}\,\bar{N}_L^1\otimes\left[\begin{pmatrix}\nu_R^\mu\\ \nu_R^\tau\end{pmatrix}\otimes\begin{pmatrix}\chi_2\\ \chi_3\end{pmatrix}\right]_{1'} + f'_{22}\,\bar{N}_L^2\otimes\left[\begin{pmatrix}\nu_R^\mu\\ \nu_R^\tau\end{pmatrix}\otimes\begin{pmatrix}\chi_2\\ \chi_3\end{pmatrix}\right]_{1'}\\
&\quad + f_{33}\,\bar{N}_L^3\otimes\left[\begin{pmatrix}\nu_R^\mu\\ \nu_R^\tau\end{pmatrix}\otimes\begin{pmatrix}\chi_2\\ \chi_3\end{pmatrix}\right]_1.
\end{aligned}
\tag{10}
$$

In writing (10), we have used the notation $\hat{\Phi}^* = i\tau_2\Phi^* = (\phi^0, \phi^-)^{\mathrm{T}}$. Here, y_α are the Yukawa couplings of the charged leptons whereas f_{ij}, g_{ij} and M_{ij} denote the dimensionless coupling constants between the leptons and the heavy fermions.

At this point we like to remark that in $\mathcal{L}_Y$ there is still a freedom to redefine a few fields (i.e. the pairs $N_L^1 - N_L^2$, $N_R^1 - N_R^2$ and $e_R - \mu_R$) in a way that certain couplings can be made equal to zero. For the sake of later convenience we choose to use this freedom of field redefinition to make $f'_{11} = M'_{12} = y'_{21} = 0$. Moreover, we relabel the remaining non-zero couplings of these redefined fields as $f'_{ij} \to f_{ij}, g'_{ij} \to g_{ij}, M'_{ij} \to M_{ij}$.

After symmetry breaking the scalar fields get VEVs $\langle\phi^0\rangle = v$, $\langle\chi_i\rangle = u_i$; $i = 2, 3$. Then the mass matrix relevant to charged leptons is given by

$$
\mathcal{M}_l = v\begin{pmatrix} y_e & y_{12} & 0\\ 0 & y_\mu & 0\\ 0 & 0 & y_\tau \end{pmatrix}. \tag{11}
$$

This mass matrix can be readily diagonalized by bi-unitary transformation. In the limit of $y_e \ll y_\mu$ we get

$$
\begin{aligned}
\theta_{12}^l &\approx \tan^{-1}\left(\frac{-y_{12}}{y_\mu}\right); & m_e &\approx v\,y_e\cos\theta_{12}^{\mathrm{l}}\\
m_\mu &\approx v\left(y_\mu\cos\theta_{12}^{\mathrm{l}} - \mathrm{y}_{12}\sin\theta_{12}^{\mathrm{l}}\right); & m_\tau &\approx v y_\tau.
\end{aligned}
\tag{12}
$$

If $y_{12} = y_\mu$ then maximal mixing is achieved, i.e. $\theta_{12}^l = -\frac{\pi}{4}$, with $m_\mu = \sqrt{2}v y_\mu$. Also, the 6×6 mass matrix spanning $(\bar{\nu}_L^e, \bar{\nu}_L^\mu, \bar{\nu}_L^\tau, \bar{N}_L^1, \bar{N}_L^2, \bar{N}_L^3)$ and

$(\nu_R^e, \nu_R^\mu, \nu_R^\tau, N_R^1, N_R^2, N_R^3)^{\mathrm{T}}$ of neutrinos and the heavy fermions is given by

$$\mathcal{M}_{\nu,N} = \begin{pmatrix} 0 & 0 & 0 & g_{11}v^* & g_{12}v^* & 0 \\ 0 & 0 & 0 & g_{21}v^* & g_{22}v^* & 0 \\ 0 & 0 & 0 & 0 & 0 & g_{33}v^* \\ 0 & f_{12}u_3 & -f_{12}u_2 & M_{11} & 0 & 0 \\ \frac{f_{21}}{\Lambda}u_2^*u_3^* & f_{22}u_3 & -f_{22}u_2 & M_{21} & M_{22} & 0 \\ 0 & f_{33}u_3 & f_{33}u_2 & 0 & 0 & M_3 \end{pmatrix}. \tag{13}$$

As remarked earlier, the mass terms M_{ij} between the heavy fermions can be naturally large, so we can block diagonalize the mass matrix assuming that $f_{ij}, g_{ij} \ll M_{ij}$. The block diagonalized mass matrix of light neutrinos is given by

$$\begin{aligned} \mathcal{M}_\nu &= m_{N_L\nu_R}\left(M_{N_LN_R}\right)^{-1} m_{L^\alpha N_R} \\ &= v^* \begin{pmatrix} \frac{(g_{21}M_{11}-g_{11}M_{21})f_{12}u_2}{M_{11}M_{22}} & \frac{(g_{22}M_{11}-g_{12}M_{21})f_{12}u_2}{M_{11}M_{22}} & \frac{-f_{12}g_{33}u_3}{M_{33}} \\ \frac{(g_{21}M_{11}-g_{11}M_{21})f_{22}u_2+f_{21}g_{11}M_{22}u_6}{M_{11}M_{22}} & \frac{(g_{22}M_{11}-g_{12}M_{21})f_{22}u_2+f_{21}g_{12}M_{22}u_6}{M_{11}M_{22}} & \frac{-f_{22}g_{33}u_3}{M_{33}} \\ \frac{(g_{21}M_{11}-g_{11}M_{21})f_{33}u_2}{M_{11}M_{22}} & \frac{(g_{22}M_{11}-g_{12}M_{21})f_{33}u_2}{M_{11}M_{22}} & \frac{f_{33}g_{33}u_3}{M_{33}} \end{pmatrix} \end{aligned} \tag{14}$$

where we have written $u_6 = \frac{u_2^*u_3^*}{\Lambda}$. Also, the 3×3 mass matrices $m_{L^\alpha N_R}$, $M_{N_LN_R}$ and $m_{N_L\nu_R}$ are obtained from the terms $\mathcal{L}_{L^\alpha N_R}$, $\mathcal{L}_{N_LN_R}$ and $\mathcal{L}_{N_L\nu_R}$ respectively. This light neutrino mass matrix can be further diagonalized by the bi-unitary transformation.

The neutrino masses and the mixing angles so obtained will be dependent on the specific values of the coupling constants f_{ij}, g_{ij}, M_{ij} as well as the VEVs v, u_i; $i = 2, 3$. In the simplifying case of $g_{ij} = g$ and $M_{ij} = M$ we get

$$\mathcal{M}_\nu = \frac{gv^*}{M}\begin{pmatrix} 0 & 0 & -f_{12}u_3 \\ f_{21}u_6 & f_{21}u_6 & -f_{22}u_3 \\ 0 & 0 & f_{33}u_3 \end{pmatrix}. \tag{15}$$

Diagonalizing the mass matrix we have

$$\begin{aligned} &\theta_{12}^\nu \approx 0; \quad \theta_{13}^\nu \approx \tan^{-1}\left(\frac{f_{12}}{f_{33}}\right); \quad \theta_{23}^\nu \approx \tan^{-1}\left(\frac{f_{22}}{\sqrt{f_{12}^2+f_{33}^2}}\right) \\ &m_1^\nu \approx 0; \qquad m_2^\nu \approx \frac{\sqrt{2(f_{12}^2+f_{33}^2)}f_{21}g|v|}{M\sqrt{f_{12}^2+f_{22}^2+f_{33}^2}}|u_6|; \\ &m_3^\nu \approx \frac{\sqrt{f_{12}^2+f_{22}^2+f_{33}^2}\,g|v|}{M}|u_3|. \end{aligned} \tag{16}$$

Since, $u_6 \ll u_3$, we have a normal hierarchy pattern with two nearly massless neutrinos and one relatively heavy neutrino. Moreover, the massless neutrino will also gain small mass, if any of the M_{ij}'s or g_{ij}'s are not equal to M or g respectively. Also, if they deviate significantly from these values then one can possibly recover degenerate or inverted hierarchy patterns also. Now, if U_l and U_ν are the mixing matrices of the charged leptons and neutrinos respectively, then the PMNS mixing matrix is given by

$$U_{\rm PMNS} = U_l^\dagger U_\nu. \tag{17}$$

Taking $y_{12} = y_\mu$, $f_{12} = -\frac{f_{33}}{2}$ and $f_{22} = \sqrt{f_{12}^2 + f_{33}^2}$ in (12), (16) we get $\theta_{23}^\nu = -\theta_{12}^l = \frac{\pi}{4}$ and $\theta_{13}^\nu = \tan^{-1}(-\frac{1}{2})$ which gives PMNS mixing angles consistent with present $3-\sigma$ limits of global fits obtained from experiments.[19]

In our minimal model with only one doublet scalar, the quark sector can be accommodated in a simple way if both the left-handed quark doublets $Q_L^i = (u_L^i, d_L^i)^{\rm T}$, $i = 1, 2, 3$ and the right-handed quark singlets u_R^i, d_R^i; $i = 1, 2, 3$ transform as 1 of S_3. A better understanding of the quark sector can be obtained if, to our minimal model, we add more doublet scalars transforming non-trivially under S_3. One such example for quark sector, albeit in context of a different model for lepton sector, has already been worked out in Refs. 17 and 18. We are currently working on a similar extension of our minimal model.

5. Inverse Seesaw

Apart from the Dirac neutrinos, other possibilities can also arise depending on the particle content.[4,15] For example instead of adding the singlet scalar $\chi_3 \sim 3$ under $U(1)_{B-L}$, one can add two complex scalar fields $\chi_2 \sim 2$ and $\chi_6 \sim 6$ under $U(1)_{B-L}$. In this case, ν_L is not connected to $\nu_{R1,R2} \sim -4$. It is connected however to $N_{L,R}$ through the mass matrix spanning $(\bar\nu_L, \bar N_R^c, \bar N_L)$ as follows:

$$M_{\nu,N} = \begin{pmatrix} 0 & m_0 & 0 \\ m_0 & m_2' & M \\ 0 & M & m_2 \end{pmatrix} \tag{18}$$

where m_2 and m_2' come from the Yukawa couplings with χ_2. This leads to an inverse seesaw,[20–22] i.e. $m_\nu \simeq m_0^2 m_2/M^2$. In the case of $\nu_{R3} \sim +5$, the corresponding mass matrix spanning $(\bar\nu_L, \bar N_R^c, \bar N_L, \bar\nu_{R3}^c)$ is given by

$$M_{\nu,N} = \begin{pmatrix} 0 & m_0 & 0 & 0 \\ m_0 & m_2' & M & 0 \\ 0 & M & m_2 & m_6 \\ 0 & 0 & m_6 & 0 \end{pmatrix} \tag{19}$$

where m_6 comes from the Yukawa coupling with χ_6. Thus ν_{R3} also gets an inverse seesaw mass $\simeq m_6^2 m_2'/M^2$ which is the 4×4 analog of the 3×3 lopsided seesaw discussed in Ref. 23. In this scheme, ν_L and ν_{R3} get small masses via inverse seesaw mechanism. Also, $N_{1,2,3}$ become heavy pseudo-Dirac fermions. However, $\nu_{R1,R2}$ remain massless. They can be given mass by adding extra scalars, e.g. by adding a third scalar $\chi_8 \sim 8$ under $B - L$.

6. Conclusion and Future Work

The idea that $B - L$ should be a gauge symmetry has been around for some time. However most of the work on gauged $B - L$ symmetry has been done for the case of the three right-handed neutrinos transforming as -1 under $U(1)_{B-L}$. In this work we looked at another possible anomaly free solution for gauged $B - L$ interaction with the three right-handed neutrinos transforming as $(+5, -4, -4)$ under $U(1)_{B-L}$. We showed how these assignments can be used to obtain seesaw Dirac neutrino masses, as well as inverse seesaw Majorana neutrino masses. We then showed that imposition of S_3 flavour symmetry to the first case can lead to realistic neutrino and charged-lepton mass matrices with a mixing pattern consistent with experiments. In our model the $B - L$ symmetry breaking scale can be as low as in TeV range. This raises the possibility of testing it in future runs of LHC or in other future colliders. We are planning to look for the phenomenological consequences of our model. The phenomenology of the Z' boson would be of particular interest. The cosmological implications of our model will also be interesting.

Acknowledgments

RS will like to thank the organizers for inviting and giving opportunity to present this work at International Conference on Massive Neutrinos, 2015, Singapore. EM is supported in part by the U.S. Department of Energy under Grant No. DE-SC0008541.

References

1. M. Auger *et al.* [EXO Collaboration], *Phys. Rev. Lett.* **109**, 032505 (2012) [arXiv:1205.5608 [hep-ex]].
2. M. Agostini *et al.* [GERDA Collaboration], *Phys. Rev. Lett.* **111**, 122503 (2013) [arXiv:1307.4720 [nucl-ex]].
3. A. Gando *et al.* [KamLAND-Zen Collaboration], *Phys. Rev. Lett.* **110**, no. 6, 062502 (2013), arXiv:1211.3863.
4. E. Ma and R. Srivastava, *Phys. Lett. B* **741**, 217 (2015) [arXiv:1411.5042 [hep-ph]].
5. E. Ma, *Mod. Phys. Lett. A* **17**, 535 (2002) [hep-ph/0112232].
6. E. Ma, *Phys. Rev. Lett.* **89**, 041801 (2002) [hep-ph/0201083].
7. M. C. Chen, A. de Gouvea and B. A. Dobrescu, *Phys. Rev. D* **75**, 055009 (2007) [hep-ph/0612017].
8. E. C. G. Stueckelberg, *Helv. Phys. Acta* **11**, 225 (1938).
9. D. Feldman, P. Fileviez Perez and P. Nath, *JHEP* **1201**, 038 (2012) [arXiv:1109.2901 [hep-ph]].
10. J. Heeck, *Phys. Lett. B* **739**, 256 (2014) [arXiv:1408.6845 [hep-ph]].
11. J. Heeck and W. Rodejohann, *Europhys. Lett.* **103**, 32001 (2013) [arXiv:1306.0580 [hep-ph]].
12. J. Heeck, *Phys. Rev. D* **88**, 076004 (2013) [arXiv:1307.2241 [hep-ph]].
13. J. C. Montero and V. Pleitez, *Phys. Lett. B* **675**, 64 (2009) [arXiv:0706.0473 [hep-ph]].

14. A. C. B. Machado and V. Pleitez, *Phys. Lett. B* **698**, 128 (2011) [arXiv:1008.4572 [hep-ph]].
15. A. C. B. Machado and V. Pleitez, *J. Phys. G* **40**, 035002 (2013) [arXiv:1105.6064 [hep-ph]].
16. P. Roy and O. U. Shanker, *Phys. Rev. Lett.* **52**, 713 (1984) [Erratum-ibid. **52**, 2190 (1984)].
17. S. L. Chen, M. Frigerio and E. Ma, *Phys. Rev. D* **70**, 073008 (2004) [Erratum-*ibid.* **70**, 079905 (2004)] [hep-ph/0404084].
18. E. Ma and B. Melic, *Phys. Lett. B* **725**, 402 (2013) [arXiv:1303.6928 [hep-ph]].
19. F. Capozzi, G. L. Fogli, E. Lisi, A. Marrone, D. Montanino and A. Palazzo, *Phys. Rev. D* **89**, 093018 (2014) [arXiv:1312.2878 [hep-ph]].
20. D. Wyler and L. Wolfenstein, *Nucl. Phys. B* **218**, 205 (1983).
21. R. N. Mohapatra and J. W. F. Valle, *Phys. Rev. D* **34**, 1642 (1986).
22. E. Ma, *Phys. Lett. B* **191**, 287 (1987).
23. E. Ma, *Mod. Phys. Lett. A* **24**, 2161 (2009) [arXiv:0904.1580 [hep-ph]].

Searching for Radiative Neutrino Mass Generation at the LHC

Raymond R. Volkas

ARC Centre of Excellence for Particle Physics at the Terascale, School of Physics, The University of Melbourne, Victoria, 3010, Australia
raymondv@unimelb.edu.au

In this talk, I describe the general characteristics of radiative neutrino mass models that can be probed at the LHC. I then cover the specific constraints on a new, explicit model of this type.

Keywords: Neutrino; radiative mass generation; effective operators; flavor-violation constraints; LHC constraints.

1. See-saw versus Radiative Neutrino Mass Generation

Many models of radiative neutrino mass are possible, some of which have been analysed in depth, while others have been examined only briefly, with quite a number yet to be explicitly written down let alone analysed. A useful organising principle for Majorana neutrino mass models is to use standard model (SM) effective operators that violate lepton number conservation by two units ($\Delta L = 2$ operators) as a base from which to construct new theories.[1–7]

Restricting ourselves to operators containing SM fermions and a single Higgs doublet, they occur at odd mass dimensions, with the lowest order one occurring at $d = 5$. That operator is the famous Weinberg operator, which has the schematic structure $LLHH$, where L is a lepton doublet and H the Higgs doublet. After electroweak symmetry breaking, this operator (actually a set of operators because of the family structure) directly induces a Majorana neutrino mass for left-handed (LH) neutrinos given by the see-saw formula,

$$m_\nu \sim \frac{\langle H \rangle^2}{M}, \tag{1}$$

where M is the scale of the new physics that gives rise to the operator in the low-energy limit.

Underlying renormalisable theories yielding $LLHH$ are constructed by "opening up" the operator. The type-1, -2 and -3 see-saw models are the minimal, tree-level ways to open up $LLHH$.[8–19] By starting with the effective operator, one may systematically construct all of the minimal underlying models.

Other $\Delta L = 2$ effective operators have mass dimension seven or higher and feature other lepton and quark fields in their expression, except for those in the generalised Weinberg class $LLHH(\bar{H}H)^n$.[3,5,7] These additional particles have to be closed off in loops to produce a Majorana neutrino mass self-energy graph from the operator. Thus, theories based on $d \geq 7$ operators (apart from the generalised Weinberg class) necessarily produce radiative neutrino mass generation. Underlying renormalisable theories can then be systematically constructed by opening up the operators in all possible ways, subject in practice to minimality assumptions, just as the three see-saw models may be derived from the Weinberg operator.

A list of all gauge invariant $\Delta L = 2$ operators at $d = 5, 7, 9, 11$, constructed from SM fermions and one Higgs doublet, was produced by Babu and Leung (BL).[1] The ninth operator in their list, $O_9 = LLe^cLe^c$ (which has $d = 9$), is the basis of the historically important Zee–Babu model of neutrino mass generation.[20,21] Two exotic scalars are introduced: an isosinglet, singly-charged state h coupling to LL and a doubly-charged, isosinglet k coupling to e^ce^c. Both carry two units of lepton number. The cubic term hhk combines with those interactions to induce $\Delta L = 2$ and produce Majorana neutrino mass at 2-loop order. ATLAS has searched for k through the same-sign dilepton channels ee, $e\mu$ and $\mu\mu$, deriving a lower bound of about 320 GeV on the mass[22] (see also Ref. 23).

It is nice that the $\Delta L = 2$ operator perspective places the radiative neutrino mass models at one end of a systematic list of Majorana mass models, bookended by the tree-level see-saw models at $d = 5$. From the phenomenological viewpoint, radiative models are interesting because the mass scale of the new physics is lower than the favoured very high scale for see-saw models, and has more chance of encroaching into the LHC regime.

It is very much worth noting that upper bounds on the see-saw scales can be derived from naturalness considerations, because the new particles and interactions will destabilise the electroweak scale if the new physics occurs at too high a scale. The type-1 upper bound was first computed by Vissani[24] to be 3×10^7 GeV, if the Higgs self-energy graph from neutrino Yukawa interactions is not to produce a correction to the μ^2 parameter in the Higgs potential greater than $(1\ \text{TeV})^2$. This one-family result has recently been generalised to the realistic three-family case,[25] producing the bounds

$$M_{N_1} \lesssim 4 \times 10^7\ \text{GeV}\,, \quad M_{N_2} \lesssim 7 \times 10^7\ \text{GeV}\,, \quad M_{N_3} \lesssim 3 \times 10^7\ \text{GeV}\left(\frac{0.05\ \text{eV}}{m_{\min}}\right), \tag{2}$$

where $N_{1,2,3}$ are the three heavy neutral lepton mass eigenstates with $M_{N_1} \leq M_{N_2} \leq M_{N_3}$, and $m_{\min}$ is the minimum light neutrino mass. It is interesting that these bounds imply that standard thermal, hierarchical leptogenesis,[26] which requires sufficiently massive heavy neutral leptons, must involve some level of fine-tuning for the minimal type-1 see-saw model. This is true for all three cases: N_1, N_2 and N_3 leptogenesis. Naturalness, if successful leptogenesis is desired, may be restored by extending the theory. An obvious example is the supersymmetric extension, and another is to add extra Higgs doublets, to divorce the leptogenesis parameter space from neutrino mass generation (one version of this is discussed in Ref. 27).

2. Opening Up $d = 7$ Operators

Let us examine the $\Delta L = 2$, $d = 7$ operators from the BL list as the next most complicated cases after the Weinberg operator. The flavour contents are

$$O_2 = LLLe^cH\,, \quad O_3 = LLQd^cH\,, \quad O_4 = LL\bar{Q}\bar{u}^cH\,, \quad O_8 = L\bar{e}^c\bar{u}^cd^cH\,, \tag{3}$$

where the BL numbering scheme has been used, with Q the quark doublet, d the RH down quark and u the RH up quark. Flavour structures $O_{3,4}$ each yield two independent operators once weak-isospin index contraction possibilities are taken into account. Operator O_3 is the basis of the pioneering 1-loop Zee model,[28] while models constructed from O_3 and O_8 have been analysed in depth by Babu and Julio.[29,30] In addition, $d = 7$ contains the Weinberg-operator generalisation $O_1' = LL\bar{H}HHH$.[3,7]

We adopt the minimality assumption that the underlying renormalisable theories obtained from opening up these operators contain only exotic heavy scalars and vector-like fermions. Table 1 lists the quantum numbers under $SU(3)\times SU(2)\times U(1)$ for scalar-only and scalar-fermion extensions.[31] The example in boldface will be analysed in detail below. The diagram topologies are given in Fig. 1.

It is clear that searches for radiative neutrino mass models are included in general searches for exotic scalars and fermions. Each specific model has a particular scalar–scalar or scalar–fermion pair. In addition, the decay branching ratios for the exotics are typically quite constrained by the requirement to reproduce the correct neutrino masses and mixing angles. This additional information must be used when interpreting the experimental constraints, and it constitutes a "smoking gun" for exotics that are responsible for radiative neutrino mass generation.

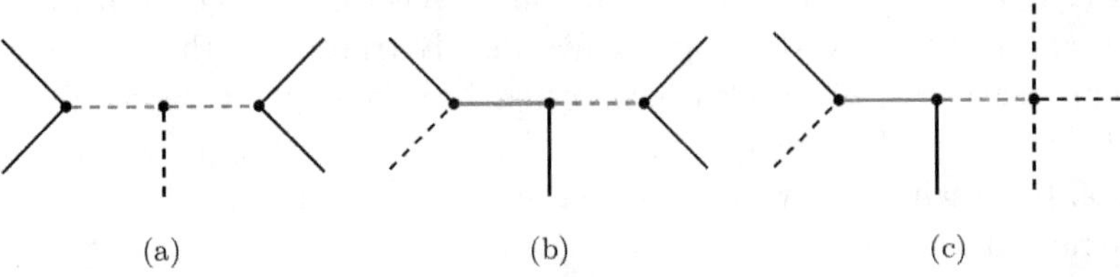

(a) (b) (c)

Fig. 1. Diagram topologies for opening up $d = 7$ operators. (a) Scalar-only. (b) Scalar plus fermion for $O_{2,3,4,8}$. (c) Scalar plus fermion for O_1'.

Table 1. Quantum numbers of exotic scalars and fermions in underlying theories for the $d = 7$ operators.

Scalar	Scalar	Dirac fermion	Operator
$(1, 2, 1/2)$	$(1, 1, 1)$		$O_{2,3,4}$
$(3, 2, 1/6)$	$(3, 1, -1/3)$		$O_{3,8}$
$(3, 2, 1/6)$	$(3, 3, -1/3)$		O_3
$(1, 1, 1)$		$(1, 2, -3/2)$	O_2
$(1, 1, 1)$		$(3, 2, -5/6)$	O_3
$(1, 1, 1)$		$(3, 1, 2/3)$	O_3
$(3, 2, 1/6)$		$(3, 2, -5/6)$	O_3
$\mathbf{(3, 1, -1/3)}$		$\mathbf{(3, 2, -5/6)}$	$\boldsymbol{O_{3,8}}$
$(3, 3, -1/3)$		$(3, 2, -5/6)$	O_3
$(3, 2, 1/6)$		$(3, 3, 2/3)$	O_3
$(1, 1, 1)$		$(3, 2, 7/6)$	O_4
$(1, 1, 1)$		$(3, 1, -1/3)$	O_4
$(3, 2, 1/6)$		$(3, 2, 7/6)$	O_8
$(3, 2, 1/6)$		$(1, 2, -1/2)$	O_8
$(1, 4, 3/2)$		$(1, 3, -1)$	O'_1

3. Collider Constraints on a New Radiative Model

We now look at the boldfaced case in Table 1.[31] Our theory will produce O_3, with a subdominant O_8. The exotic scalar ϕ and exotic fermion χ have the quantum numbers

$$\phi \sim (\bar{3}, 1, 1/3)\,, \quad \chi \sim (3, 2, -5/6)\,. \tag{4}$$

The Lagrangian is

$$-\mathcal{L} = m_\phi^2 \phi^\dagger \phi + m_\chi \bar{\chi}\chi + \left(Y_{ij}^{\bar{e}\bar{u}\phi} \bar{e}_i \bar{u}_j \phi^\dagger + \text{h.c.}\right) + \left(Y_{ij}^{LQ\phi} L_i Q_j \phi + Y_i^{L\bar{\chi}\phi} L_i \bar{\chi}\phi^\dagger + Y_i^{\bar{d}\chi H} \bar{d}_i \chi H + \text{h.c.}\right), \tag{5}$$

where we use two-component notation for the fermions and i, j are family indices. The Yukawa coupling constants $Y^{\bar{e}\bar{u}\phi}$ play no role in neutrino mass generation and are thus set to zero for simplicity. We also impose baryon-number conservation to forbid the $QQ\phi^\dagger$ and $\bar{d}\bar{u}\phi$ interactions permitted by the gauge symmetry.

The diagram responsible for neutrino mass generation is given in Fig. 2. It is proportional to the down-type quark masses. Barring unaesthetic hierarchies in the coupling constants, the terms proportional to the b-quark mass will dominate, which is what we will assume is the case. Associated with that, for simplicity we switch off the mixing of χ with first- and second-generation quarks.

The neutrino mass matrix is then

$$(m_\nu)_{ij} = \frac{3}{16\pi^2}\left(Y_{i3}^{LQ\phi} Y_j^{L\bar{\chi}\phi} + (i \leftrightarrow j)\right) m_{bB} \frac{m_b m_B}{m_\phi^2 - m_B^2} \ln \frac{m_B^2}{m_\phi^2}\,. \tag{6}$$

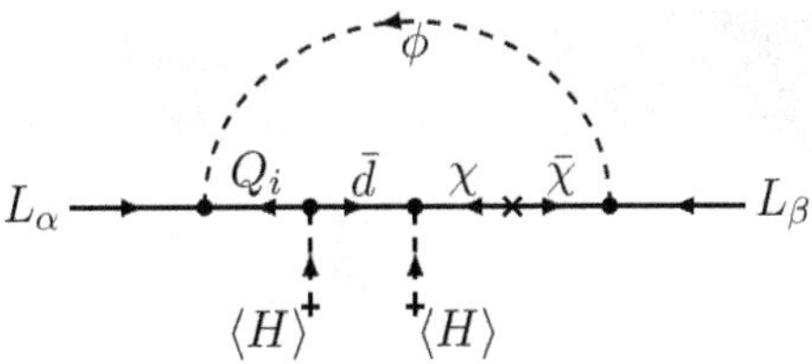

Fig. 2. Feynman diagram responsible for 1-loop generation of neutrino masses and mixings.

The bottom quark mixes with the isospin $+1/2$ component of χ to form two mass eigenstates, b and B, with masses m_b and m_B respectively. The parameter $m_{bB} = Y_3^{d\chi H} v/\sqrt{2}$, where $\langle H^0\rangle = v/\sqrt{2}$. At this level of approximation, there are two massive neutrinos and a massless one. In reality, the lightest neutrino will pick up a small mass, but its value is unimportant and will be neglected. CP violating phases will also be set to zero for simplicity.

The rank-2 mass matrix may be expressed as the symmetrised outer product of two vectors $a_\pm$,

$$m_\nu = a_+ a_-^T + a_- a_+^T\,, \tag{7}$$

where

$$\begin{aligned} a_\pm^{\rm NO} &= \frac{\zeta^{\pm 1}}{\sqrt{2}}\left(\sqrt{m_2}u_2^* \pm i\sqrt{m_3}u_3^*\right), \\ a_\pm^{\rm IO} &= \frac{\zeta^{\pm 1}}{\sqrt{2}}\left(\sqrt{m_1}u_1^* \pm i\sqrt{m_2}u_2^*\right), \end{aligned} \tag{8}$$

with NO and IO denoting normal and inverted ordering, respectively. The vectors $u_{1,2,3}$ are the columns of the PMNS matrix, and the complex number ζ is a Casas–Ibarra-like parameter, not determined by low-energy experiments. Equating (6) and (7) constrains the parameter space to the region compatible with the neutrino oscillation results.

The next set of constraints on the parameter space come from the lepton flavour violating processes $\mu \to e\gamma$, $\mu \to eee$ and $\mu N \to eN$.[31] An example of the results is presented in Fig. 3. The green dash-dotted, blue dotted and red-dashed lines bound the regions corresponding to $\mathrm{BR}(\mu \to e\gamma) < 5.7 \times 10^{-13}$, $\mathrm{BR}(\mu \to eee) < 10^{-12}$ and $\mathrm{BR}(\mu\,\mathrm{Au} \to e\,\mathrm{Au}) < 7 \times 10^{-13}$, respectively. The grey region is thus excluded. The magenta dashed line shows the reach of the $\mu\,\mathrm{Ti} \to e\,\mathrm{Ti}$ Mu2E and COMET experiments. The allowed region is divided into B and T regions in which $\mathrm{BR}(\phi \to b\nu) = 1$ and $\mathrm{BR}(\phi \to b\nu) < 1$, respectively.

The vector-like fermion doublet χ consists of the charge $-1/3$ quark B' which mixes with the bottom quark, and an exotic charge $-4/3$ quark Y. CMS have searched for B-like particles, whose dominant decay modes are $B \to Zb$ and $B \to Hb$ in our model with branching ratios that are strongly constrained by the need to fit the neutrino oscillation data. Taking that into account, the CMS bound is $m_B \gtrsim 620$ GeV.[32–36] No Y search has been performed.

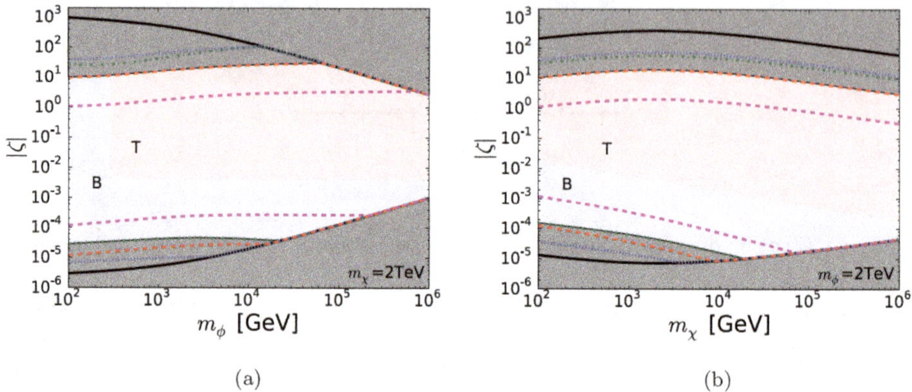

Fig. 3. Lepton flavour violation bounds. (a) Allowed region in the $|\zeta|$, m_ϕ plane for $m_\chi = 2$ TeV. (b) Allowed region in the $|\zeta|$, m_χ plane for $m_\phi = 2$ TeV.

The other collider constraint comes from searches for leptoquark scalar ϕ. It is pair-produced at the LHC from gluon–gluon fusion and $q\bar{q}$ annihilation. Since the production is through its colour charge, the rate depends only on m_ϕ. For example, for $m_\phi = 500(600)$ GeV, the production cross-section $\sigma(pp \to \phi\phi)$ is 82(23.5) fb.

The main decay modes are into Lt and $b\nu$, where $L = (e, \mu, \tau)$. In the parameter region $m_{Y,B} \gg m_\phi$, so that LY and $B\nu$ final states are kinematically forbidden, the partial decay rates are,

$$
\begin{aligned}
\Gamma(\phi \to Lt) &= \frac{m_\phi}{8\pi}\left|Y_{L3}^{LQ\phi}\right|^2 f(m_\phi, m_L, m_t)\,, \\
\Gamma(\phi \to \nu_L b) &\simeq \frac{m_\phi}{8\pi}\left(\left|Y_{L3}^{LQ\phi}\cos\theta_2\right|^2 + \left|Y_L^{L\bar{\chi}\phi}\sin\theta_1\right|^2\right) f(m_\phi, m_L, m_t)\,,
\end{aligned}
\tag{9}
$$

where f is given in Eqs. (4.16) and (4.21) of Ref. 31, and the mixing angles $\theta_{1,2}$ are defined through

$$
\sin\theta_1 = \frac{m_{bB} m_\chi}{m_\chi^2 - m_{b'}^2}\,, \quad \sin\theta_2 = \frac{m_{bB} m_{b'}}{m_\chi^2 - m_{b'}^2}\,, \tag{10}
$$

where $m_{b'} = y_b v/\sqrt{2}$ is the bottom quark mass in the absence of mixing with B'. The decay rates depend on the same Yukawa coupling constants that also contribute to neutrino mass generation, so are quite constrained, as well as $|\zeta|$.

Recall that the allowed regions B and T in Fig. 3 correspond to regimes where $\mathrm{BR}(\phi \to b\nu) = 1$ and $\mathrm{BR}(\phi \to b\nu) < 1$, respectively. In the B region, the main signature is thus $pp \to \phi\phi \to bb +$ missing E_T, where sbottom pair searches apply. The bound in this case is $m_\phi \gtrsim 730$ GeV at 95% C.L.[37,38] The analysis for region T requires recasting stop searches as well and is summarised in Fig. 4, which plots the various branching ratios as functions of m_ϕ, as well as the limits from ATLAS and CMS.[37–41] In numbers: The m_ϕ lower limit from CMS in the bb+missing E_T channel is in the range 520–600 GeV. From the $(e, \mu)+$missing $E_T + (b-)$ jets, ATLAS sets

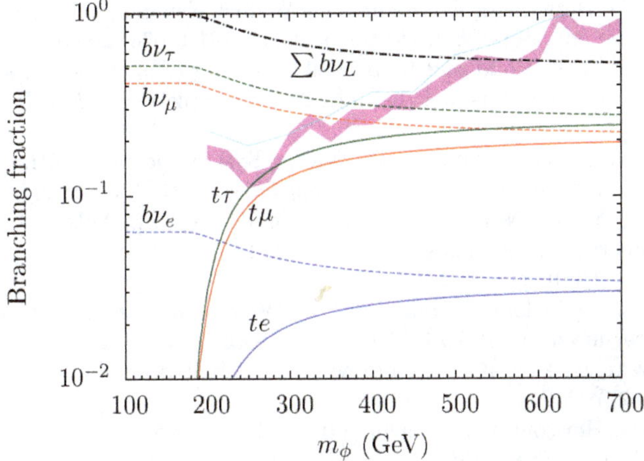

Fig. 4. Decay branching ratios of ϕ as a function of m_ϕ in region T. The curves with positive slope running to the top right show the ATLAS (thin line) and CMS (thick band) limits.

a limit of approximately 580 GeV. Finally, the $(e, \mu)^+(e, \mu)^- +$ missing $E_T +$ jets channel yields about a 620 GeV limit from ATLAS data.

4. Final Remarks

A systematic procedure for constructing radiative Majorana neutrino mass models has been developed from standard model $\Delta L = 2$ effective operators containing leptons, quarks and the Higgs doublet. The LHC is a useful tool for searching for the exotic scalars and fermions that appear in these models, and specific bounds for a new model were summarised in this talk. The 13 TeV LHC will extend the reach of past searches. Some naturalness concerns for hierarchical, thermal leptogenesis in the minimal type-1 see-saw model were also mentioned in passing.

Acknowledgments

This work was supported in part by the Australian Research Council. I thank my coauthors P. W. Angel, Y. Cai, J. D. Clarke, R. Foot, N. L. Rodd and M. A. Schmidt. I also thank K. S. Babu, A. de Gouvêa and W. Winter for discussions over the last few years on radiative neutrino mass generation.

References

1. K. S. Babu and C. N. Leung, Classification of effective neutrino mass operators, *Nucl. Phys. B* **619**, 667 (2001).
2. A. de Gouvea and J. Jenkins, A survey of lepton number violation via effective operators, *Phys. Rev. D* **77**, 013008 (2008).

3. F. Bonnet, D. Hernandez, T. Ota and W. Winter, Neutrino masses from higher than $d = 5$ effective operators, *J. High Energy Phys.* **0910**, 076 (2009).
4. P. W. Angel, N. L. Rodd and R. R. Volkas, Origin of neutrino masses at the LHC: $\Delta L = 2$ effective operators and their ultraviolet completions, *Phys. Rev. D* **87**, 073007 (2013).
5. F. Bonnet, M. Hirsch, T. Ota and W. Winter, Systematic study of the $d = 5$ Weinberg operator at one-loop order, *J. High Energy Phys.* **1207**, 153 (2012).
6. P. W. Angel, Y. Cai, N. L. Rodd, M. A. Schmidt and R. R. Volkas, Testable two-loop radiative neutrino mass model based on an $LLQd^cQd^c$ effective operator, *J. High Energy Phys.* **1310**, 118 (2013).
7. M. B. Krauss, D. Meloni, W. Porod and W. Winter, Neutrino mass from a $d = 7$ effective operator in an SU(5) SUSY-GUT framework (January 2013), arXiv:1301.4221.
8. P. Minkowski, $\mu \to e\gamma$ at a rate of one out of 1-billion muon decays?, *Phys. Lett. B* **67**, 421 (1977).
9. T. Yanagida, Horizontal gauge symmetry and masses of neutrinos, in *Proceedings of the Workshop on The Unified Theory and the Baryon Number in the Universe*, eds. O. Sawada and A. Sugamoto (1979).
10. S. L. Glashow, The future of elementary particle physics, in *Proceedings of the 1979 Cargèse Summer Institute on Quarks and Leptons*, eds. M. Lévy, J.-L. Basdevant, D. Speiser, J. Weyers, R. Gastmans and M. Jacob (Plenum Press, New York, 1980).
11. M. Gell-Mann, P. Ramond and R. Slansky, Complex spinors and unified theories, in *Supergravity*, eds. P. van Nieuwenhuizen and D. Z. Freedman (North Holland, Amsterdam, 1979).
12. R. N. Mohapatra and G. Senjanović, Neutrino mass and spontaneous parity violation, *Phys. Rev. Lett.* **44**, 912 (1980).
13. M. Magg and C. Wetterich, Neutrino mass problem and gauge hierarchy, *Phys. Lett. B* **94**, 61 (1980).
14. J. Schechter and J. Valle, Neutrino masses in $SU(2) \times U(1)$ theories, *Phys. Rev. D* **22**, 2227 (1980).
15. C. Wetterich, Neutrino masses and the scale of $B - L$ violation, *Nucl. Phys. B* **187**, 343 (1981).
16. G. Lazarides, Q. Shafi and C. Wetterich, Proton lifetime and fermion masses in an SO(10) model, *Nucl. Phys. B* **181**, 287 (1981).
17. R. N. Mohapatra and G. Senjanovic, Neutrino masses and mixings in gauge models with spontaneous parity violation, *Phys. Rev. D* **23**, 165 (1981).
18. T. Cheng and L.-F. Li, Neutrino masses, mixings and oscillations in $SU(2) \times U(1)$ models of electroweak interactions, *Phys. Rev. D* **22**, 2860 (1980).
19. R. Foot, H. Lew, X. He and G. C. Joshi, Seesaw neutrino masses induced by a triplet of leptons, *Z. Phys. C* **44**, 441 (1989).
20. A. Zee, Quantum numbers of Majorana neutrino masses, *Nucl. Phys. B* **264**, 99 (1986).
21. K. Babu, Model of 'calculable' Majorana neutrino masses, *Phys. Lett. B* **203**, 132 (1988).
22. G. Aad *et al.*, Search for doubly-charged Higgs bosons in like-sign dilepton final states at $\sqrt{s} = 7$ TeV with the ATLAS detector, *Eur. Phys. J. C* **72**, 2244 (2012).
23. S. Chatrchyan *et al.*, A search for a doubly-charged Higgs boson in *pp* collisions at $\sqrt{s} = 7$ TeV, *Eur. Phys. J. C* **72**, 2189 (2012).
24. F. Vissani, Do experiments suggest a hierarchy problem?, *Phys. Rev. D* **57**, 7027 (1998).
25. J. D. Clarke, R. Foot and R. R. Volkas, Electroweak naturalness in three-flavour type I

see-saw and implications for leptogenesis (2015), arXiv:1502.01352 [hep-ph].
26. M. Fukugita and T. Yanagida, Baryogenesis without grand unification, *Phys. Lett. B* **174**, 45 (1986).
27. H. Davoudiasl and I. M. Lewis, Right-handed neutrinos as the origin of the electroweak scale, *Phys. Rev. D* **90**, 033003 (2014).
28. A. Zee, A theory of lepton number violation, neutrino Majorana mass, and oscillation, *Phys. Lett. B* **93**, 389 (1980).
29. K. Babu and J. Julio, Two-loop neutrino mass generation through leptoquarks, *Nucl. Phys. B* **841**, 130 (2010).
30. K. Babu and J. Julio, Radiative neutrino mass generation through vector-like quarks, *Phys. Rev. D* **85**, 073005 (2012).
31. Y. Cai, J. D. Clarke, M. A. Schmidt and R. R. Volkas, Testing radiative neutrino mass models at the LHC, *J. High Energy Phys.* **1502**, 161 (2015).
32. CMS Collab., Search for vector-like b' pair production with multilepton final states in pp collisions at $\sqrt{s} = 8$ TeV, Tech. Rep. CMS-PAS-B2G-13-003 (2013).
33. CMS Collab., Search for pair-produced vector-like quarks of charge $-1/3$ in lepton+jets final state in pp collisions at $\sqrt{s} = 8$ TeV, Tech. Rep. CMS-PAS-B2G-12-019 (2012).
34. CMS Collab., Search for pair-produced vector-like quarks of charge $-1/3$ in dilepton+jets final state in pp collisions at $\sqrt{s} = 8$ TeV, Tech. Rep. CMS-PAS-B2G-12-021 (2013).
35. S. Chatrchyan *et al.*, Inclusive search for a vector-like T quark with charge $\frac{2}{3}$ in pp collisions at $\sqrt{s} = 8$ TeV, *Phys. Lett. B* **729**, 149 (2014).
36. S. Chatrchyan *et al.*, Search for top-quark partners with charge 5/3 in the same-sign dilepton final state, *Phys. Rev. Lett.* **112**, 171801 (2014).
37. G. Aad *et al.*, Search for direct third-generation squark pair production in final states with missing transverse momentum and two b-jets in $\sqrt{s} = 8$ TeV pp collisions with the ATLAS detector, *J. High Energy Phys.* **1310**, 189 (2013).
38. CMS Collab., Search for direct production of bottom squark pairs, Tech. Rep. CMS-PAS-SUS-13-018, CERN (Geneva, 2014).
39. ATLAS Collab., Search for direct top squark pair production in final states with one isolated lepton, jets, and missing transverse momentum in $\sqrt{s} = 8$ TeV pp collisions using 21 fb^{-1} of ATLAS data, Tech. Rep. ATLAS-CONF-2013-037, CERN (Geneva, 2013).
40. S. Chatrchyan *et al.*, Search for top-squark pair production in the single-lepton final state in pp collisions at $\sqrt{s} = 8$ TeV, *Eur. Phys. J. C* **73**, 2677 (2013).
41. G. Aad *et al.*, Search for direct top-squark pair production in final states with two leptons in pp collisions at $\sqrt{s} = 8$ TeV with the ATLAS detector (2014), arXiv:1403.4853 [hep-ex].

Lepton-Flavor Violating Signatures in Supersymmetric $U(1)'$ Seesaw

Eung Jin Chun

Korea Institute for Advanced Study, Seoul 130-722, Republic of Korea
ejchun@kias.re.k

In a supersymmetric $U(1)'$ seesaw model, a right-handed sneutrino can be a good thermal dark matter candidate if the extra gaugino $\tilde{Z}'$ is light enough to provide an appropriate annihilation cross-section through a t-channel diagram. We first discuss how right thermal relic density of the right-handed sneutrino dark matter can arise and then explore lepton number and flavor violating signatures followed by cascade production of $\tilde{Z}'$ from the third generation squarks at the LHC.

Keywords: Supersymmetry; seesaw; Z prime; sneutrino dark matter.

1. Introduction

The most popular way of generating the observed neutrino masses and mixing would be to introduce right-handed neutrinos. Although they are completely neutral under the Standard Model (SM) gauge group, their presence might be associated with an extra gauge symmetry $U(1)'$ whose spontaneous breaking sets the mass scales of an extra heavy gauge boson Z' and the right-handed neutrinos.[1] This scheme leads to exotic collider signatures of di-jet/di-lepton resonances from the decays $Z' \to q\bar{q}, l\bar{l}$ which are actively searched for at the LHC. The current limit on various Z' masses is pushed up to around 3 TeV.[2] An important consequence of the model is that the Majorana nature of a right-handed neutrino N can be revealed by the observation of same-sign di-leptons (SSD) caused by the process:

$$pp \to Z' \to NN \to l^{\pm}l^{\pm}W^{\mp}W^{\mp}\,.$$

In the (R-parity conserving) supersymmetric version of a $U(1)'$ seesaw model, there appear two additional dark matter (DM) candidates, $\tilde{Z}'$ and $\tilde{N}$, the superpartner of Z' and N, respectively. While $\tilde{Z}'$ has a DM property similar to the bino, $\tilde{N}$ has a characteristically different feature.[3] After introducing a representative supersymmetric $U(1)'$ seesaw model in Sec. 2, we will show in Sec. 3 how $\tilde{N}$ becomes a good thermal DM candidate when $\tilde{Z}'$ is relatively light and the neutrino Yukawa coupling is not too small. Such light $\tilde{Z}'$ can be copiously produced from cascade decays and then pair-produced $\tilde{Z}'$ can lead to SSD signals associated with $\tilde{N}$:[4]

$$pp \to \tilde{Z}'\tilde{Z}' \to NN\tilde{N}\tilde{N} \to l^\pm l^\pm W^\mp W^\mp \tilde{N}\tilde{N}\,.$$

In this process the right-handed neutrino decay is generically flavor-dependent reflecting the flavor structure of the neutrino Yukawa coupling. The LHC phenomenology of the resulting lepton flavor and number violation associated with missing energy will be discussed in Sec. 4.

2. A Supersymmetric $U(1)'$ Seesaw Model

As a representative model of $U(1)'$ seesaw, we will take the $U(1)_\chi$ model which has the following particle content:[1]

$SU(5)$	10_F	$\bar{5}_F$	$1(N)$	5_H	$\bar{5}_H$	$1(X)$	$1(S_1)$	$1(S_2)$
$2\sqrt{10}Q'$	-1	3	-5	2	-2	0	10	-10

(1)

where the $SU(5)$ notation is used to show the $U(1)'$ charges of the SM fermions $(10_F, \bar{5}_F)$, Higgs bosons $(5_H, \bar{5}_H)$, and additional singlet fields $(N, X, S_{1,2})$. Note that unconventional singlets $S_{1,2}$, vector-like under $U(1)'$, are added to break $U(1)'$ and generate the Majorana mass of N. The gauge invariant superpotential in the seesaw sector is given by

$$W_{\text{seesaw}} = y^\nu_{ij} L_i H_u N_j + \frac{\lambda_{N_i}}{2} S_1 N_i N_i\,, \tag{2}$$

where L_i and H_u denote the lepton and Higgs doublet superfields, respectively. After the $U(1)'$ breaking by the vacuum expectation value $\langle S_1 \rangle$, the right-handed neutrinos obtain the mass $m_{N_i} = \lambda_{N_i}\langle S_1\rangle$ and induce the seesaw mass for the light neutrinos:

$$\tilde{m}^\nu_{ij} = -y^\nu_{ik} y^\nu_{jk} \frac{\langle H^0_u \rangle^2}{m_{N_k}}\,. \tag{3}$$

The atmospheric neutrino mass scale $\tilde{m}_\nu = 0.05$ eV sets the neutrino Yukawa coupling $y_\nu \sim 4 \times 10^{-7}$ with the right-handed neutrino mass $m_N = 100$ GeV. For $U(1)'$ breaking, one may introduce the μ' term:[5]

$$W' = \mu' S_1 S_2\,, \tag{4}$$

and arrange radiative $U(1)'$ breaking with a large Yukawa coupling, say λ_{N_3}, similarly to the radiative electroweak symmetry breaking. Another possibility is to consider

$$W' = \lambda X S_1 S_2 + \frac{\kappa}{3} X^3 \tag{5}$$

as in the nonminimal Higgs sector.[6]

Given the vacuum expectation values $\langle S_{1,2}\rangle$, the Z' gauge boson mass is set to be $M_{Z'}^2 = 2g'^2 \sum_i Q_{S_i}'^2 \langle S_i\rangle^2$ where g' denotes the $U(1)'$ gauge coupling constant. We will take the reference value of $g' = \sqrt{5/3}g_2 \tan\theta_W \approx 0.46$ in our analysis. The $U(1)'$ gaugino and dark matter masses are important input parameters for our discussion. Defining $\tan\beta' = \langle S_2\rangle/\langle S_1\rangle$, the $U(1)'$ gaugino-Higgsino mass matrix in the basis of $[\tilde{Z}', \tilde{S}_1, \tilde{S}_2]$ is given by

$$\mathcal{M} = \begin{bmatrix} m_M & M_{Z'}c_{\beta'} & -M_{Z'}s_{\beta'} \\ M_{Z'}c_{\beta'} & 0 & \mu' \\ -M_{Z'}s_{\beta'} & \mu' & 0 \end{bmatrix}, \tag{6}$$

where m_M denotes a soft supersymmetry breaking mass. In the limit of $\mu' \gg m_M, M_{Z'}$, the lightest state has the mass $m_{\tilde{Z}'}$ given by $m_{\tilde{Z}'} \approx m_M + M_{Z'}^2 s_{2\beta'}/\mu'$. As a light $\tilde{Z}'$ is favored for our dark matter relic density, we will work in this limit taking $m_{\tilde{Z}'}$ as a free parameter. Among three right-handed neutrino superfields, let us take the lightest one N suppressing the flavor index in the following discussion. Denoting its supersymmetric mass by m_N, the right-handed sneutrino $\tilde{N}$ has the mass term:

$$V_{\text{mass}} = \left(m_N^2 + m_{\tilde{N}}^2 - \frac{1}{4}m_{Z'}^2 c_{2\beta'}\right)|\tilde{N}|^2 - \frac{1}{2}B_N m_N(\tilde{N}\tilde{N} + \tilde{N}^*\tilde{N}^*), \tag{7}$$

where $B_N m_N = -\lambda_N A_N \langle S_1\rangle$. Due to the lepton number violating (Majorana) mass term of $B_N m_N$ which is assumed to be positive, the real and imaginary components of the sneutrino, $\tilde{N} = (\tilde{N}_1 + i\tilde{N}_2)/\sqrt{2}$, get a mass splitting and their masses are given by $m_{\tilde{m}_{1,2}}^2 = m_N^2 + m_{\tilde{N}}^2 - \frac{1}{4}m_{Z'}^2 \mp B_N m_N$. The real scalar field $\tilde{N}_1$ is taken to the lightest supersymmetric particle being the DM candidate.

3. Right-Handed Sneutrino as Thermal Dark Matter

While the neutrino Yukawa coupling ($y_\nu \sim 10^{-6-7}$) is too small to thermalize the DM $\tilde{N}_1$, it can be kept in thermal equilibrium by the $U(1)'$ gauge interaction. As the gauge interaction mediated by Z' is too weak to generate an appropriate freeze-out density of $\tilde{N}_1$, one may ask whether the DM annihilation, $\tilde{N}_1\tilde{N}_1 \to NN$, through the t-channel $\tilde{Z}'$ exchange shown in Fig. 1(a) can be useful. As will be shown in detail, the DM relic density is strongly depends on whether the right-handed neutrino N remains sufficiently in thermal equilibrium. The gauge interaction of N coupling to Z' in Fig. 1(b) is again too weak to maintain thermal equilibrium long enough due to too heavy Z' in the multi-TeV region. It is then crucial to include the decay and inverse-decay effect of the right-handed neutrino caused by the Yukawa interaction in Fig. 1(c) and 1(d). Although the neutrino Yukawa coupling is small, it plays a crucial role in maintaining N in thermal equilibrium and thus in determining the freeze-out density of $\tilde{N}_1$.

In Fig. 2 we show the number densities of $\tilde{N}_1$ and N varying the effective neutrino mass $\tilde{m}_\nu \equiv |y_\nu|^2\langle H_u^0\rangle^2/m_N$. For $\tilde{m}_\nu \gtrsim 10^{-3}$ eV, the N decay effect is strong enough to dominate other interaction terms of N before the annihilation effect

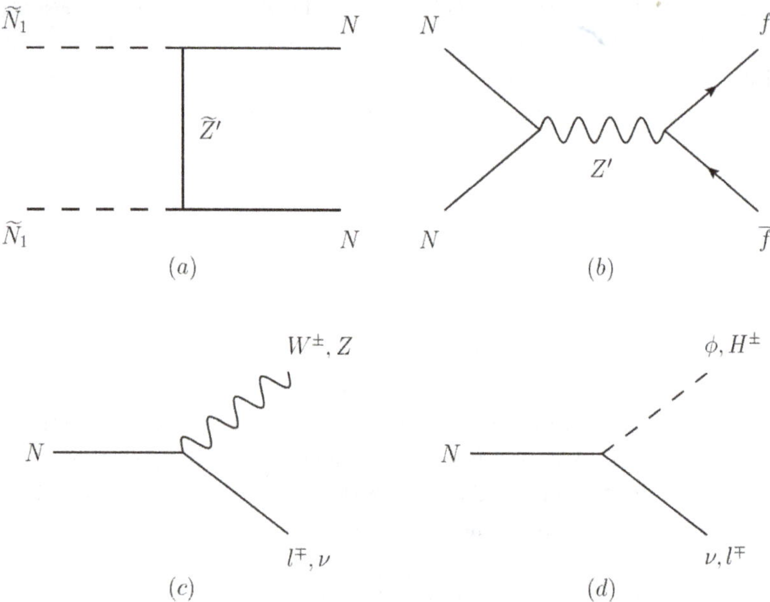

Fig. 1. Annihilation and decay channels of $\tilde{N}_1$ and N. In panel (d), $\phi = h$, H and A.

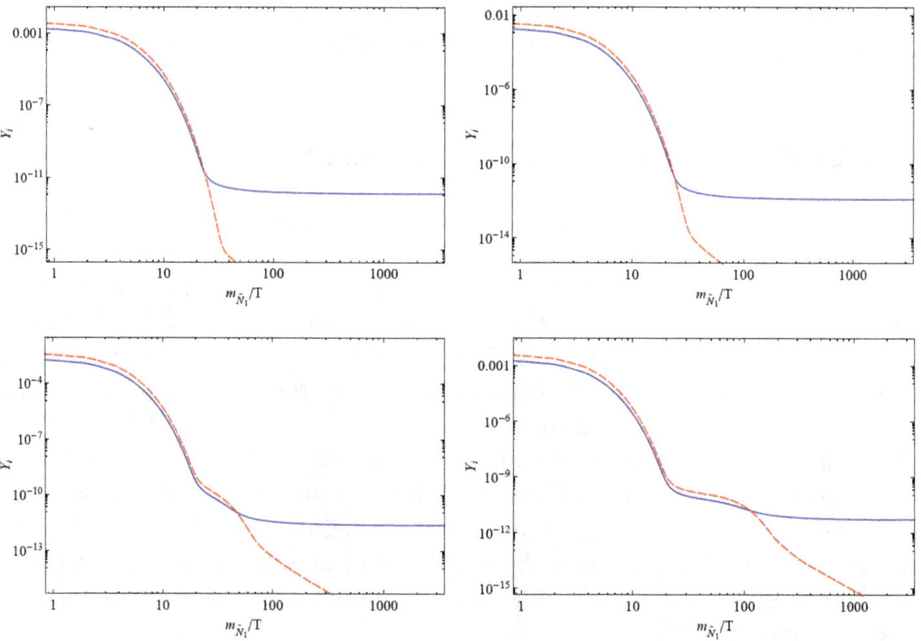

Fig. 2. (Color online) The actual number of $\tilde{N}_1$ and N per comoving volume. The panels correspond to $\tilde{m}_\nu = 10^{-2}, 10^{-3}, 10^{-5}$ and 10^{-6} eV respectively from left to right and top to bottom. Blue solid and red dashed lines show $Y_{\tilde{N}_1} \equiv n_{\tilde{N}_1}/s$ and $Y_N \equiv n_N/s$.

of N becomes weaker than the dilution effect due to the expansion of the Universe. Therefore, the decay effect keeps N in thermal equilibrium for a longer time compared with the case that N is stable, and N can continuously remain in the thermal bath before $\tilde{N}_1$ is decoupled from the thermal bath as shown in the top two panels of Fig. 2. On the other hand, for $\tilde{m}_\nu \lesssim 10^{-3}$ eV, the annihilation effect of N becomes weaker than the dilution effect before the N decay effect dominates other reactions. As a result, a retarded behavior appears in the number density evolution of $\tilde{N}_1$ and N as shown in the bottom two panels of Fig. 2.

Figure 3 shows the countour lines of the DM relic density depending on various model parameters. As expected from the previous discussion, $\tilde{Z}'$ is required to be lighter for heavier Z', smaller $\tilde{m}_\nu$ and smaller g'. In the following section, we will discuss LHC phenomenology of such light $\tilde{Z}'$ assuming that the decay channel of $\tilde{Z}' \to N\tilde{N}_1$ is open.

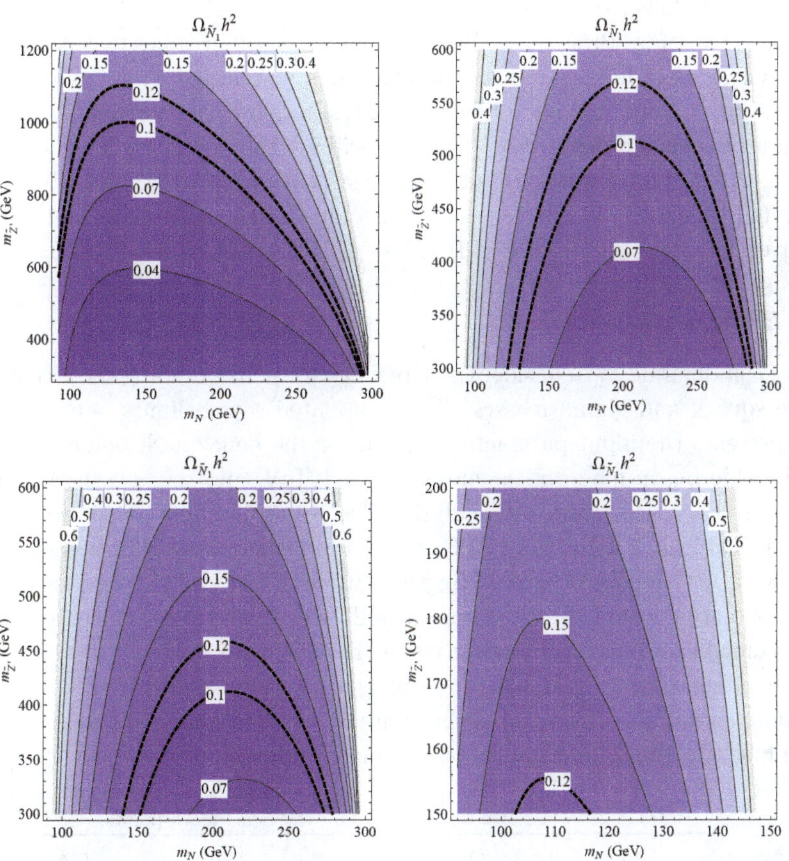

Fig. 3. Contour plots for the relic abundance of the right-handed sneutrino dark matter $\tilde{N}_1$ in the $m_N - m_{\tilde{Z}'}$ plane. Each panel shows the case with $(m_{\tilde{N}_1}, M_{Z'}, \tilde{m}_\nu, g')$ set to be (300 GeV, 1.2 TeV, 10^{-3} eV, 0.46) (upper-left), (300 GeV, 1.2 TeV, 10^{-5} eV, 0.46) (upper-right), (300 GeV, 4 TeV, 10^{-3} eV, 0.46) (lower-left), (150 GeV, 1.2 TeV, 10^{-3} eV, 0.2) (lower-right).

Table 1. Input parameters (masses in GeV) for the benchmark points.

	$M_{\tilde{q}_{1,2}}$	$M_{\tilde{Q}_3}$	$M_{\tilde{t}_R}$	$M_{\tilde{b}_R}$	$\tan\beta$	μ	M_1	M_2	M_3	$m_{\tilde{Z}'}$	A_t	A_b
BP1	1000	700	800	650	20	−730	700	750	1400	300	1600	1500
BP2	1000	700	800	650	15	−130	230	400	1400	220	1625	1500
BP3	2000	800	800	700	20	−730	700	750	2000	300	1600	1500

4. LHC Signatures: Lepton Number and Flavor Violation

The extra gaugino $\tilde{Z}'$ can be pairly produced mainly through third generation squark cascades and then decay to N and $\tilde{N}_1$. Due to the Majorana nature of N, it can decay to either sign leptons, $N \to l^\pm W^\mp$, leading to the lepton number violating signature of same-sign dilepton (SSD) events in addition to the usual opposite-sign dilepton (OSD) events. Furthermore, the neutrino Yukawa couplings are generically flavor-dependent and thus lead to lepton flavor violating decays, e.g. $\mathrm{Br}(N \to e^\pm W^\mp) \gg \mathrm{Br}(N \to \mu^\pm W^\mp)$. It is then an interesting task to investigate the LHC prospect for detecting such signatures manifested by a flavor difference, e.g. $2e - 2\mu$, in both SSD and OSD final states. Similar phenomenon should appear also in a tri-lepton difference $3e - 3\mu$. We will see that 5σ discovery is expected in the $2e - 2\mu$ SSD final states for an optimistic benchmark point even with $\sim 2\ \mathrm{fb}^{-1}$ integrated luminosity at the very early stage of LHC14. A similar conclusion could be driven for the opposite case of $\mathrm{Br}(N \to \mu^\pm W^\mp) \gg \mathrm{Br}(N \to e^\pm W^\mp)$.

4.1. *Benchmark points*

For the collider study we consider three benchmark points evading the recent bounds on the squark and gluino masses[8–10] (for updated search limits, see Ref. 7). Table 1 presents the input parameters chosen for the benchmark points. The heavy pseudo-scale boson mass m_A is chosen to be 1 TeV, and thus all the heavy Higgs bosons are decoupled from the analysis. The resulting SUSY particle spectrum is shown in Table 2. For BP1 and BP3, $\tilde{Z}'$ is NLSP, whereas for BP2 it is next to next LSP (NNLSP) and NLSP is the Higgsino. In BP3 first two generations of squarks and gluino are decoupled having masses ~ 2 TeV. In all three benchmark points, a right-handed sneutrino $\tilde{N}_1$ is taken to be the LSP with $m_{\tilde{N}_1} = 110$ GeV, and the corresponding right-handed neutrino mass is $m_N = 100$ GeV. Table 3 shows the production cross-sections for third generation squarks and gluinos. The cross-sections of the first two generation squark pairs which are less than 10 fb are not shown.

Table 2. Mass spectra (in GeV) for the benchmark points.

	$m_{\tilde{t}_1}$	$m_{\tilde{t}_2}$	$m_{\tilde{b}_1}$	$m_{\tilde{b}_2}$	$m_{\tilde{g}}$	$m_{\tilde{\chi}_1^0}$	$m_{\tilde{\chi}_1^\pm}$	$m_{\tilde{Z}'}$
BP1	561.5	910.0	634.3	730.8	1400	671.1	693.1	300.0
BP2	547.4	904.0	656.2	711.3	1400	117.3	129.5	220.0
BP3	543.5	900.0	622.7	760.9	2000	677.2	699.6	300.0

Table 3. Cross-sections in fb at LHC14.

	$\tilde{t}_1\tilde{t}_1$	$\tilde{t}_2\tilde{t}_2$	$\tilde{b}_1\tilde{b}_1$	$\tilde{b}_2\tilde{b}_2$	$\tilde{g}\tilde{g}$
BP1	176.29	10.0	88.63	38.70	3.68
BP2	200.00	10.05	82.66	45.86	3.68
BP3	213.6	6.19	65.78	19.03	1.52

Table 4. Decay branching fraction of squarks to $\tilde{Z}'$.

	$\tilde{u}_L$	$\tilde{u}_R$	$\tilde{d}_L$	$\tilde{d}_R$	$\tilde{t}_1$	$\tilde{t}_2$	$\tilde{b}_1$	$\tilde{b}_2$
BP1	0.117	0.347	0.120	0.679	1.00	0.009	1.00	0.034
BP2	0.04	0.15	0.04	0.42	0.017	0.005	0.11	0.01
BP3	0.04	0.18	0.04	0.47	1.00	0.01	1.00	0.018

Let us now look at the $\tilde{Z}'$ production rates from the SUSY cascade decay. Table 4 gives the decay branching fractions of the various squarks to $\tilde{Z}'$. For BP1 and BP3 where $\tilde{Z}'$ is the NLSP, $\mathrm{Br}(\tilde{t}_1 \to t\tilde{Z}')$ and $\mathrm{Br}(\tilde{b}_1 \to b\tilde{Z}')$ are 100%. The right-handed neutrino having the lepton flavor violating Yukawa coupling is supposed to have the following decay modes:

$$\begin{aligned} N &\to e^{\pm}W^{\mp} \quad (79\%) \\ &\to \nu_e Z \qquad (21\%) \\ &\to \nu_e h \qquad (0\%)\,, \end{aligned} \tag{8}$$

where in the parentheses are shown the branching ratios for $m_N = 100$ GeV.

The final state coming from the sbottom production and decay will have two b-jet and four non-b-jet demanding both Ws to decay hadronically:

$$\begin{aligned} \tilde{b}_{1,2} &\to b\tilde{Z}' \to bN\tilde{N}_1 \to beW\not{p}_T\,, \\ \tilde{b}_{1,2}\tilde{b}^*_{1,2} &\to 2e + 2b + 4q + \not{p}_T\,. \end{aligned} \tag{9}$$

Similarly, for $\tilde{t}_{1,2}$ we have

$$\begin{aligned} \tilde{t}_{1,2} &\to t\tilde{Z}' \to bWN\chi^0_1 \to b + 2W + e + \not{p}_T\,, \\ \tilde{t}_{1,2}\tilde{t}^*_{1,2} &\to 2e + 2b + 8q + \not{p}_T\,. \end{aligned} \tag{10}$$

Thus, it will be our primary interest to look for the lepton flavor violating final state:

$$(2e - 2\mu) + 2b + n_q + \not{p}_T \tag{11}$$

with $n_q \geq 4$ for both same-sign or opposite-sign e or μ. In addition, $3l$ and $4l$ signatures coming from leptonic decays of W are also promising to look for the signal events.

Table 5. Number of $(2e-2\mu)$ events at LHC14 with an integrated luminosity of 50 fb^{-1}.

14 TeV/50 fb^{-1}		Signal			Background			
$(2e-2\mu)+n_j \geq 6$		BP1	BP2	BP3	$t\bar{t}Z$	$t\bar{t}W$	$t\bar{t}b\bar{b}$	$t\bar{t}$
	OSD	712.72	37.31	784.54	−7.52	−1.24	−0.59	98.55
	SSD	606.07	30.86	629.71	−0.66	1.56	−0.59	−14.78
Significance	OSD	25.17	3.31	26.54				
	SSD	24.91	7.62	25.39				

4.2. *2ℓ signature*

The signal signature (9) involves the decay $N \to e\,W(jj)$ leading to $2e + 2b + 4j + \not{p}_T$ in the final state. In the process of hadronization and jet formation with ISR/FSR, more number of jets are produced. To extract out the lepton flavor asymmetry of electron and muon we look for final states with $2e/2\mu + n_j \geq 6$ $(n_b \geq 2)$. Background events come mainly from $t\bar{t}Z$, $t\bar{t}W$ and $t\bar{t}b\bar{b}$ producing same numbers of OSD e^+e^- and $\mu^+\mu^-$, and thus can be efficiently suppressed by looking for flavor asymmetric SSD events: $e^\pm e^\pm - \mu^\pm \mu^\pm$. Table 5 shows the event numbers in the $(2e-2\mu) + n_j \geq 5$ $(n_b \geq 2)$ final state for all the benchmark points and the backgrounds. For BP1 and BP3 we can have around 25σ signal significance for OSD and SSD flavor difference. It is encouraging to see that the significance of BP3 can also reach to about 7σ for SSD. Note that the missing energy cut is not imposed as the requirement of lepton flavor/number violation is powerful enough.

4.3. *3ℓ signature*

The $3e$ final state is possible if one of the Ws from the right-handed neutrino decays leptonically. The resulting final state consists of $3e + n_j \geq 4$ $(n_b \geq 2)$ from the production and decay of $\tilde{b}_1\tilde{b}_1^*$. Here, we also impose a missing energy cut $\not{p}_T \geq 100$ GeV to reduce the SM backgrounds. Table 6 shows the number of the $3e - 3\mu$ final states whose observation will provide a consistent check of the model prediction.

4.4. *4ℓ signature*

Finally 4ℓ final states, can appear when two Ws from the decays of the right-handed neutrinos, decay leptonically. Here we do not distinguish the flavors of the charged lepton and consider both e and μ in the final state. Of course, among these 4ℓ, two of them are electrons coming the flavor violating decays of the right-handed neutrino, $N \to eW$. Table 7 presents the number of events for the $4\ell + n_j \geq 3$ $(n_b \geq 2)$ final state.

Table 6. Number of $(3e-3\mu)$ events at LHC14 with an integrated luminosity of 50 fb^{-1}.

14 TeV/50 fb^{-1}	Signal			Background			
$(3e-3\mu)+n_j \geq 4$ $(n_b \geq 2) + \not{p}_T \geq 100$ GeV	BP1	BP2	BP3	$t\bar{t}Z$	$t\bar{t}W$	$t\bar{t}b\bar{b}$	$t\bar{t}$
$\tilde{t}_1\tilde{t}_1^*$	37.54	0.92	35.55	−2.24	0.00	0.00	0.00
$\tilde{b}_1\tilde{b}_1^*$	46.26	3.46	31.64				
$\tilde{b}_2\tilde{b}_2^*$	0.86	0.35	0.24				
$\tilde{t}_2\tilde{t}_2^*$	0.04	0.03	0.04				
Total	84.71	4.15	67.47	−2.24			
Significance	9.33	3.00	8.35				

Table 7. Number of 4ℓ events at LHC14 with an integrated luminosity of 50 fb^{-1}.

14 TeV/50 fb^{-1}	Signal			Background			
$4\ell + n_j \geq 3$ $(n_b \geq 2)$	BP1	BP2	BP3	$t\bar{t}Z$	$t\bar{t}W$	$t\bar{t}b\bar{b}$	$t\bar{t}$
$\tilde{t}_1\tilde{t}_1^*$	122.15	2.47	143.26	7.45	0.00	0.00	0.00
$\tilde{b}_1\tilde{b}_1^*$	35.14	5.92	24.26				
$\tilde{b}_2\tilde{b}_2^*$	0.53	0.20	0.14				
$\tilde{t}_2\tilde{t}_2^*$	0.06	0.04	0.05				
Total	157.88	8.64	167.71	7.45			
Significance	12.28	0.25	12.67				

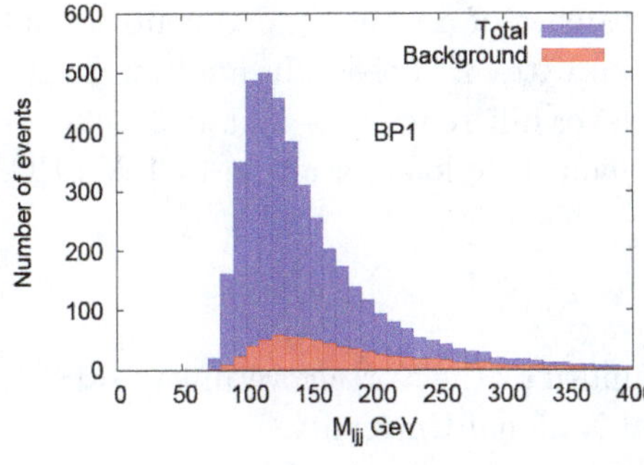

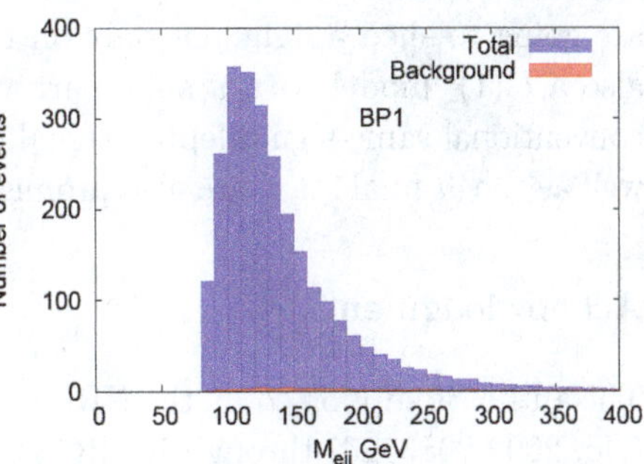

Fig. 4. $M_{\ell jj}$ distribution (left) for $n_\ell \geq 3 + n_j \geq 4(n_b \geq 2) + \not{p}_T \geq 100$ GeV and M_{ejj} (right) $n_\ell \geq 3(n_e \geq 2) + n_j \geq 4(n_b \geq 2) + \not{p}_T \geq 100$ GeV for final state at an integrated luminosity of 50 fb^{-1}.

4.5. *Reconstruction of the right-handed neutrino*

For the $2e$ or 3ℓ final states, an invariant mass of ℓjj will give the right-handed neutrino mass. In Fig. 4 (left) we demonstrate the invariant mass distribution of one electron and two jets coming from W: M_{ejj} reconstructing the decay $N \to eW$. To control the dominant SM background $t\bar{t}$, we have chosen final states with $n_\ell \geq 3 + n_j \geq 4(n_b \geq 2) + \not{p}_T \geq 100$ GeV. The demand of additional jets and b-jets reduces the SM backgrounds substantially. We can see that the signal peaked around ~ 100 GeV, which is the right-handed neutrino mass, m_N. Clearly it has more than 60σ signal significance a 50 fb^{-1} luminosity.

If we use the flavor violating decay of the right-handed neutrino and demand that two of the three leptons are electrons, the backgrounds can be efficiently suppressed. Figure 4 (right) shows the corresponding invariant mass distribution of ejj for the final state of $n_\ell \geq 3(n_e \geq 2) + n_j \geq 4(n_b \geq 2) + \not{p}_T \geq 100$ GeV. One can see that the signal stands out over the backgrounds much clearly.

5. Conclusion

A $U(1)'$ supersymmetric seesaw model with R-parity can be motivated by simultaneous explanation of the observed neutrino masses and mixing, the existence of dark matter, and the stabilization of the Higgs boson mass assuming TeV-scale SUSY breaking scale. In this scheme, a right-handed sneutrino $\tilde{N}_1$ becomes a good thermal dark matter candidate whose annihilation cross-section is in the right range when the extra gaugino $\tilde{Z}'$ is relatively light, and the right relic density can arise if the decay and inverse-decay effect of the right-handed neutrino N and thus its Yukawa coupling is large enough.

Considering stop and sbottom below TeV, we showed that $\tilde{Z}'$ produced from third generation SUSY cascades can lead to significant lepton number and flavor violating signatures in final states with multi-lepton accompanied by multi-jet (+missing energy) through the decay chain of $\tilde{Z}' \to N\tilde{N}_1 \to e^\pm W^\mp \tilde{N}_1$. These signatures are going to shed a light not only on the existence of a right-handed neutrino but also a $U(1)'$ model with a superpartner of an extra Z' boson. In addition to the conventional same-sign dilepton signal, the flavor differences $2e - 2\mu$ and $3e - 3\mu$, as well as the $4l$ final state are also promising channels to look for at the 14 TeV LHC.

Acknowledgment

The author is supported by the NRF grant funded by the Korea government (MSIP) (No. 2009-0083526) through KNRC at Seoul National University.

References

1. P. Langacker, *Rev. Mod. Phys.* **81**, 1199 (2009), arXiv:0801.1345 [hep-ph].

2. ATLAS Collab. (G. Aad *et al.*), *Phys. Rev. D* **90**, 052005 (2014), arXiv:1405.4123 [hep-ex].
3. P. Bandyopadhyay, E. J. Chun and J. C. Park, *J. High Energy Phys.* **1106**, 129 (2011), arXiv:1105.1652 [hep-ph].
4. P. Bandyopadhyay and E. J. Chun, arXiv:1412.7312 [hep-ph].
5. S. Khalil and A. Masiero, *Phys. Lett. B* **665**, 374 (2008), arXiv:0710.3525 [hep-ph].
6. U. Ellwanger, C. Hugonie and A. M. Teixeira, *Phys. Rep.* **496**, 1 (2010), arXiv: 0910.1785 [hep-ph].
7. https://twiki.cern.ch/twiki/bin/view/AtlasPublic/SupersymmetryPublicResults; https://twiki.cern.ch/twiki/bin/view/CMSPublic/PhysicsResultsSUS.
8. ATLAS Collab., ATLAS-CONF-2013-007.
9. CMS Collab., CMS-PAS-SUS-13-007.
10. CMS Collab. (S. Chatrchyan *et al.*), *Eur. Phys. J. C* **73**, 2568 (2013), arXiv:1303.2985 [hep-ex].

From Electromagnetic Neutrinos to New Electromagnetic Radiation Mechanism in Neutrino Fluxes

Ilya Balantsev*,‡ and Alexander Studenikin*,†,§

**Department of Theoretical Physics, Faculty of Physics, Lomonosov Moscow State University, Moscow, 119991, Russia*

†*Joint Institute for Nuclear Research, Dubna, Moscow Region, 141980, Russia*

‡*balantsev@physics.msu.ru*

§*studenik@srd.sinp.msu.ru*

A massive neutrino has nonzero magnetic moment and is involved in the electromagnetic interactions with external fields and photons. The electromagnetic neutrino moving in matter can emit the spin light ($SL\nu$) in the process of transition between two quantum states in matter. In quite resembling way an electron can emit spin light in moving background composed of neutrinos, that is "the spin light of an electron in neutrino flux" (SLe_ν). In this paper we obtain the exact solution for the wave function and energy spectrum for an electron moving in a neutrino flux and consider the SLe_ν as the transition process between two electron quantum states in the background. The SLe_ν radiation rate, power and emitted photon energy are calculated. Notably, the energy spectrum of the emitted SLe_ν photons can span up to gamma-rays. We argue that the considered SLe_ν can be of interest for astrophysical applications, for supernovae processes in particular.

Keywords: Electromagnetic radiation; neutrino fluxes; Dirac equation exact solutions; supernovae.

1. Introduction

The confirmation that neutrinos are massive particles, obtained in many neutrino oscillation experiments, no doubt is, together with the observation of the Higgs boson, among two most important recent discoveries in particle physics. The later finally confirms the foundation of the Standard Model, whereas the former opens a window to new physics. It was established long ago that a massive neutrino should have nonvanishing magnetic moment[1] that provides a room for neutrinos electromagnetic interactions (see Refs. 2–4 for a review).

In the year 2015, that has been claimed by the United Nations the Year of Light, it is relevant to discuss how in the studies of neutrino electromagnetic properties (or electromagnetic neutrinos) we have arrived to a prediction[5,6] for a new mechanism of electromagnetic radiation that can be emitted by an electron in a dense neutrino flux.

In our studies[7] of neutrino electromagnetic properties we have predicted a new mechanism of electromagnetic radiation that can be emitted by a neutrino propagating in dense matter and have termed it "the spin light of neutrino" in matter ($SL\nu$). At first the $SL\nu$ was investigated within the quasiclassical treatment (see also Refs. 8 and 9) based on use of the generalized Bargmann–Michel–Telegdi equation for the neutrino spin evolution in the background environment.[10,11]

The quantum theory of the $SL\nu$ was first revealed in our studies[12,13] (see also Ref. 14) within implication of the so-called "method of wave equations exact solutions" that implies use of exact solutions of modified Dirac equations that contain the corresponding effective potentials accounting for the matter influence on neutrinos.[12,13,15–22] Different aspects related to the proposed the $SL\nu$ have been discussed and investigated recently. For instance, the importance of the plasma influence on the proposed new mechanism of electromagnetic radiation as well as corrections to the effective matter density due to nonlocality of neutrino interaction with particles of the environment have been considered in the subsequent series of papers dedicated to the $SL\nu$ (see Refs. 23–25). The $SL\nu$ mechanism was also considered for the case when in addition to matter a gravitational field is also present,[9] or the model describing neutrino interactions with the environment permits of the Lorentz and CPT invariance violation.[26]

The main properties of the $SL\nu$ are summarized[19] in the following way: (1) a neutrino with nonzero magnetic moment when moving in dense matter can emit electromagnetic waves; (2) the $SL\nu$ radiation rate and power depend on the neutrino magnetic moment and energy, and also on the matter density; (3) for a wide range of matter densities the radiation is beamed along the neutrino momentum; (4) the emitted photon energy is also essentially dependent on the neutrino energy and matter density; (5) in the most interesting for possible astrophysical and cosmology applications case of ultrahigh energy neutrinos, the average energy of the $SL\nu$ photons equals a reasonable part of the initial neutrino energy so that the $SL\nu$ spectrum can span up to gamma-rays.

In spite of the listed above notable properties of the $SL\nu$ its possible role and impact in astrophysical processes is constraint due to the fact that the rate of the process is proportional to the neutrino magnetic moment squared that is in fact a very small quantity for the most of theories beyond the Standard Model (see, for instance, Ref. 2). Other constraints on possible visualization of the $SL\nu$ are imposed by the above-mentioned effects of the background plasma.

To avoid the suppression of the radiation produced by the spin light mechanism in the case of a neutrino, we considered the electromagnetic radiation by an electron moving in matter (the "spin light of electron in matter," SLe). From the order-of-magnitude estimation,[15] we predicted that the ratio of rates of the SLe and the

$SL\nu$ in matter is

$$R = \frac{\Gamma_{SLe}}{\Gamma_{SL\nu}} \sim \frac{e^2}{\omega^2\mu^2}, \quad (1)$$

that gives $R \sim 10^{20}$–10^{14} for the value of the neutrino magnetic moment $\mu \sim 10^{-11}\mu_0$ and the predicted wide range SLe photon's energies $\omega \sim 5$ MeV–5 GeV. This estimation was confirmed by the direct evaluation[27] of the SLe properties based on the exact solutions of the modified Dirac equation for the electron in matter. We expected that in certain cases the SLe in matter would be more effective than the $SL\nu$. However, the possibility of phenomenological consequences of the SLe in astrophysics is quite not obvious.

In this paper we continue studies of the spin light mechanism of the electromagnetic radiation in a dense environment and consider a new possible realization of this mechanism. The predicted new mechanism implies the electromagnetic radiation emitted by an electron in a dense flux of ultrarelativistic neutrinos. We term this mechanism "the spin light of an electron in dense neutrino fluxes," SLe_ν. This new realization of the spin light provides a possibility to avoid two suppression factors, peculiar to the $SL\nu$, in the radiation rate and power, the discussed above suppression due to smallness of a neutrino magnetic moment and one due to the plasma effects. It is also shown that the relativistic motion of the radiating electron can significantly influence characteristics of the SLe_ν in respect to the SLe_ν by an electron at rest. In particular, the SLe_ν rate, radiation power and the energy of the emitted photon can be reasonably increased. Therefore, the predicted new realization of the spin light mechanism can be of interest for physics of cosmic rays[28,29] and description of phenomenology in different astrophysical settings such as stellar core-collapse and supernova explosion.[30–32]

2. Method of Exact Solutions in Studies of Neutrino and Electron Moving in External Environment

An effective tool in studies of particle interactions in presence of external electromagnetic fields and/or dense matter is based on implication of exact solutions of quantum field equations of motion. This method was first established and applied in quantum electrodynamics for studies of motion and radiation of an electron in a magnetic field, i.e. in the quantum theory of the synchrotron radiation (see, for instance, Ref. 33), and also for studies of the electrodynamics and weak interactions in different configurations of external electromagnetic fields.[34,35] Initially this method was based on the Furry representation[36] of quantum electrodynamics. Recently it has been shown that the method of exact solutions can be also applied for the problem of neutrinos and electron motion in presence of dense matter (see Refs. 15, 17 and 19 for a review on this topic). In Refs. 12, 13, 18, 37 the detailed derivation of the exact solution for the modified Dirac equation for a neutrino moving in matter is presented. In Refs. 19 and 27 the corresponding exact solution for an electron moving in matter is obtained. Analogous problem for electron in magnetized matter is solved in Ref. 20. The problem of neutrino propagation in transversally moving matter was first solved in Refs. 38 and 39 we considered neutrino propagation in a rotating matter accounting for the effect of nonzero neutrino mass.

In this paper we consider an electron propagating in matter composed of neutrinos (the neutrino flux) within the method of exact solutions and then study the accompanying electromagnetic radiation by an electron in the neutrino flux.

3. Modified Dirac Equation

We consider a beam of electrons moving inside of the neutrino flux composed of three flavors ν_e, ν_μ and ν_τ with number densities n_i (in the laboratory rest frame) moving in the same direction. Following the previous discussions,[5,6] we introduce the average value n of the neutrino number density and the parameter δ_e,

$$n = \frac{n_e + n_\mu + n_\tau}{3}\,, \quad \delta_e = \frac{n_\mu + n_\tau - n_e}{n}\,, \tag{2}$$

and obtain the modified Dirac equation for an electron in the neutrino flux,

$$\left\{\gamma_\mu p^\mu + \gamma_\mu \frac{c + \delta_e \gamma^5}{2} f^\mu - m\right\}\Psi(x) = 0\,, \tag{3}$$

where m and p^μ are the electron mass and momentum, $c = \delta_e - 12\sin^2\theta_W$, $G = \frac{G_F}{\sqrt{2}}$, and G_F is the Fermi constant. The speeds of relativistic neutrinos are $\beta^\mu_{\nu_i} \simeq (1,0,0,1)$, thus the effective neutrino potential in (2) is $f^\mu = G(n,0,0,n)$. We suppose here that the neutrino flux propagates along the direction of z axis.

4. Exact Solution

Equation (3) can be solved exactly (see Refs. 5 and 6) and for the electron energy spectrum we get

$$E^\varepsilon_s(\boldsymbol{p}) = \varepsilon\sqrt{m^2 + \boldsymbol{p}_\perp^2 + (p_3 + A)^2} - A\,, \tag{4}$$

where $A = \frac{Gn}{2}(c - s\delta)$, $\delta = |\delta_e|$, p_3 is the electron momentum in the direction of the neutrino flux propagation and $\boldsymbol{p} = (\boldsymbol{p}_\perp, p_3)$ is the total electron momentum. Comparing (4) with corresponding spectra of a neutrino[12,13] or an electron[27] in nonmoving matter we conclude that the number $s = \pm 1$ distinguishes two possible spin states of the electron. Two particular electron energy branches $E^\varepsilon_s(\boldsymbol{p})_{|\varepsilon=+1} = E_s(\boldsymbol{p})$ with $s = \pm 1$ as functions of the momentum $\boldsymbol{p}$ are plotted in Fig. 1.

It is possible to show that always $E_+(\boldsymbol{p}) > E_-(\boldsymbol{p})$ (for any values of $\boldsymbol{p}$). Two exact solutions of Eq. (3) characterized by $E_+(\boldsymbol{p})$ and $E_-(\boldsymbol{p})$ are given by[5,6]

$$\psi_i(\boldsymbol{r}, t) = e^{i(-E_+ t + \boldsymbol{p}\boldsymbol{r})}\tilde{\psi}_i\,, \tag{5}$$

$$\psi_f(\boldsymbol{r}, t) = e^{i(-E_- t + \boldsymbol{p}\boldsymbol{r})}\tilde{\psi}_f\,, \tag{6}$$

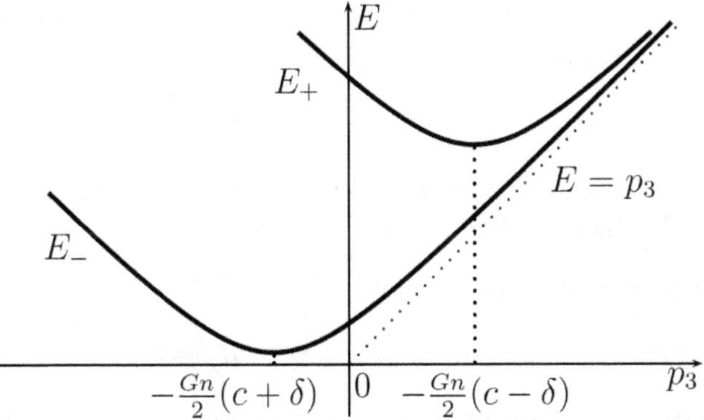

Fig. 1. The dependence of the electron energies in two different spin states, $E_+(\boldsymbol{p})$ and $E_-(\boldsymbol{p})$, on the momentum component p_3.

where

$$\tilde{\psi}_i = \frac{1}{L^{\frac{3}{2}}C_+}\begin{pmatrix} 0 \\ m \\ p_\perp e^{-i\phi} \\ E_+ - p_3 \end{pmatrix}, \quad \tilde{\psi}_f = \frac{1}{L^{\frac{3}{2}}C_-}\begin{pmatrix} E_- - p_3 \\ -p_\perp e^{i\phi} \\ m \\ 0 \end{pmatrix}, \tag{7}$$

here L is the normalization length and

$$C_\pm = \sqrt{m^2 + p_\perp^2 + (E_\pm - p_3)^2} \tag{8}$$

are the normalization coefficients.

5. Spin Light of Relativistic Electron in Dense Neutrino Flux

Now we consider the radiative transition of an electron from the state $\psi_i(\boldsymbol{r}, t)$ to the state $\psi_f(\boldsymbol{r}, t)$ with emission of a photon in the presence of the relativistic flux of neutrinos. This process we have termed the *spin light of electron in a neutrino flux* (SLe_ν). The element of S-matrix defining the process amplitude is given by (see Refs. 19 and 27):

$$S_{fi}^{(\lambda)} = -e\sqrt{4\pi}\int d^4x\,\bar{\psi}_f(x)(\boldsymbol{\gamma}\boldsymbol{e}^{(\lambda)*})\frac{e^{ikx}}{\sqrt{2\omega L^3}}\psi_i(x)\,, \tag{9}$$

where e is the electron charge, $\psi_i(x)$ and $\psi_f(x)$ are the wave functions of the initial and final electron states in the background neutrino flux given by (5) and (6), $k = (\omega, \boldsymbol{k})$ and $\boldsymbol{e}^{(\lambda)}$ ($\lambda = 1, 2$) are the momentum and polarization vectors of the emitted photon.

In general case the considered electron can also move in respect to an observer. The electron rest frame in moving neutrino background is defined as one where the electron energy E_+ gets its minimum, $\frac{\partial E_+}{\partial \boldsymbol{p}} = 0$ (see Refs. 12, 40–42): $p_3 =$

$-\frac{Gn}{2}(c-\delta)$, $\boldsymbol{p}_\perp = 0$. Thus the initial value of the electron momentum third component can be represent as

$$p_3 = -\frac{Gn}{2}(c-\delta) + \tilde{p}_3\,, \tag{10}$$

where $\tilde{p}_3$ is an "access" of the momentum component over its (minimum) value in the rest frame. In the following we distinguish two cases:

(1) for the nonmoving electron we have

$$p_3 = -\frac{Gn}{2}(c-\delta)\,, \quad \boldsymbol{p}_\perp = 0\,, \tag{11}$$

(2) for the relativistic electron moving in the opposite direction to the neutrino flux propagation we have

$$|\tilde{p}_3| \gg m\,, \quad |\tilde{p}_3| Gn\delta \ll m^2\,, \quad \text{and} \quad \tilde{p}_3 < 0\,. \tag{12}$$

As for the supernova environment $\frac{Gn}{m} \sim 10^{-8}$, the electron momentum in this case should be within the range $1 \ll \frac{|\tilde{p}_3|}{m} \ll 10^8$.

From the energy–momentum conservation for the emitted photon energy in the case of the nonmoving initial electron we obtain[6]

$$\omega = \frac{m}{1 - \cos\theta + \frac{m}{Gn\delta}}\,, \tag{13}$$

whereas for the case of the relativistic initial electron we get[5]

$$\omega = \frac{2Gn\delta}{1 + \cos\theta + \frac{1}{2}\frac{m^2}{\tilde{p}_3^2}}\,, \tag{14}$$

where θ is the angle between the direction of the SLe_ν and neutrino flux propagation.

From Eqs. (13) and (14) it follows that in general the emission is possible in all directions. For the most realistic case, when $Gn\delta \ll m$, the photon energy in the nonmoving case simplifies to

$$\omega = Gn\delta\,. \tag{15}$$

Thus, in the nonmoving case for an initial charged particle with rather large mass or for the case of the background environment with enough small density the emitted photon energy does not depend on the direction of radiation and is determined only by the density of the environment.

For the relativistic case, it is interesting to compare the emitted photon energies in the case of the considered here SLe_ν by relativistic electrons and one produced by electrons at rest. Taking into account that in the case of nonmoving electrons $\omega = Gn\delta$, for the photons energy ratio (in the case of electron motion against the neutrino flux propagation, $\theta = \pi$) we get

$$\frac{\omega(|\tilde{p}_3| \gg m)}{\omega(|\tilde{p}_3| \ll m)} = 4\frac{\tilde{p}_3^2}{m^2} \gg 1\,. \tag{16}$$

It follows[5] that there is a reasonable increase of the emitted photon energy in case of the relativistic motion of the emitters (the electrons) with respect to nonmoving case.

Using expressions for the amplitude (9) and the wave functions of the initial and final electrons (7), and also for the emitted photon energy (13) and (14) we get for the SLe_ν total rate and power in the two discussed cases:

(1) for the nonmoving electron[6]

$$\Gamma = \frac{4}{3}e^2 m\left(\frac{Gn\delta}{m}\right)^3, \quad I = \frac{4}{3}e^2 m^2\left(\frac{Gn\delta}{m}\right)^4, \tag{17}$$

(2) for the relativistic electron[5]

$$\Gamma = \frac{16}{3}e^2 m\left(\frac{Gn\delta}{m}\right)^3\left(\frac{|\tilde{p}_3|}{m}\right)^2, \quad I = 16e^2 m^2\left(\frac{Gn\delta}{m}\right)^4\left(\frac{|\tilde{p}_3|}{m}\right)^4. \tag{18}$$

Comparing the corresponding characteristics (17), (18) of the SLe_ν obtained for two different cases, we get that

$$\frac{\Gamma(|\tilde{p}_3| \gg m)}{\Gamma(|\tilde{p}_3| \ll m)} = 4\left(\frac{|\tilde{p}_3|}{m}\right)^2, \quad \frac{I(|\tilde{p}_3| \gg m)}{I(|\tilde{p}_3| \ll m)} = 12\left(\frac{|\tilde{p}_3|}{m}\right)^4.$$

For the case of relativistic electrons $\frac{|\tilde{p}_3|}{m} \gg 1$. Thus there is a reasonable amplification of the SLe_ν rate and power by the relativistic motion of the initial radiating electron.

6. Effect of Plasma

The electromagnetic wave propagation in the background environment is influenced by the plasma effects. For the $SL\nu$ in matter these effects have been discussed in details in Refs. 13, 23, 24. In Ref. 5 we have shown that the effect of nonzero emitted photon mass (the plasmon mass m_γ) in the case of SLe_ν is not important, $\frac{m_\gamma}{Gn\delta} \ll 1$. The Debye screening of electromagnetic waves (another possible plasma effect) could be important for the SLe_ν radiation propagation if electron number density $N_e < 10^{35}$ cm^{-3}. However, the electron matter with $N_e \sim 10^{19}$ cm^{-3} considered here is quite transparent for the SLe_ν.

7. Conclusions and Indications for Possible Phenomenology

It is interesting to apply the considered SLe_ν of nonmoving and relativistic electrons in dense fluxes of neutrinos to an environment peculiar to a supernova phenomena. As it is shown in Ref. 29 one can estimate the effective neutrino matter density to be $n \sim 10^{35}$ cm^{-3}, thus the characteristic parameter $\frac{Gn}{m} \sim 10^{-8}$. As it is discussed in Refs. 31 and 32, the surrounding interstellar medium can contain regions with reasonably high electron density relativistically moving towards the neutrino flux.

Under these conditions, the spin light can be emitted by the relativistic electrons in the quantum transition from the energy states E_+ to the states E_-.

From Eq. (13) it follows for nonmoving initial electron the SLe_ν photon energy

$$\omega \sim 1 \text{ eV} . \qquad (19)$$

From (17) we also find for the SLe_ν rate and power

$$\Gamma \sim 10^{-19} \text{ eV} \sim 10^{-4} \text{ s}^{-1} , \quad I \sim 10^{-7} \text{ eV s}^{-1} . \qquad (20)$$

The corresponding characteristic time of the SLe_ν process is rather big, $\tau \sim 10^4$ s. It means that SLe_ν from a single electron is hardly observable.

From (14) and (18) for the relativistic electrons characterized by $\frac{|\tilde{p}_3|}{m} = 10^7$ we get the following estimations for the SLe_ν photon energy, rate and power, respectively,

$$\omega \sim 10^{14} \text{ eV} , \quad \Gamma \sim 10^{10} \text{ s}^{-1} , \quad I \sim 10^{21} \text{ eV s}^{-1} . \qquad (21)$$

The electron number density at the distance $R = 10$ km from the star center can be of order $N_e \sim 10^{19}$ cm^{-3}. Thus, the amount of SLe_ν flashes per second from 1 cm^3 of the electron matter under the influence of a dense neutrino flux is $N \sim 10^{28}$ cm^{-3} s^{-1}. For the energy release of 1 cm^3 per one second, we get

$$\frac{\delta E}{\delta t \delta V} = I N_e \sim 10^{40} \text{ eV cm}^{-3} \text{ s}^{-1} . \qquad (22)$$

Now let us also estimate the efficiency of the energy transfer from the total neutrino flux to the electromagnetic radiation due to the proposed SLe_ν mechanism. The total neutrino energy in the neutrino flux (characterized by $n \sim 10^{35}$ cm^{-3} and $\langle E \rangle \sim 10^7$ eV) is

$$\frac{\delta E_\nu}{\delta V} \sim \langle E \rangle n \sim 10^{42} \text{ eV cm}^{-3} . \qquad (23)$$

It follows that each second a considerable part of neutrino flux energy transforms into gamma-rays by the SLe_ν mechanism. The performed studies illustrates an increase of the efficiency of such energy transfer mechanism in the case when the emitting electrons are moving with relativistic speed against the neutrino flux propagation in comparison with the case of nonmoving initial electrons. We predict that this may have important consequences in astrophysics and for the supernova process in particular.

Acknowledgments

We are thankful to Alexander Grigoriev, Alexey Lokhov and Alexei Ternov for many fruitful discussions on the considered phenomenon. This study has been partially supported by the Russian Foundation for Basic Research (grants No. 14-22-03043-ofi and No. 15-52-53112). One of the authors (A. Studenikin) is thankful to Harald Fritzsch and Zhi-Zhong Xing for the invitation to attend the International

Conference on Massive Neutrinos and to Phua Kok Khoo for the kind hospitality provided at the Institute for Advanced Studies of the Nanyang Technological University in Singapore.

References

1. K. Fujikawa and R. Shrock, *Phys. Rev. Lett.* **45**, 963 (1980).
2. C. Giunti and A. Studenikin, Neutrino electromagnetic interactions: A window to new physics, to appear in *Rev. Mod. Phys.* **87** (2015).
3. C. Broggini, C. Giunti and A. Studenikin, *Adv. High Energy Phys.* **2012**, 459526 (2012).
4. C. Giunti and A. Studenikin, *Phys. Atom. Nucl.* **72**, 2089 (2009).
5. I. Balantsev and A. Studenikin, Spin light of electron in dense neutrino fluxes, arXiv:1405.6598.
6. I. A. Balantsev and A. I. Studenikin, Spin light of relativistic electrons in neutrino fluxes, arXiv:1502.05346.
7. A. Lobanov and A. Studenikin, *Phys. Lett. B* **564**, 27 (2003).
8. A. Lobanov and A. Studenikin, *Phys. Lett. B* **601**, 171 (2004).
9. M. Dvornikov, A. Grigoriev and A. Studenikin, *Int. J. Mod. Phys. D* **14**, 309 (2005).
10. A. Lobanov and A. Studenikin, *Phys. Lett. B* **515**, 94 (2001).
11. M. Dvornikov and A. Studenikin, *J. High Energy Phys.* **0209**, 016 (2002).
12. A. Studenikin and A. Ternov, *Phys. Lett. B* **608**, 107 (2005).
13. A. Grigorev, A. Studenikin and A. Ternov, *Phys. Lett. B* **622**, 199 (2005).
14. A. Lobanov, *Phys. Lett. B* **619**, 136 (2005).
15. A. Studenikin, *J. Phys. A* **39**, 6769 (2006).
16. A. Grigoriev, A. Studenikin and A. Ternov, *Phys. Atom. Nucl.* **69**, 1940 (2006).
17. A. I. Studenikin, *Ann. Fond. Louis de Broglie* **31**, 289 (2006).
18. A. Studenikin, *Phys. Atom. Nucl.* **70**, 1275 (2007).
19. A. I. Studenikin, *J. Phys. A* **41**, 164047 (2008).
20. I. Balantsev, Y. Popov and A. Studenikin, *J. Phys. A* **44**, 255301 (2011).
21. I. Balantsev, A. Studenikin and I. Tokarev, *Phys. Atom. Nucl.* **76**, 489 (2013).
22. A. Studenikin and I. Tokarev, *Nucl. Phys. B* **884**, 396 (2014).
23. A. Kuznetsov and N. Mikheev, *Int. J. Mod. Phys. A* **22**, 3211 (2007).
24. A. Grigoriev, A. Lokhov, A. Studenikin and A. Ternov, *Phys. Lett. B* **718**, 512 (2012).
25. A. Kuznetsov, N. Mikheev and A. Shitova, *Phys. Atom. Nucl.* **76**, 1359 (2013).
26. S. Kruglov, *Int. J. Mod. Phys. A* **29**, 1450031 (2014).
27. A. Grigoriev, S. Shinkevich, A. Studenikin, A. Ternov and I. Trofimov, *Grav. Cosmol.* **14**, 248 (2008).
28. G. Raffelt, *Stars as Laboratories for Fundamental Physics: The Astrophysics of Neutrinos, Axions, and Other Weakly Interacting Particles* (University of Chicago Press, 1996), ISBN 0-226-70272-3.
29. M. Kachelriess, Lecture notes on high energy cosmic rays, arXiv:0801.4376.
30. H. Bethe, *Rev. Mod. Phys.* **62**, 801 (1990).
31. C. Frohlich *et al.*, *Astrophys. J.* **637**, 415 (2006).
32. H.-T. Janka, K. Langanke, A. Marek, G. Martinez-Pinedo and B. Mueller, *Phys. Rep.* **442**, 38 (2007).
33. A. Sokolov and I. Ternov, *Synchrotron Radiation* (Elsevier, 1969), ISBN 0-08-012945-5.

34. V. Ritus, *Issues in Intense-Field Quantum Electrodynamics*, Proceedings of the Lebedev Physics Institute, Vol. 111, ed. V. Ginzburg (Nova Science Publishers, 1979).
35. A. Nikishov, *Issues in Intense-Field Quantum Electrodynamics*, Proceedings of the Lebedev Physics Institute, Vol. 111, ed. V. Ginzburg (Nova Science Publishers, 1979).
36. W. Furry, *Phys. Rev.* **81**, 115 (1951).
37. A. Grigoriev, A. Studenikin and A. Ternov, *Grav. Cosmol.* **11**, 132 (2005).
38. A. Grigoriev, A. Savochkin and A. Studenikin, *Russ. Phys. J.* **50**, 845 (2007).
39. I. Balantsev, Y. Popov and A. Studenikin, *Nuovo Cimento C* **32**, 53 (2009).
40. L. N. Chang and R. Zia, *Phys. Rev. D* **38**, 1669 (1988).
41. J. T. Pantaleone, *Phys. Lett. B* **268**, 227 (1991).
42. J. T. Pantaleone, *Phys. Rev. D* **46**, 510 (1992).

Lepton Number Violation and the Baryon Asymmetry of the Universe

Julia Harz and Wei-Chih Huang

Department of Physics and Astronomy, University College London, London WC1E 6BT, UK

Heinrich Päs

Fakultät für Physik, Technische Universität Dortmund, D-44221 Dortmund, Germany

heinrich.paes@tu-dortmund.de

Neutrinoless double beta decay, lepton number violating collider processes and the Baryon Asymmetry of the Universe (BAU) are intimately related. In particular, lepton number violating processes at low energies in combination with sphaleron transitions will typically erase any preexisting BAU. In this contribution, we briefly review the tight connection between neutrinoless double beta decay, lepton number violating processes at the LHC and constraints from successful baryogenesis. We argue that far-reaching conclusions can be drawn unless the baryon asymmetry is stabilized via some newly introduced mechanism.

Keywords: Lepton number violation; baryogenesis; neutrinos.

1. Introduction

The discovery of neutrino masses is typically understood as a hint for physics beyond the Standard Model (SM). Intimately related to this link is the question whether lepton number is conserved or broken. After all, neutrino masses can be realized in two different ways, either as Majorana masses $\overline{\nu_L^C}\,\nu_L$ or as Dirac masses $\overline{\nu_L}\,\nu_R + \overline{\nu_R}\,\nu_L$. In the first case, lepton number is broken. In the latter case the newly introduced right-handed neutrino is an SM singlet so that a Majorana mass $\overline{\nu_R^C}\,\nu_R$ is allowed by the SM symmetry. So either lepton number is broken again or this operator has to be forbidden by a new symmetry. In this sense the problem of how neutrino masses are related to physics beyond the Standard Model boils down to the question whether the accidental lepton number conservation in the Standard Model is enforced by a new symmetry or violated by lepton number violation (LNV) operators.

LNV can be searched for directly for example in neutrinoless double beta decay

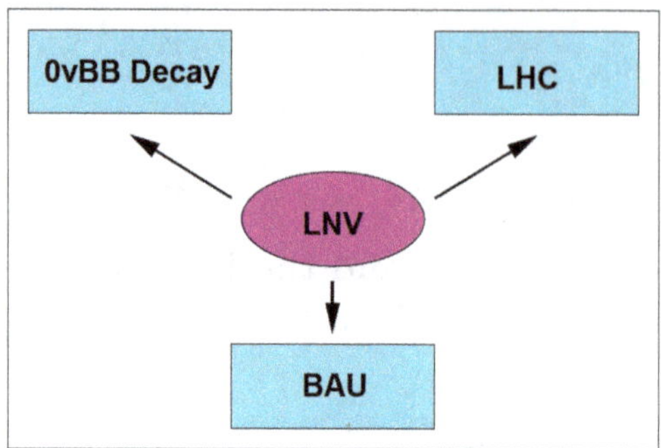

Fig. 1. Neutrinoless double beta decay, lepton number violating processes at the LHC and the generation and survival of the BAU are closely interrelated.

or at colliders. Moreover, lepton number violating interactions can be important in cosmology where they can both wash out or create the baryon asymmetry of the Universe (BAU). These apparently different phenomena are thus closely related (see Fig. 1) as will be discussed in the following.

2. Probing Lepton Number Violation with Neutrinoless Double Beta Decay

A sensitive probe of low energy LNV is neutrinoless double beta decay ($0\nu\beta\beta$), the simultaneous transition of two neutrons into two protons and two electrons, without emission of any antineutrinos:

$$2n \to 2p + 2e^- . \tag{1}$$

While the most prominent decay mode is triggered by a massive Majorana neutrino being exchanged between Standard Model (SM) $V-A$ vertices, providing a bound on the effective Majorana neutrinos mass

$$\langle m_\nu \rangle = \sum_j U_{ej}^2 m_j \equiv m_{ee} , \tag{2}$$

in the sub-eV range, in principle any operator violating lepton number by two units and transforming two neutrons into two protons, two electrons and nothing else will induce the decay.

As discussed in detail in Refs. 4 and 5, the most general operator triggering the decay can be parametrized in terms of effective couplings ϵ as shown in Fig. 2. The diagram depicts the exchange of a light Majorana neutrino between two SM vertices (contribution (a)), the exchange of a light Majorana neutrino between an SM vertex and an effective operator which is pointlike at the nuclear Fermi momentum scale $\mathcal{O}(100 \text{ MeV})$ (contribution (b)) and a short-range contribution triggered by a single dimension-9 operator being pointlike at the Fermi momentum scale (contribution (d)). Contribution (c) which contains two non-SM vertices can be neglected

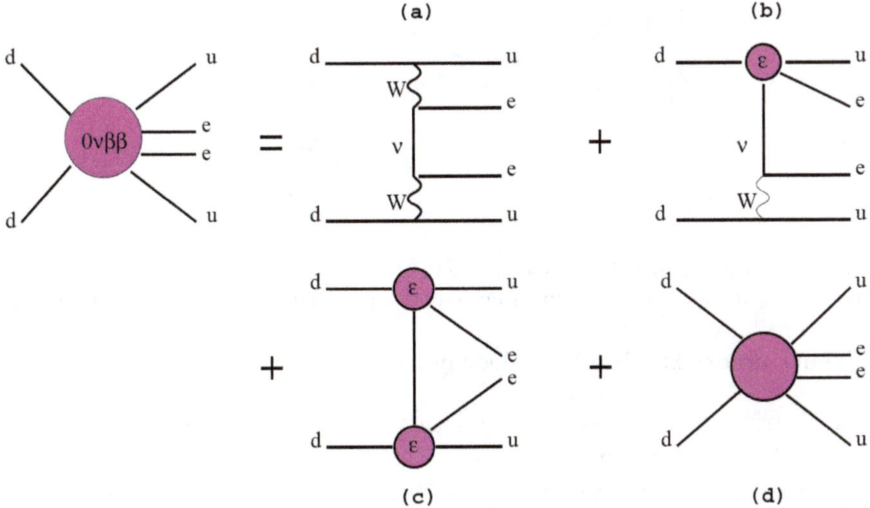

Fig. 2. Mechanisms for neutrinoless double beta decay: the most general effective operator triggering the decay can be decomposed into diagrams with SM vertices and effective vertices being pointlike at the nuclear Fermi scale (from Ref. 4).

when compared to contribution (b). The most general decay rate contains all combinations of leptonic and hadronic currents induced by the operators

$$\mathcal{O}_{V\mp A} = \gamma^\mu(1 \mp \gamma_5), \quad \mathcal{O}_{S\mp P} = (1 \mp \gamma_5), \quad \mathcal{O}_{T_{L/R}} = \frac{i}{2}[\gamma_\mu, \gamma_\nu](1 \mp \gamma_5), \quad (3)$$

allowed by Lorentz invariance.

Examples for contribution (b) are the Leptoquark and SUSY accompanied decay modes, examples for contribution (d) are decay modes where only SUSY particles or heavy neutrinos and gauge bosons in left–right-symmetric models are exchanged between the decaying nucleons, for a recent overview see Ref. 2. Present experiments have a sensitivity to the effective couplings of

$$\epsilon < (\text{few}) \times (10^{-7}\text{–}10^{-10})\,. \quad (4)$$

For the $d = 9$ operator triggering the contribution (d) it can be estimated that an observation of $0\nu\beta\beta$ decay with present-day experiments would involve TeV scale particles and thus would offer good chances to see new physics associated with LNV at the LHC. A crucial prerequisite for such a conclusion is of course a possibility to discriminate among the various mechanisms which may be responsible for the decay. This is a difficult task but may be possible at least for some of the mechanisms by observing neutrinoless double beta decay in multiple isotopes[6,7] or by measuring the decay distribution, for example in the SuperNEMO experiment.[8] Another possibility to discriminate between various short range contributions to neutrinoless double beta decay at the LHC itself is to identify the invariant mass peaks of particles produced resonantly in the intermediate state or to analyze the charge asymmetry between final states involving particles and/or antiparticles.[9,10]

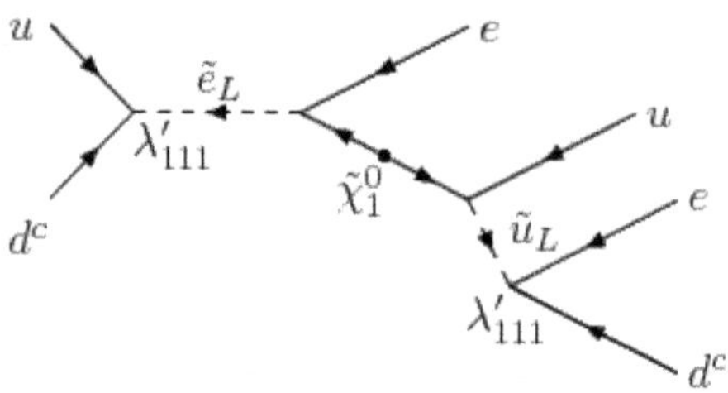

Fig. 3. Neutrinoless double beta decay at the LHC: the case for R-Parity violation. Two quarks in the initial state are converted into a same-sign di-lepton signal and two jets (from Ref. 11).

3. Neutrinoless Double Beta Decay at the LHC

While neutrinoless double beta decay is the prime probe for massive Majorana neutrinos, LNV in general can be searched for also in collider processes. Indeed, as has been discussed for example for the special cases of left–right symmetric models[13,14] and R-parity violating supersymmetry[11,12] the short range contribution (d) can easily be crossed into a diagram with two quarks in the initial state where resonant production of a heavy particle leads to a same-sign dilepton signature plus two jets at the LHC, see Fig. 3. In order to discuss the LHC bounds in a model-independent way similar to the effective field theory approach of Refs. 4 and 5, it is necessary to specify, which particles are produced in the process which requires a decomposition of the $d = 9$ operator. Such a decomposition has been worked out in Ref. 15 where two different topologies (topology 1 with two fermions and a boson in the internal lines and topology 2 with an internal 3-boson-vertex) have been specified. This decomposition was applied to the LHC analogue of $0\nu\beta\beta$ decay and first results for topology 1 have been derived in Refs. 9 and 10. The conclusion reached was that with the exception of leptoquark exchange, the LHC was typically more sensitive than $0\nu\beta\beta$ decay on the short range operators. Thus one could infer that typically and with some exceptions:

- *Either* an observation of $0\nu\beta\beta$ decay would imply an LHC signal of LNV as well. In turn, no sign of LNV at the LHC would exclude an observation of $0\nu\beta\beta$ decay.
- *Or* $0\nu\beta\beta$ decay would be triggered by a long-range mechanism (a) or (b).

4. Baryon Asymmetry Washout

An observation of LNV at low energies has important consequences for a preexisting lepton asymmetry in the Universe. For example, the prominent leptogenesis scenario for a generation of the BAU assumes a lepton number (or $B-L$) and CP asymmetry created in the decays of heavy Majorana neutrinos in the early Universe, which later on is converted into a baryon asymmetry by the nonperturbative $B+L$ violating sphaleron transitions present in the SM. Obviously, such a lepton asymmetry can be washed out by lepton number violating interactions, and indeed in Ref. 16 it has been pointed out that any observation of LNV at the LHC will falsify high-scale leptogenesis.

The basic argument is that the observation of LNV at the LHC will yield a lower bound on the washout factor for the lepton asymmetry in the early Universe. It is easy to see that this argument can be extended even further.

Just like the combination of $B-L$ violating ν_R decays in leptogenesis with $B+L$ violating sphaleron processes can produce a baryon asymmetry, $B-L$ violation observed, e.g. at the LHC or elsewhere in combination with $B+L$ violating sphaleron processes will lead to a washout of any preexisting baryon asymmetry, irrespective of the concrete mechanism of baryogenesis.

Combining this argument with the results of Refs. 9 and 10 discussed above, one can argue that an observation of short-range $0\nu\beta\beta$ decay will typically imply that LNV processes should be detected at the LHC as well, and this in turn will falsify leptogenesis and in general any high-scale scenario of baryogenesis.

Indeed, such arguments are not new. They have been first discussed in Ref. 17 and later on used, e.g. to constrain neutrino Majorana masses,[18] light lepton number violating sneutrinos[19] or Majorana mass terms for 4th generation neutrino states.[20] However, only quite recently it has been realized in Ref. 21 that the argument can be shown to apply for all short range contributions (d) and also for the long-range contribution (b) in Fig. 2. It has been shown that the $\Delta L = 2$ processes induced by the operator $\mathcal{O}_D$ can be considered to be in equilibrium and the washout of the lepton asymmetry is effective if

$$\frac{\Gamma_W}{H} = c'_D \frac{\Lambda_{\rm Pl}}{\Lambda_D}\left(\frac{T}{\Lambda_D}\right)^{2D-9} \gtrsim 1\,, \tag{5}$$

where Λ_D is the scale of the associated effective operator (assumed to be generated at tree level) from Eq. (3) and c'_D being a prefactor of order $\mathcal{O}(10^{-3}-1)$.

Thus the far-reaching and strong conclusion can be drawn that the observation of *any* new physics mechanism (i.e. not the mass mechanism) of neutrinoless double beta decay will typically exclude *any* high-scale generation of the BAU.

Even more recently further studies have been published which analyze the relation of LNV and the BAU in concrete models such as left–right symmetry or low energy seesaw models.[23,24]

5. Loopholes

Of course these arguments are rather general and various loopholes exist in specific models. These include:

- Scenarios where LNV is confined to a specific flavor sector only. For example, $0\nu\beta\beta$ decay probes $\Delta L_e = 2$ LNV, only. It may be possible for example that lepton number could still be conserved in the τ flavor which is not necessarily in equilibrium with the e and μ flavors in the early Universe.[16] It has been discussed in Ref. 21, however, that an observation of LFV decays such as $\tau \to \mu\gamma$

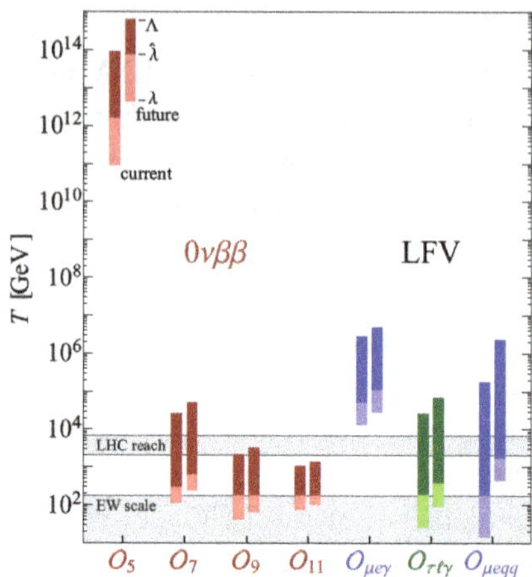

Fig. 4. Temperature intervals where the given LNV and LFV operators are in equilibrium, assumed that the corresponding process is observed at the current (future) experimental sensitivity (from Ref. 21).

may require LFV couplings large enough to wash out such a flavor specific lepton asymmetry when combined with LNV observed in a different flavor sector. In Fig. 4, the temperature intervals are shown where two individual flavor number asymmetries are equilibrated by LFV processes. When this interval overlaps with the $\Delta L = 2$ washout interval of one net flavor number (i.e. electron number if $0\nu\beta\beta$ is observed), the net number of the other flavor will be efficiently washed out as well. As can be seen, if $\tau \to \ell\gamma$ or $\mu - e$ conversion in nuclei was observed, the involved flavors would be equilibrated around the same temperatures as the washout from the LNV operators.

- Models with hidden sectors, new symmetries and/or conserved charges may stabilize a baryon asymmetry against LNV washout as suggested for the example of hypercharge in Ref. 22.
- Lepton number may be broken at a scale below the electroweak phase transition where sphalerons are no longer active.

It should be realized though that in general an observation of low energy LNV would invalidate any high-scale generation of the baryon asymmetry and that the aforementioned protection mechanisms should be addressed explicitly in any model combining low-scale LNV with high-scale baryogenesis.

6. Conclusions

By simply combining the arguments made above, we can conclude as follows.

If neutrinoless double beta decay is observed, it is:

- *Either* due to a long-range mechanism, e.g. a light Majorana neutrino mass.

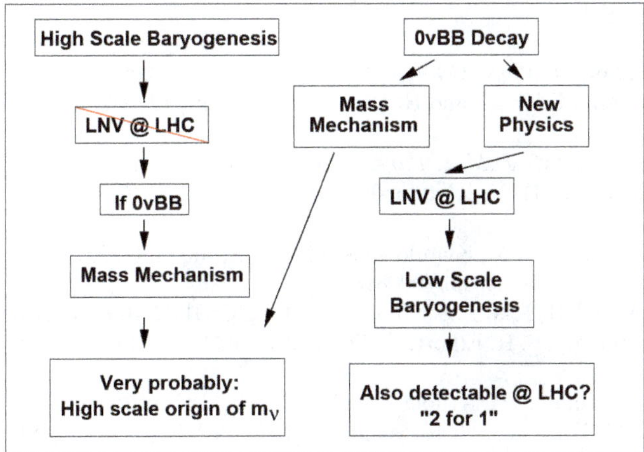

Fig. 5. Conclusions as a logic tree: a discovery of LNV at low energies will have far-reaching consequences for the origin of the BAU. On the other hand, if the Baryon Asymmetry is generated by a high-scale mechanism of baryogenesis interesting consequences for the search of low energy LNV and the origin of neutrino masses can be deduced.

- *Or* due to a short-range mechanism. In this case it is very probable that lepton number is observed at the LHC. This, however, implies that baryogenesis is a low-scale phenomenon which also may be observable at the LHC. In this case there thus may well be a "two-for-one" deal at the LHC.

If, on the other hand, the BAU is generated at a high-scale, there will be no LNV at the LHC. If, in this case, neutrinoless double beta decay is observed, it thus will be typically due to a long-range mechanism. In combination with the assumption that we do not have a hint for LNV at a low-scale in this case and on the other hand a mechanism for the generation of the BAU at a high-scale, this will probably point towards a high-scale origin of the neutrino mass as well, such as a vanilla-type seesaw mechanism in combination with leptogenesis.

Thus, in summary, an observation of neutrinoless double beta decay will typically (see Fig. 5):

- *Either* imply LNV at the LHC and low-scale baryogenesis and thus a possible observation of both processes in the near future.
- *Or* very probably a high-scale origin of both neutrino masses and baryogenesis.

We thus think that even if possible loopholes to these arguments may exist, it is important to stress these relations to make both model builders and experimentalists aware of the tight connections between neutrinoless double beta decay, the search for LNV at the LHC and the origin of the BAU.

References

1. W. Rodejohann, *J. Phys. G* **39**, 124008 (2012), arXiv:1206.2560 [hep-ph].
2. F. F. Deppisch, M. Hirsch and H. Päs, *J. Phys. G* **39**, 124007 (2012), arXiv:1208.0727 [hep-ph].
3. W. Rodejohann, *Int. J. Mod. Phys. E* **20**, 1833 (2011), arXiv:1106.1334 [hep-ph].
4. H. Päs, M. Hirsch, H. V. Klapdor-Kleingrothaus and S. G. Kovalenko, *Phys. Lett. B* **453**, 194 (1999).
5. H. Päs, M. Hirsch, H. V. Klapdor-Kleingrothaus and S. G. Kovalenko, *Phys. Lett. B* **498**, 35 (2001), arXiv:hep-ph/0008182.
6. F. Deppisch and H. Päs, *Phys. Rev. Lett.* **98**, 232501 (2007), arXiv:hep-ph/0612165.
7. V. M. Gehman and S. R. Elliott, *J. Phys. G* **34**, 667 (2007) [Erratum: *ibid.* **35**, 029701 (2008)].
8. SuperNEMO Collab. (R. Arnold *et al.*), *Eur. Phys. J. C* **70**, 927 (2010).
9. J. C. Helo, M. Hirsch, S. G. Kovalenko and H. Päs, *Phys. Rev. D* **88**, 011901 (2013), arXiv:1303.0899 [hep-ph].
10. J. C. Helo, M. Hirsch, H. Päs and S. G. Kovalenko, *Phys. Rev. D* **88**, 073011 (2013), arXiv:1307.4849 [hep-ph].
11. B. C. Allanach, C. H. Kom and H. Päs, *Phys. Rev. Lett.* **103**, 091801 (2009), arXiv:0902.4697 [hep-ph].
12. B. C. Allanach, C. H. Kom and H. Päs, *J. High Energy Phys.* **0910**, 026 (2009), arXiv:0903.0347 [hep-ph].
13. W. Y. Keung and G. Senjanovic, *Phys. Rev. Lett.* **50**, 1427 (1983).
14. V. Tello, M. Nemevsek, F. Nesti, G. Senjanovic and F. Vissani, *Phys. Rev. Lett.* **106**, 151801 (2011), arXiv:1011.3522 [hep-ph].
15. F. Bonnet, M. Hirsch, T. Ota and W. Winter, *J. High Energy Phys.* **1303**, 055 (2013) [*J. High Energy Phys.* **1404**, 090 (2014)], arXiv:1212.3045 [hep-ph].
16. F. F. Deppisch, J. Harz and M. Hirsch, *Phys. Rev. Lett.* **112**, 221601 (2014), arXiv:1312.4447 [hep-ph].
17. M. Fukugita and T. Yanagida, *Phys. Rev. D* **42**, 1285 (1990).
18. G. Gelmini and T. Yanagida, *Phys. Lett. B* **294**, 53 (1992).
19. H. V. Klapdor-Kleingrothaus, S. Kolb and V. A. Kuzmin, *Phys. Rev. D* **62**, 035014 (2000), arXiv:hep-ph/9909546.
20. S. Hollenberg, H. Pas and D. Schalla, arXiv:1110.0948 [hep-ph].
21. F. F. Deppisch, J. Harz, M. Hirsch, W. C. Huang and H. Päs, arXiv:1503.04825 [hep-ph].
22. A. Antaramian, L. J. Hall and A. Rasin, *Phys. Rev. D* **49**, 3881 (1994), arXiv:hep-ph/9311279.
23. M. Dhuria, C. Hati, R. Rangarajan and U. Sarkar, arXiv:1503.07198 [hep-ph].
24. P. S. Bhupal Dev, C. H. Lee and R. N. Mohapatra, arXiv:1503.04970 [hep-ph].

Status of the MAJORANA DEMONSTRATOR: A Search for Neutrinoless Double-Beta Decay

Yu. Efremenko,[*,1] N. Abgrall,[2] I. J. Arnquist,[3] F. T. Avignone III,[4,5] C. X. Baldenegro-Barrera,[5] A. S. Barabash,[6] F. E. Bertrand,[5] A. W. Bradley,[2] V. Brudanin,[7] M. Busch,[8,9] M. Buuck,[10] D. Byram,[11] A. S. Caldwell,[12] Y.-D. Chan,[2] C. D. Christofferson,[12] C. Cuesta,[10] J. A. Detwiler,[10] H. Ejiri,[13] S. R. Elliott,[14] A. Galindo-Uribarri,[5] T. Gilliss,[9,15] G. K. Giovanetti,[9,15] J. Goett,[14] M. P. Green,[5] J. Gruszko,[10] I. Guinn,[10] V. E. Guiseppe,[4] R. Henning,[9,15] E. W. Hoppe,[3] S. Howard,[12] M. A. Howe,[9,15] B. R. Jasinski,[11] K. J. Keeter,[16] M. F. Kidd,[17] S. I. Konovalov,[6] R. T. Kouzes,[3] B. D. LaFerriere,[3] J. Leon,[10] J. MacMullin,[9,15] R. D. Martin,[11] S. J. Meijer,[9,15] S. Mertens,[2] J. L. Orrell,[3] C. O'Shaughnessy,[9,15] A. W. P. Poon,[2] D. C. Radford,[5] J. Rager,[9,15] K. Rielage,[14] R. G. H. Robertson,[10] E. Romero-Romero,[1,5] B. Shanks,[9,15] M. Shirchenko,[7] N. Snyder,[11] A. M. Suriano,[12] D. Tedeschi,[4] J. E. Trimble,[9,15] R. L. Varner,[5] S. Vasilyev,[7] K. Vetter,[2] K. Vorren,[9,15] B. R. White,[5] J. F. Wilkerson,[5,9,15] C. Wiseman,[4] W. Xu,[14] E. Yakushev,[7] C.-H. Yu,[5] V. Yumatov[6] and I. Zhitnikov[7]

(The MAJORANA Collaboration)

[1]*Department of Physics and Astronomy, University of Tennessee, Knoxville, TN, USA*

[2]*Nuclear Science Division, Lawrence Berkeley National Laboratory, Berkeley, CA, USA*

[3]*Pacific Northwest National Laboratory, Richland, WA, USA*

[4]*Department of Physics and Astronomy, University of South Carolina, Columbia, SC, USA*

[5]*Oak Ridge National Laboratory, Oak Ridge, TN, USA*

[6]*Institute for Theoretical and Experimental Physics, Moscow, Russia*

[7]*Joint Institute for Nuclear Research, Dubna, Russia*

[8]*Department of Physics, Duke University, Durham, NC, USA*

[9]*Triangle Universities Nuclear Laboratory, Durham, NC, USA*

[10]*Center for Experimental Nuclear Physics and Astrophysics, and Department of Physics, University of Washington, Seattle, WA, USA*

[11]*Department of Physics, University of South Dakota, Vermillion, SD, USA*

[12]*South Dakota School of Mines and Technology, Rapid City, SD, USA*

[13]*Research Center for Nuclear Physics and Department of Physics, Osaka University, Ibaraki, Osaka, Japan*

[14]*Los Alamos National Laboratory, Los Alamos, NM, USA*

[15]*Department of Physics and Astronomy, University of North Carolina,*

Chapel Hill, NC, USA
[16]*Department of Physics, Black Hills State University, Spearfish, SD, USA*
[17]*Tennessee Tech University, Cookeville, TN, USA*
**yefremen@utk.edu*

If neutrinos are Majorana particles, i.e. fermions that are their own antiparticles, then neutrinoless double-beta ($0\nu\beta\beta$) decay is possible. In such a process, two neutrons can simultaneously decay into two protons and two electrons without emitting neutrinos. Neutrinos being Majorana particles would explicitly violate lepton number conservation, and might play a role in the matter–antimatter asymmetry in the universe. The MAJORANA DEMONSTRATOR experiment is under construction at the Sanford Underground Research Facility in Lead, SD and will search for the neutrinoless double-beta ($0\nu\beta\beta$) decay of the ^{76}Ge isotope. The goal of the experiment is to demonstrate that it is possible to achieve a sufficiently low background rate in the 4 keV region of interest (ROI) around the 2039 keV Q-value to justify building a tonne-scale experiment. In this paper, we discuss the physics and design of the MAJORANA DEMONSTRATOR, its approach to achieving ultra-low background and the status of the experiment.

Keywords: Neutrinoless double beta decay; germanium detector; Majorana.

1. Introduction

Since the discovery and confirmation of neutrino oscillations by Super-Kamiokande,[1] the Sudbury Neutrino Observatory (SNO),[2] KamLAND,[3] and other experiments,[4] a nonzero neutrino mass has been recognized as an indication of physics beyond the Standard Model of particle physics. If neutrinos are Majorana particles, neutrinoless double-beta ($0\nu\beta\beta$) decay should take place in some even–even nuclei where single-beta decay is energetically forbidden or highly suppressed. In this process, two neutrons simultaneously decay into two protons and two electrons without emitting neutrinos. This would explicitly violate lepton number conservation.[5–7] $0\nu\beta\beta$ decay experiments are the only way to unambiguously establish the Majorana or Dirac nature of neutrinos and the observation of $0\nu\beta\beta$ decay would have a profound impact on the paradigm of particle and nuclear physics.

$0\nu\beta\beta$ decay requires neutrinos to be massive and their own antiparticles. A direct relationship can be established between $0\nu\beta\beta$ decay half life and the effective Majorana neutrino mass as

$$\left(T_{1/2}^{0\nu}\right)^{-1} = G^{0\nu}|M_{0\nu}|^2\left(\frac{\langle m_{\beta\beta}\rangle}{m_e}\right)^2, \tag{1}$$

where $G^{0\nu}$ is the phase space of the decay, $M_{0\nu}$ is the nuclear matrix element and $\langle m_{\beta\beta}\rangle = |\Sigma_{i=1}^{3}U_{ei}^2 m_i|$ is the effective Majorana neutrino mass with U_{ei} being the Pontecorvo–Maki–Nakagawa–Sakata mixing matrix.[4] Therefore, measurement of the $0\nu\beta\beta$ decay half life is a complementary way to probe the absolute neutrino mass scale. For example, the inverted mass hierarchy has an effective neutrino

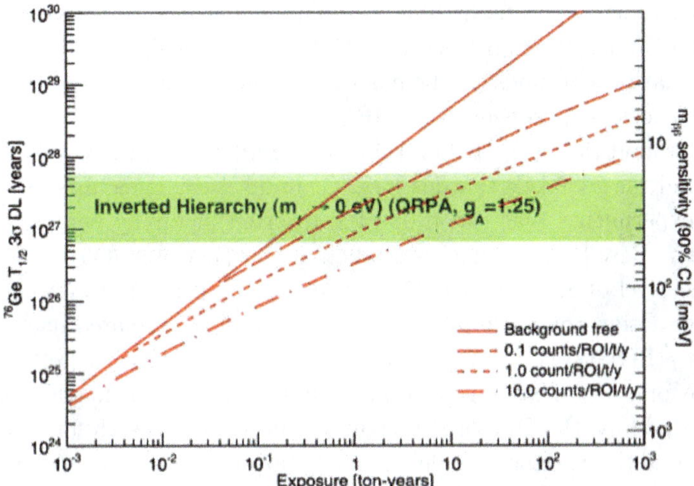

Fig. 1. Sensitivity to the $0\nu\beta\beta$ half life for various assumptions for the background level in the ROI of 4 keV.

mass between about 15 meV and 50 meV, and the lower boundary corresponds to a ^{76}Ge $0\nu\beta\beta$ decay half life longer than 10^{27} years.

In recent decades, many experiments have searched for $0\nu\beta\beta$ decay in several isotopes including: ^{76}Ge, ^{136}Xe, ^{130}Te, and others. For recent reviews of $0\nu\beta\beta$ decay experiments, see Refs. 7 and 9. All of these isotopes also undergo two-neutrino double-beta ($2\nu\beta\beta$) decay with typical half-lives on the order of 10^{19} to 10^{24} years.[9–13] Similar to a single-beta decay spectrum, the sum energy of the two electrons in $2\nu\beta\beta$ decay is a continuum extending up to the Q-value ($Q_{\beta\beta}$) of the decay. In contrast, due to the absence of neutrinos, the two electrons in $0\nu\beta\beta$ decay carry almost all of the energy released in the decay, and their total kinetic energy is peaked at the Q-value. Compared to other detection systems, High-Purity Germanium (HPGe) detectors have the critical advantage of excellent energy resolution, which eliminates the otherwise almost irreducible $2\nu\beta\beta$ decay background and improves the discovery potential due to the clean line signature.

To have high sensitivity in the search for $0\nu\beta\beta$ decay, an experiment should be virtually background free in a narrow window (the region of interest) around the Q-value. Figure 1 shows the large effect on discovery potential of even a few background events in the region of interest (ROI).

2. The MAJORANA DEMONSTRATOR

The MAJORANA Collaboration is constructing the MAJORANA DEMONSTRATOR (MJD) in class-100 clean rooms located on the 4850 feet underground level at the Sanford Underground Research Facility (SURF)[a] located in Lead, SD. The collabo-

[a]http://www.sanfordlab.org; also see Ref. 14.

ration plans to deploy a total of 40-kg Germanium detectors into two modules, with 30 kg of the Ge detectors enriched to 87% ^{76}Ge. The goal of the DEMONSTRATOR is to show that backgrounds can be made low enough to justify feasibility of a large double beta decay experiment using HPGe.

A background rate at or below 1.0 count/(ROI·t·y) in the 4 keV region of interest around the 2039 keV Q-value for ^{76}Ge $0\nu\beta\beta$ decay is required for tonne-scale Ge-based experiments that can probe the inverted hierarchy parameter space for $0\nu\beta\beta$ decay. Based on simulation studies, if the DEMONSTRATOR can achieve a background level of 3.0 counts/(ROI·t·y) or lower, then a tonne-scale experiment with similar design and material is expected to have the required background level of 1.0 count/(ROI·t·y) or better, thanks to various effects such as self-shielding and longer time for decay of cosmogenic backgrounds such as ^{68}Ge. In addition to searching for $0\nu\beta\beta$ decay, the DEMONSTRATOR will be used to search for a range of new physics beyond the Standard Model, including low mass dark matter and axions.[8,15]

A modular approach was chosen for the MJD for its natural expandability to larger scale experiments. In the current approach, four or five HPGe detectors are stacked together to form a string, and seven strings are mounted into a single cryostat with dedicated cryogenic and vacuum systems, together referred to as a module. The construction of the DEMONSTRATOR is organized in three phases. First, a prototype module with three strings of natural HPGe detectors was constructed in 2014 and has been taking commissioning data since August of 2014. By the middle of 2015, the first production module with more than half of the total enriched HPGe detectors and some natural detectors, with a total mass of 20 kg of Ge, will be completed. By late 2015, the second production module with the rest of HPGe detectors will be constructed, also with a total mass of 20 kg of Ge. Thanks to separate support systems, those modules can be individually operated and maintained, and data-taking will begin as soon as each module is ready.

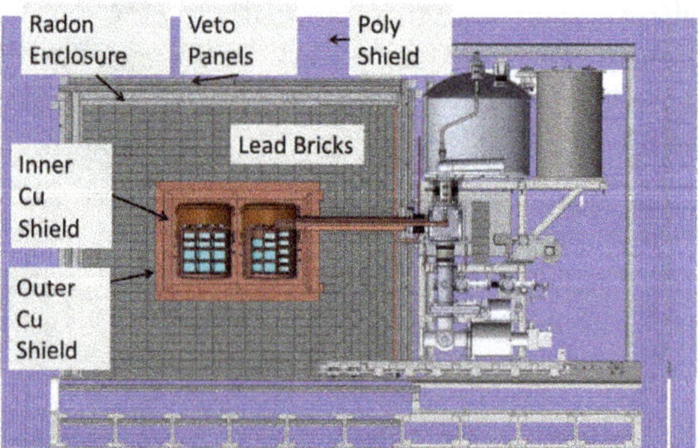

Fig. 2. Configuration of the MAJORANA DEMONSTRATOR. The experimental apparatus is sitting on the top of a stainless-steel over-floor with orthogonal channels, allowing HDPE panels (poly shield) and veto panels underneath the apparatus. Detectors located in two cryostats are surrounded by several layers of passive and active shields. See the text for detailed discussions of the shield. Overall height of the assembly is about two meters.

2.1. *Shield*

MJD utilizes both active and passive shielding, as shown in Fig. 2. The outermost component of the shield configuration is comprised of layers of high density polyethylene (HDPE) panels, two layers of which are borated. Upon completion, the poly shield will enclose the entire apparatus, including the support systems of each module. Inside the poly shield, there are two layers of active veto panels on all six sides of the apparatus. Veto panels are made out of one inch thick plastic scintillator, and read out by photo-multiplier tubes (PMTs). Each panel is read out by a single half-inch PMT via wavelength shifting fibers. Detailed design of veto panels is discussed at.[18] A total of 32 veto panels will be installed for the completed shield. At present, veto panels have been installed and commissioned on four sides of the shield. Constant monitoring of the muon flux at the 4850' level has been implemented by the MJD Data-Acquisition (DAQ) system and the veto data has been combined with the HPGe detector data-stream. Inside the veto panels, a Radon exclusion box made of aluminum has been constructed to provide a nearly Rn-free environment. A nitrogen gas delivery system that constantly purges the Rn box with boil-off nitrogen gas was implemented. Further improvements are currently being implemented to use a cold, clean charcoal trap to filter out the remaining Rn in the nitrogen purge gas. Inside the Rn box is a 45 cm thick lead shield composed of $5.1 \times 10.2 \times 20.3$ cm^3 lead bricks, which are carefully stacked in a pattern that eliminates any direct path for photons originating from outside the shield. The main body of the lead shield has already been completed, as shown in the left side of Fig. 3. In the photo, keyed structures can be seen on two sides of the lead shield.

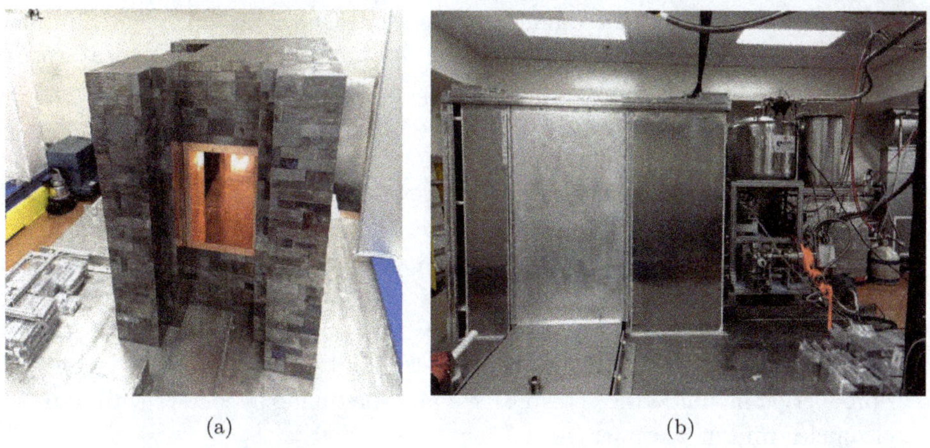

(a) (b)

Fig. 3. (a): A photograph of the main body of the lead shield and outer copper shield taken before it was covered by the radon exclusion box. (b): A photograph of the nearly completed shield, inside which the prototype module has been taking commissioning data. The support systems of the prototype module can be seen in the photo on the right of the shield. The current shielding includes four sides of veto panels, a radon exclusion box purged with nitrogen gas, a complete lead shield, and an outer copper shield.

When cryostats are inserted into the shield, additional lead shielding matching the keyed structures are also installed, completing the lead shield. Inside the lead shield, an outer copper shield made of Oxygen-Free High thermal Conductivity (OFHC) copper has already been installed. An inner copper shield made of electroformed copper will be installed to directly enclose the cryostats. For more details of the shield design, see Ref. 8. Since August 2014, the MJD prototype module has been taking commissioning data in the nearly completed shield described here, as shown in the photo on the right side of Fig. 3.

2.2. *Detectors, strings, and modules*

The collaboration has chosen to use P-type Point Contact (P-PC) HPGe detectors.[16,17] As compared to the more conventional coaxial HPGe detectors, P-PC detectors have several important advantages for a rare-event search. Their small readout contact produces a highly localized "weighting potential," and this combined with a larger range of charge drift times allows for excellent pulse-shape discrimination sensitivity to distinguish between multi-site and single-site events. This suppresses the main background of Compton-scattered gamma rays. Both P-PC and p-type coaxial detectors also have a thick outer lithium contact that fully absorbs alpha particles, further reducing the background. Lastly, the small capacitance of PPC detectors greatly improves the energy resolution at low energies, lowering the energy thresholds and allowing for low-mass dark matter searches.

A total of 14 production strings of HPGe detectors will be constructed, seven for each production module. The construction of strings for the first module is complete. In addition, three strings with a total of ten natural HPGe detectors of both Broad Energy Germanium Detectors (BEGe) from CANBERRA[19] and ORTEC[20] types have been constructed for the prototype module and are being used for commissioning the experimental apparatus.

Each MJD module consists of a cryostat with detector strings, cryogenic and vacuum systems, parts of the shield and its own calibration system. Each module can be moved as a whole inside the clean room on top of a movable air bearing table with minimum vibrations, and together they are referred to as a monolith. The cryogenic system is a two-phase closed-loop thermosyphon using nitrogen for thermal mass transport. The nitrogen is condensed at one end by a heat reservoir of liquid nitrogen (LN), and is thermally connected to a cold plate inside the cryostat dewar where heat is transported from the detectors by evaporating the nitrogen.[23] For production modules, the cryostat itself and the tubing in contact with it are all made of electroformed copper.

To validate the apparatus design and to debug issues in the construction and operation of modules and the DAQ systems, a prototype module was constructed and has been commissioned. The prototype module is identical to production modules, except it is made of commercial OFHC copper. The vacuum system of the prototype module will be reused for the second production module. Ten natural

HPGe detectors of both BEGe and ORTEC types are arranged into three strings and mounted onto the cold plate inside the prototype module cryostat, as shown in the left panel of Fig. 4.

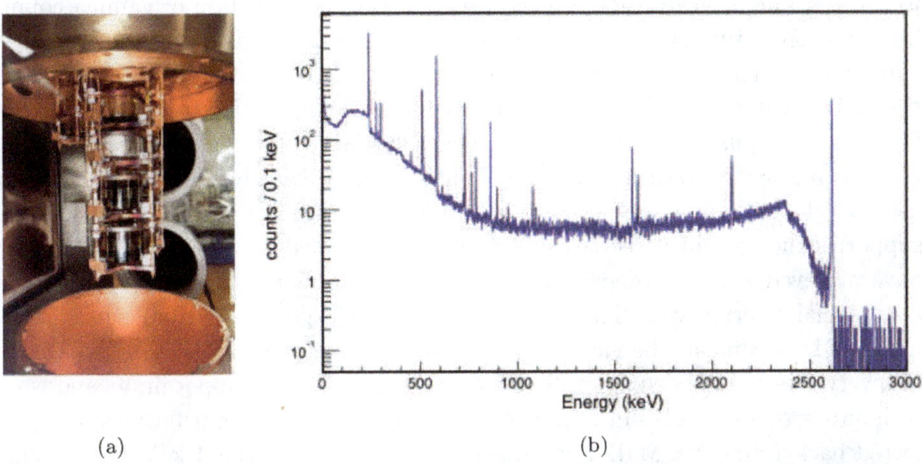

Fig. 4. (a) A photograph inside the prototype cryostat, which is opened up to allow the mounting of three strings of HPGe detectors. The string in the front has two Canberra BEGe detectors on the top and two ORTEC detectors at the bottom. (b) A spectrum taken by one of the HPGe detectors in the prototype cryostat during a calibration run with a ^{228}Th line source.

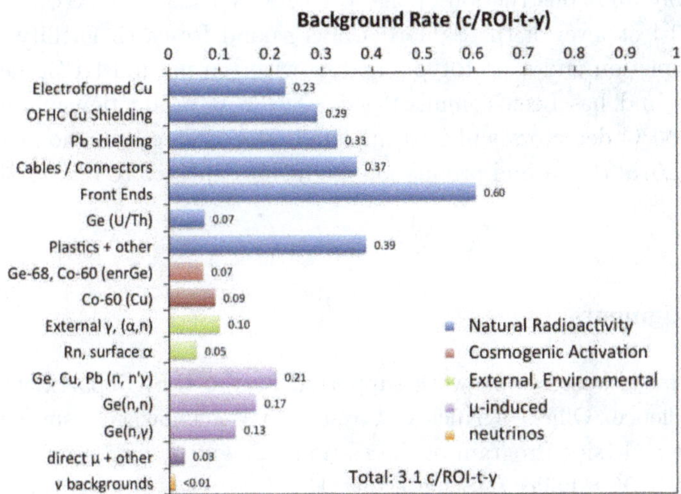

Fig. 5. Estimated background contributions in the $0\nu\beta\beta$ decay ROI for the MJD. The estimations come from simulation studies based on material assay results as well as existing measurements of various physics processes, and some of the estimations are upper limits. The contributions sum to $\leq$ 3.1 counts/(ROI-t-y) in the DEMONSTRATOR.

2.3. *Background estimation*

A vigorous program to assay the natural radioactivity in almost all materials used in the MJD experiment has provided a foundation for accurate background projections, summarized in Fig. 5. For assay we use a combination of gamma counting, neutron activation analysis, inductively coupled plasma mass spectrometry, and glow discharge mass spectrometry. The radioactivity of components close to the active HPGe detector volumes can contribute significantly to the background level. In particular, ultra pure copper electroformed underground has both much reduced primordial radioactivity and limited cosmogenically produced ^{60}Co. By using electroformed copper at some locations closest to the detectors, such as the support structures in detector units and strings as well as the cryostat itself, we have achieved several orders-of-magnitude background reduction over the use of commercial alternatives. The biggest projected background is from natural uranium and thorium in the electronics front-end, cables and connectors, which are also very close to the detectors. Some of the estimations are upper limits and work to improve them are still in progress. Summarizing all the contributions, the projected background for MJD is $\leq$ 3.1 counts/(ROI-t-y) in the 4 keV ROI around 2039 keV. More details about the MJD background model can be found in Refs. 8 and 25.

3. Summary

The assembly and construction of the MAJORANA DEMONSTRATOR is proceeding at the 4850 feet level of the Sanford Underground Research Facility. The shield is near completion and a prototype module with ten natural HPGe detectors was constructed and has been commissioned. The first production module with 20-kg HPGe P-PC detectors will be completed and enter commissioning before the summer of 2015; the second production module is expected to be completed in late 2015.

Acknowledgments

This material is based upon work supported by the U.S. Department of Energy, Office of Science, Office of Nuclear Physics. We acknowledge support from the Particle Astrophysics Program of the National Science Foundation. This research uses these US DOE Office of Science User Facilities: the National Energy Research Scientific Computing Center and the Oak Ridge Leadership Computing Facility. We acknowledge support from the Russian Foundation for Basic Research. We thank our hosts and colleagues at the Sanford Underground Research Facility for their support.

References

1. Super-Kamiokande Collab. (Y. Fukuda *et al.*), *Phys. Rev. Lett.* **81**, 1562 (1998).
2. SNO Collab. (Q. R. Ahmad *et al.*), *Phys. Rev. Lett.* **87**, 071301 (2001).
3. KamLAND Collab. (K. Eguchi *et al.*), *Phys. Rev. Lett.* **90**, 021802 (2002).
4. Particle Data Group (K. A. Olive *et al.*), *Chin. Phys. C* **38**, 090001 (2014).
5. F. T. Avignone, III, S. R. Elliott and J. Engel, *Rev. Mod. Phys.* **80**, 481 (2008).
6. L. Camilleri, E. Lisi and J. F. Wilkerson, *Annu. Rev. Nucl. Part. Sci.* **58**, 343 (2008).
7. S. R. Elliott, *Mod. Phys. Lett. A* **27**, 1230009 (2012).
8. N. Abgrall *et al.*, *Adv. High Energy Phys.* **2014**, 365432 (2014).
9. B. Schwingenheuer, *Ann. Phys. (Berlin)* **525**, 4 (2013).
10. A. S. Barabash, *Nucl. Phys. A* **935**, 52 (2015).
11. M. Goeppert-Mayer, *Phys. Rev.* **48**, 512 (1935).
12. S. R. Elliott, A. A. Hahn and M. K. Moe, *Phys. Rev. Lett.* **59**, 18 (1987).
13. R. Saakyan, *Annu. Rev. Nucl. Part. Sci.* **63**, 503 (2013).
14. J. Heise, *AIP Conf. Proc.* **1604**, 331 (2014).
15. G. K. Giovanetti *et al.*, A dark matter search with MALBEK, arXiv:1407.2238.
16. P. N. Luke, F. S. Goulding, N. W. Madden and R. H. Pehl, *IEEE Trans. Nucl. Sci.* **36**, 926 (1989).
17. P. S. Barbeau, J. I. Collar and O. Tench, *J. Cosmol. Astropart. Phys.* **09**, 009 (2007).
18. W. Bugg, Yu. Efremenko and S. Vasilyev, *NIM A* **758**, 91 (2014).
19. Canberra Industries, Meriden, CN, USA, 2009.
20. ORTEC, Oak Ridge, TN, USA, 2009.
21. W. Xu *et al.*, Testing the Ge detectors for the Majorana Demonstrator, arXiv:1404.7399.
22. E. W. Hoppe *et al.*, *NIM A* **764**, 116 (2014).
23. E. Aguayo *et al.*, *NIM A* **709**, 17 (2013).
24. M. A. Howe *et al.*, *IEEE Trans. Nucl. Sci.* **51**, 878 (2004).
25. C. Cuesta *et al.*, Background model of the Majorana Demonstrator, arXiv:1405.1370.

Towards Neutrino Mass Spectroscopy Using Atoms/Molecules

M. Yoshimura

Center of Quantum Universe, Faculty of Science, Okayama University, Tsushima-naka 3-1-1 Kita-ku Okayama 700-8530 Japan

In this short talk I shall explain the most important aspects of our new project that aims at determination of remaining matrix elements of the neutrino mass and at detection of 1.9 K relic neutrino. The recent experimental result of the verified macro-coherence amplification of weak QED rate and prospects towards our final goals are also briefly touched upon.

1. Introduction

The conventional target for exploration of neutrino parameters has been nuclei. The problem of nuclei as targets is the remoteness of available energies of a few to several MeV from the expected neutrino mass of sub-eV range. Our Okayama group proposed a new experimental principle using atoms or molecules as targets, initiated R and D works, and recently succeeded in experimentally verifying the crucial mechanism of rate amplification in weak QED process: the macro-coherence. I shall cover in this talk theoretical aspects of the precision neutrino mass spectroscopy, the recent experimental result, and how to proceed further towards our goal in neutrino physics.

A great merit of atoms/molecules for a next generation of neutrino physics is a rich variety of their available energies, less than 10 eV downward to the THz range or even much less. Also, a great variety of experimental tools to manipulate their level spacings and their properties (parity, angular momentum mixing, etc.) exist: notably the magnetic field and the electric field, both completely static or rf range. The process we use is atomic de-excitation producing a photon and a neutrino pair: $|e\rangle \rightarrow |g\rangle + \gamma + \nu\bar{\nu}$. The process is called by us RENP (radiative neutrino pair emission). Since neutrino detection of this low energy is hopeless, it is important to measure the photon spectrum with precision.

A disadvantage is that atomic weak process is very much suppressed due to the small level spacing. Thus, it is crucial to amplify rates. We have proposed the macro-coherence amplification mechanism which I shall explain. The macro-coherent RENP occurs in a long target irradiated by two excitation lasers and a trigger light whose energy is set equal to that of the detected photon. When this macro-coherence amplification works, atomic de-excitation process conserves both the energy and the momentum, thereby making kinematics much like elementary particle decays. This gives in principle six mass thresholds of photon emission at $\epsilon_{eg}/2 - (m_i + m_j)^2/(2\epsilon_{eg})$ where ϵ_{eg} is the level spacing and $m_i, i = 1, 2, 3$ is the ith neutrino mass value. The small spacing $(m_i + m_j)^2/(2\epsilon_{eg})$ of threshold steps can be measured with a great resolution of trigger laser frequency, and a high accuracy of photon detection energy is not required.

We shall first show what can be achieved if the macro-coherence mechanism works, and later explain how this can be achieved.

2. Sensitivity of RENP Photon Spectrum on Neutrino Parameters

We show in Figs. 1 and 2 the photon energy spectrum of Xe de-excitation from ~ 8.4 eV level as a whole and in the threshold regions, respectively, calculated according to the nuclear monopole contribution.[2] Only Z-boson exchange contributes in the nuclear monopole diagram, hence there are three neutrino mass thresholds as is clear in this figure. Rates of the nuclear monopole contribution is much larger than the valence electron spin contribution, because RENP amplitude is proportional to the electroweak charge of nuclei, $Q_w = N - (1 - 4\sin^2\theta_w)Z \sim N$ (N, Z are the neutron and the proton number of a nucleus) and is large for heavy atoms.

The overall photon spectrum has a feature similar to the decay spectrum of an elementary particle such as the muon decay into three nearly massless particles $\mu \to e + \nu_\mu \bar{\nu}_e$. A high resolution spectrum near thresholds however exhibits remarkable three-step thresholds. With sufficient statistics of data, one should be able to determine all three neutrino masses with precision. Moreover, it is relatively easy to determine the hierarchical mass pattern, the normal vs the inverted mass hierarchy.

The next easier physics item is distinction of the Majorana from the Dirac neutrino. The principle of the distinction is existence or absence of effect of identical fermion. This causes difference of predicted spectrum shape and from fitting with experimental data one should be able to tell which mass type is favored by nature. In practice, a small level spacing is required for this distinction, as indicated in Ref. 3.

A further challenge awaits us for determination of CPV (CP violating) phases intrinsic to Majorana neutrinos, because one has to rely on the valence spin current contribution with much smaller rates for this measurement.

3. Detection of Relic Neutrino of 1.9 K

About 300 cm^{-3} neutrinos, all flavors added, exist on the average in the entire universe (ignoring the gravitational clustering of massive neutrino within our galaxy). It obeys a quasi-thermal Fermi–Dirac distribution $f(\vec{p}/T_\nu)$ of 1.9 K temperature as if they were massless, ignoring a small effect of possible finite chemical potential.

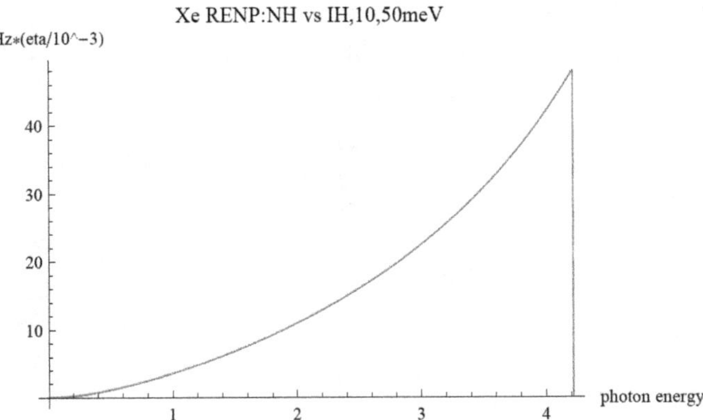

Fig. 1. Xe RENP spectrum in the whole photon energy range. $(5p^+6s^-)_{J=1}$ (hole$^+$–electron$^-$ system) of excitation energy ~ 8.4 eV is the initial state $|e\rangle$. Assumed parameters are the number density 7×10^{19} cm^{-3}, the target volume 10^2 cm^3, and the activity factor 10^{-3} expected from numerical simulations of macro-coherence development.

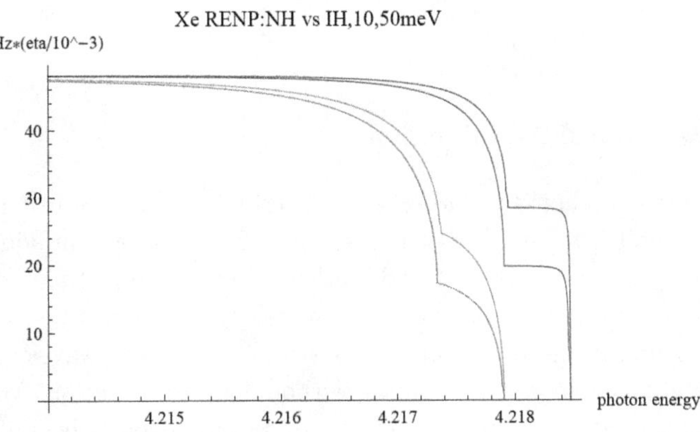

Fig. 2. Threshold region of Xe RENP corresponding to the same level and the same parameters of Fig. 1. Four different cases of normal and inverted hierarchical mass patterns of two smallest masses, 10 and 50 meV, are compared.

The ambient cosmic neutrino distorts RENP spectrum by the Pauli-blocking effect. If the momentum of emitted neutrino is zero, the reduction factor is maximal, because $1 - f = 1/2$ at the zero momentum. In practice, the energy given by the atomic level spacing is shared by three particles, the photon and the neutrino pair. The photon energy region most sensitive to the relic Pauli-blocking is the threshold region of neutrino pair emission of the smallest mass. The pair neutrino shares about a half of the level spacing, and their momenta cannot vanish simultaneously. Thus, the maximal suppression down to $1/4$ cannot occur and we need detailed spectrum calculation assuming a level spacing.

It is readily anticipated that if the atomic spacing ϵ_{eg} is close to $2m_0$ (m_0 the smallest neutrino mass), the Pauli-blocking effect becomes large. This is confirmed from Fig. 3 in which the case $\epsilon_{eg} = 2m_0 + 1$ meV is calculated.

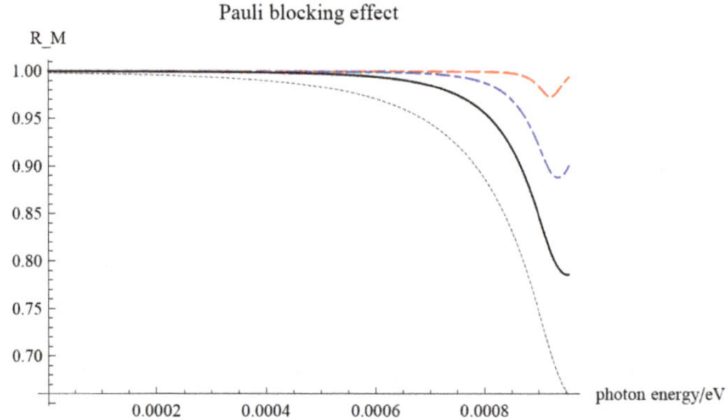

Fig. 3. (Color online) Spectral distortion caused by the Pauli-blocking due to relic neutrinos, $T_\nu = 1.9/2$ K in dashed red, 1.9 K in dash-dotted blue, 2.7 K in solid black and 1.9×2 K in dotted black, all assuming $m_0 = 5$ meV, $\epsilon_{eg} = 11$ meV and the zero chemical potential. The plotted quantity is the ratio of two rates, the one with the Pauli-blocking to the one without. The nuclear monopole contribution is used.

4. Macro-coherence Amplification

We shall now discuss how to evade the most serious problem of using atoms: small rates due to small level spacings. The small rate is true for individual and independent de-excitation of atoms. On the other hand, Dicke pointed out a long time ago that de-excitation of atoms within the wavelength size may be enhanced by the factor of N which is the total number of atoms within the wavelength, if atomic matrix elements share a common phase without phase relaxation. We pointed out that this enhancement may further be increased in multi-particle emission of nearly massless particles.[5]

Let us first explain this mechanism in an intuitive, but not completely rigorous way. A total rate of RENP from a collective body of excited atoms involves a quantum mechanical formula, $|\sum_a e^{i(\vec{k}+\vec{p}_1+\vec{p}_2)\cdot\vec{r}_a}\mathcal{M}_a|^2$, where emitted plane waves at atomic site $\vec{r}_a$ are explicitly written. We assume that atoms are distributed uniformly and take the continuum limit of this discrete sum. If the atomic phase relaxation at different atomic sites is slow in time and atomic matrix elements $\mathcal{M}_a$ are spatially slowly varying, then one may factor out this part of matrix elements, to obtain $(2\pi)^3\delta^{(3)}(\vec{k}+\vec{p}_1+\vec{p}_2)n^2V|\mathcal{M}|^2$, where n is the number density of excited atoms in a coherent volume V.

A more rigorous derivation of the continuum limit relation is based on numerical simulations of the Maxwell–Bloch equation including relaxation effects, as given in Refs. 1 and 5. The Maxwell–Bloch equation is a non-linear set of partial differential equations in 1+ 1 space–time dimension, involving the density matrix element $\rho_i(x,t), i = 1 \sim 3$ (which evolves according to the Schrödinger equation modified by relaxation terms) and light field amplitude $E(x,t)$ (which follows the Maxwell equation whose right-hand side is given by the polarization of the matter system). One space dimension is selected since laser is irradiated in this direction. The density matrix elements give the state of target atoms such as the population difference and the off-diagonal term $\rho_{eg}(x,t)$ usually called the coherence term from which one can compute the macroscopic polarization of the matter system (allowing a nearly random, hence vanishing on average in a special case). It is further coupled to two-photon fields. One can clarify both the state of target matter or the light field inside the target, and what comes out from two ends of a long target.

We have extensively performed numerical simulations in our past references, and further compared these results with our experimental outputs of PSR process, which is described in Ref. 7. Agreement of simulations with experimental results is reasonably good.

Results show that there are both explosive PSR events and PSR signals in the weak field region. In explosive events most of the stored energy in the upper level $|e\rangle$ is released in the form of short pulse, as illustrated in Fig. 4. Although the explosive event is spectacular, it may be difficult to experimentally realize this situation. In the weak trigger filed case the PSR signal has an exponential time dependence roughly of the form, $e^{cn|\rho_{eg}|t}$. This case is illustrated in Fig. 5. The signal development is terminated either by the phase relaxation time $t = T_2$ or a finite target length $t = L/c$. The important exponent factor here is the product $n|\rho_{eg}|$. Thus, it should have a large enough of $n|\rho_{eg}|$ to observe PSR signals in actual experiments.

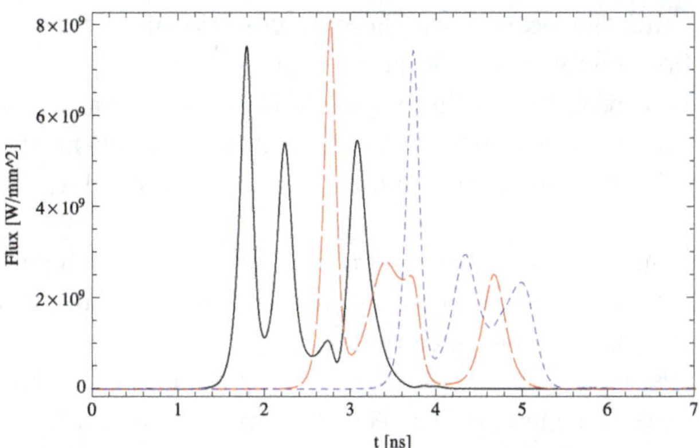

Fig. 4. (Color online) Trigger power dependence of time-evolving output flux from the symmetric trigger irradiation of the power range, $10^{-12} \sim 1$ Wmm^{-2}, under the conditions of $n = 1 \times 10^{21}$ cm^{-3}, target length = 30 cm, relaxation times $T_2 = 10, T_1 = 10^3$ ns, and the initial polarization, $r_1 = 1, r_2 = r_3 = 0$. Depicted outputs from 1 Wmm^{-2} trigger power in solid black, from 10^{-6} Wmm^{-2} in dashed red, and from 10^{-12} Wmm^{-2} in dotted blue are displaced almost equi-distantly in the first peak positions. Transition $Xv = 1 \to Xv = 0$ of pH$_2$ is considered. ~ 70 % stored energy in the initial metastable state is released in these cases. Taken from Ref. 5.

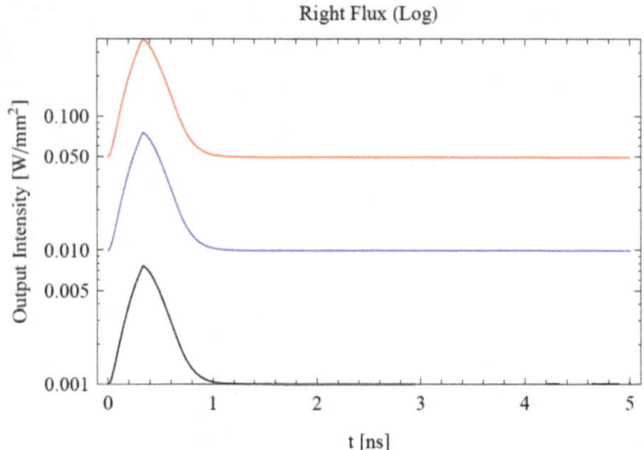

Fig. 5. (Color online) PSR signals in the linear trigger regime. Right-moving output flux from pH_2 vibrational excitation level ($Xv = 1$). The initial coherence is assumed to be spatially homogeneous ($r_1^{(0)} = 1$). The black, blue and red lines correspond to trigger laser intensities of 1 mW/mm^2, 10 mW/mm^2, and 50 mW/mm^2. Note logarithmic scale for the vertical axis. Left flux is identical to the right flux. Taken from Ref. 1.

5. PSR Experiment

We found it imperative to experimentally prove by ourselves the macro-coherence mechanism prior to actual RENP experiments. Our group in Okayama has launched the experimental project to verify the macro-coherence in weak QED process which is easier than RENP measurement, but still is not trivial.

The target and the process we chose is vibrational de-excitation of para-H_2 molecule. Its first excited state of energy ~ 0.5 eV is forbidden to decay by the electric dipole transition due to the symmetry of homo-nuclear molecule. The main radiative decay mode is two-photon emission whose spontaneous decay rate is estimated to be $1/(3 \times 10^{12}$ sec). At liquid nitrogen temperature its relaxation time is long enough.

In the first round of experiments, we used as the trigger light one of the side band lights which is generated by two excitation lasers of Raman type. Two laser frequencies, ω_1, ω_2, were chosen such that $\omega_1 - \omega_2 \approx \epsilon_{eg}$ (deviation from the equality is called detuning and may be of some interest experimentally). Since we achieved a considerable degree of coherence between two levels, $|e\rangle$ and $|g\rangle$, we could measure 13 orders of side bands, both in higher and lower energy sides than ϵ_{eg}. Simulations suggest that our achieved coherence is $\sim 6\%$ on the average over 15 cm target length.[7] Two excitation lasers are irradiated from the same direction and one expects from the momentum and the energy conservation that the Stokes side band of 4.6 μm lies within the level spacing and its partner making up the full level spacing should be emitted as a result of PSR process into the same direction. We succeeded in observation of this signal. We refer to Ref. 7 for experimental details.

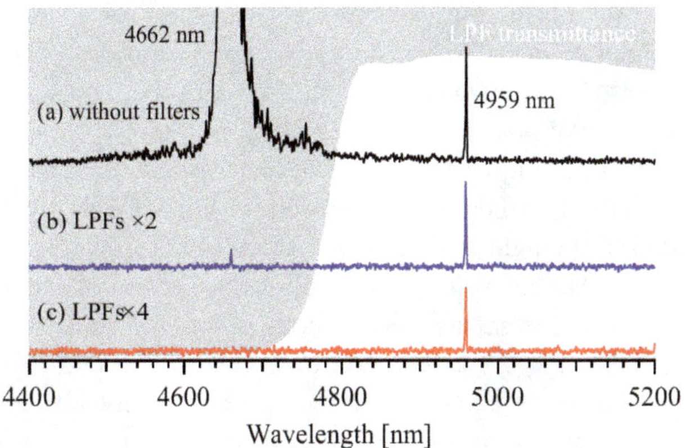

Fig. 6. Observed spectra at 60 kPa (a) without the long-pass filter (LPF), (b) with two LPFs and (c) with four LPFs. The white portion excluded by the gray hatch shows the LPF transmittance; it is ~0.85 at 4.96 μm. Taken from Ref. 7.

From the measured pulse strength of PSR signal it is concluded that the signal is stronger by $\sim 10^{15}$ than the expected event rate from the spontaneous two-photon decay.

We still need to clarify details of PSR. One of the important items is whether the PSR signal is also measured when one uses an independent trigger rather than one of the side band light. This experiment is now going on and our preliminary result shows that PSR from the external trigger is there and its strength is proportional to the irradiated trigger strength, as expected in the linear regime of PSR theory.

6. Future Prospects

Let me summarize our goals. Using atoms or molecules, we wish to determine all neutrino mass matrix elements with precision. In particular, (1) the mass hierarchy pattern, (2) the absolute neutrino mass scale and the smallest neutrino mass, (3) the neutrino mass type, Majorana or Dirac, (4) CPV phases including α, β intrinsic to Majorana neutrinos. When the smallest neutrino mass is known with some precision by RENP experiments, we can add (5) detection of relic neutrino of 1.9 K.

We only achieved the experimental verification of the macro-coherence amplification mechanism, which is however crucial for further development of our experimental project. Not only this proves the amplification mechanism, but it also gives an insight towards RENP. From simulations of PSR we anticipate that remnant states after PSR activities are nearly static objects. This appears to be static condensate of coupled matter polarization and light field. In other words, it seems that two-photon analogue of stopped light may be formed.

Indeed, one may formulate condensate formation as a kind of non-linear eigenvalue problem and identify eigenstates as soliton solutions. The important idea of soliton-condensate has been examined extensively in the usual electric dipole transition under the name of polariton. This is a coherent state of target in which light electric field is strongly coupled with atomic macroscopic polarization and forms a joint eigenstate. The light field can be stored within a target by a large amount, and in a special case the light field does not move at all, apparently: stopped light. Needless to say, this is achieved by the supporting system of atomic polarization, and one may regard the event as a continuous process of emission and absorption occurring always within a target. Our soliton-condensate is two-photon analogue of the stopped light and is yet to be discovered experimentally, although its existence has been shown even in the presence of relaxation.[8] An important outcome of the stopped light in the two-photon case is that light can escape the target only from its ends, with an exponentially suppressed rate. This feature of suppressed photon emission is very useful to reduce QED backgrounds against RENP, since RENP process can occur perturbatively from the bulk of target with no suppression at all.

Two-photon condensate can occur by irradiation of lasers in counter-propagating directions. Hence our next important experimental step would be PSR experiments using counter-propagating laser pulses and formation of soliton-condensates.

Although soliton-condensate is ideal for the background rejection, it cannot be used as a target state of RENP, because the relation of the initial and the final states is different for RENP and PSR processes: they have different relative parities. In order to convert a soliton-condensate state after PSR emission into a RENP-ready state, it is necessary either to have a mixture from different parity state or to convert the parity as a real process. This can be done either by application of static electric field or irradiation of another laser. We may generically call this process switching. Preparation of RENP-ready state by the switching is an important step towards the precision neutrino mass spectroscopy.

References

1. A. Fukumi *et al.*, *Prog. Theor. Exp. Phys.* (2012) 04D002, and earlier references cited therein.
2. M. Yoshimura and N. Sasao, *Phys. Rev. D* **89**, 053013 (2014).
3. D. N. Dinh, S. Petcov, N. Sasao, M. Tanaka, and M. Yoshimura, *Phys. Lett. B* **719**, 154 (2012).
4. M. Yoshimura, N. Sasao, and M. Tanaka, *Phys. Rev. D* **91**, 063516(2015).
5. M. Yoshimura, N. Sasao, and M. Tanaka, *Phys. Rev. A* **86**, 013812 (2012).
6. R. H. Dicke, *Phys. Rev.* **93**, 99 (1954).
7. Y. Miyamoto *et al.*, *Prog. Theor. Exp. Phys.* 113C01 (2014).
8. M. Yoshimura and N. Sasao, *Prog. Theor. Exp. Phys.* 073B02 (2014).

Detection Prospects of the Cosmic Neutrino Background

Yu-Feng Li

Institute of High Energy Physics, Chinese Academy of Sciences, P. O. Box 918, Beijing 100049, China

liyufeng@ihep.ac.cn

The existence of the cosmic neutrino background (CνB) is a fundamental prediction of the standard Big Bang cosmology. Although current cosmological probes provide indirect observational evidence, the direct detection of the CνB in a laboratory experiment is a great challenge to the present experimental techniques. We discuss the future prospects for the direct detection of the CνB, with the emphasis on the method of captures on beta-decaying nuclei and the PTOLEMY project. Other possibilities using the electron-capture (EC) decaying nuclei, the annihilation of extremely high-energy cosmic neutrinos (EHECνs) at the Z-resonance, and the atomic de-excitation method are also discussed in this review.

Keywords: CνB; direct detection; beta decay.

1. Introduction

As weakly-interacting and rather stable particles, relic neutrinos were decoupled from radiation and matter at a temperature of about one MeV and an age of one second after the Big Bang (see e.g. Ref. 1). It is quite similar to the cosmic microwave background (CMB) radiation, whose formation was at a time of around 3.8×10^5 years after the Big Bang. This cosmic neutrino background (CνB) played an important role in the evolution of the Universe, and its existence has been indirectly proved from current cosmological data on the Big Bang nucleosynthesis (BBN), large-scale structures of the cosmos and CMB anisotropies.[2]

The properties of the CνB are tightly related to the properties of the CMB. In particular, in the absence of lepton asymmetries the temperature and average number density of relic neutrinos can be expressed as[1]

$$T_\nu = \left(\frac{4}{11}\right)^{1/3} T_\gamma \approx 1.945\,\mathrm{K}\,, \quad n_\nu = \frac{9}{11} n_\gamma \approx 336\,\mathrm{cm}^{-3}\,. \tag{1}$$

As a consequence, one predicts the average three-momentum today for each species of the relic neutrino is very small (for a brief review, see Ref. 3):

$$\langle p_\nu \rangle = 3T_\nu \approx 5.8\,\mathrm{K} \approx 5 \times 10^{-4}\,\mathrm{eV}\,, \tag{2}$$

implying that at least two mass eigenstates of relic neutrinos are already non-relativistic, no matter whether the neutrino mass spectrum is the normal or inverted hierarchy.

Although cosmological observations provide the indirect evidence for the existence of the CνB, direct detection in a laboratory experiment is a great challenge to the present experimental techniques. Among several possibilities,[3] the most promising one seems to be the neutrino capture experiment using radioactive β-decaying nuclei.[4–13] The proposed PTOLEMY project[14] aims to obtain the sensitivity required to detect the CνB using 100 grams of ^{3}H as the capture target. Other interesting possibilities include the electron-capture (EC) decaying nuclei,[15–18] the annihilation of extremely high-energy cosmic neutrinos (EHECνs) at the Z-resonance,[19–22] and the atomic de-excitation method.[23]

The remaining parts of this work are organized as follows. In Sec. 2 we introduce the method of captures on the beta-decaying nuclei and calculate the β-decay energy spectrum and the relic neutrino capture rate. Section 3 is devoted to flavor effects of the relic neutrino capture spectrum. We shall present a brief description on other interesting possibilities of the CνB detection in Sec. 4, and then conclude in Sec. 5.

2. Captures on the Beta-Decaying Nuclei

In the presence of $3+n$ species of active and sterile neutrinos, the flavor eigenstates of three active neutrinos and n sterile neutrinos can be written as[1,2]

$$\begin{pmatrix} \nu_e \\ \nu_\mu \\ \nu_\tau \\ \vdots \end{pmatrix} = \begin{pmatrix} U_{e1} & U_{e2} & U_{e3} & \cdots \\ U_{\mu 1} & U_{\mu 2} & U_{\mu 3} & \cdots \\ U_{\tau 1} & U_{\tau 2} & U_{\tau 3} & \cdots \\ \vdots & \vdots & \vdots & \ddots \end{pmatrix} \begin{pmatrix} \nu_1 \\ \nu_2 \\ \nu_3 \\ \vdots \end{pmatrix}, \tag{3}$$

where ν_i is a mass eigenstate of active (for $1 \le i \le 3$) or sterile (for $4 \le i \le 3+n$) neutrinos, and $U_{\alpha i}$ stands for an element of the $(3+n) \times (3+n)$ neutrino mixing matrix.

One of the most promising method to measure the absolute electron neutrino mass is the nuclear β-decay

$$\mathcal{N}(A, Z) \rightarrow \mathcal{N}'(A, Z+1) + e^- + \bar{\nu}_e\,, \tag{4}$$

where A and Z are the mass and atomic numbers of the parent nucleus, respectively.

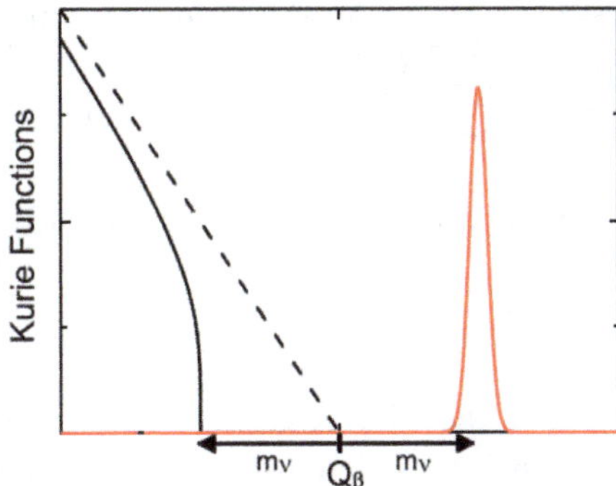

Fig. 1. Idealized electron spectra for the tritium beta decay and relic neutrino capture. The dashed and black-solid lines are shown for β-decay spectra of the massless and massive neutrinos respectively. The red-solid line with the sharp peak is for the relic neutrino signal.

The differential decay rate of a β-decay can be written as[24]

$$\frac{\mathrm{d}\lambda_\beta}{\mathrm{d}T_e} = \int_0^{Q_\beta - \min(m_i)} \mathrm{d}T'_e \left\{ \frac{G_\mathrm{F}^2 \cos^2\theta_\mathrm{C}}{2\pi^3} F(Z, E_e) |\mathcal{M}|^2 E_e \sqrt{E_e^2 - m_e^2} \right.$$
$$\left. \times (Q_\beta - T'_e) \sum_{i=1}^{4} \left[|U_{ei}|^2 \sqrt{(Q_\beta - T'_e)^2 - m_i^2}\, \Theta(Q_\beta - T'_e - m_i) \right] \right\} R(T_e, T'_e)\,, \tag{5}$$

where $T'_e = E_e - m_e$ denotes the intrinsic kinetic energy of the outgoing electron, $F(Z, E_e)$ is the Fermi function, $|\mathcal{M}|^2$ is the dimensionless nuclear matrix elements,[24] and $\theta_\mathrm{C} \simeq 13^\circ$ is the Cabibbo angle. Note that a Gaussian energy resolution function,

$$R(T_e, T'_e) = \frac{1}{\sqrt{2\pi}\,\sigma} \exp\left[-\frac{(T_e - T'_e)^2}{2\sigma^2} \right], \tag{6}$$

is implemented in Eq. (5) to include the finite energy resolution, and the theta function is adopted to ensure the kinematic requirement. The spectral shape near the β-decay endpoint represents a kinetic measurement of the absolute neutrino masses, which can be understood by comparing the dashed and black solid lines of Fig. 1. The gap between end points of two black lines stands for a measurement of the effective electron neutrino mass.

On the other hand, the threshold-less neutrino capture process,

$$\nu_e + \mathcal{N}(A, Z) \to \mathcal{N}'(A, Z+1) + e^-\,, \tag{7}$$

is located well beyond the end point of the β-decay, where the signal is characterized by the monoenergetic kinetic energy of the electron for each neutrino mass eigenstate. A measurement of the distance between the decay and capture processes will directly probe the CνB and constrain or determine the masses and mixing angles.

The differential neutrino capture rate of this process reads

$$\frac{\mathrm{d}\lambda_\nu}{\mathrm{d}T_e} = \sum_i |U_{ei}|^2 \sigma_{\nu_i} v_{\nu_i} n_{\nu_i} R(T_e, T_e^{\nu i}), \tag{8}$$

where the sum is for all the neutrino mass eigenstates and

$$n_{\nu_i} = \frac{n_{\nu_i}}{\langle n_{\nu_i} \rangle} \cdot \langle n_{\nu_i} \rangle \equiv \zeta_i \cdot \langle n_{\nu_i} \rangle, \tag{9}$$

denotes the number density of the relic neutrinos ν_i around the Earth. The standard Big Bang cosmology gives the prediction $\langle n_{\nu_i} \rangle \approx \langle n_{\bar{\nu}_i} \rangle \approx 56\ \mathrm{cm}^{-3}$ for each species of active neutrinos, and the prediction is also expected to hold for each species of sterile neutrinos if they could be fully thermalized in the early Universe. The number density of relic neutrinos around the Earth may be enhanced by the gravitational clustering effect (i.e. the factor ζ_i) when the neutrino mass is larger than 0.1 eV.[25] In Eq. (7) the capture cross-section times neutrino velocity can be written as

$$\sigma_{\nu_i} v_{\nu_i} = \frac{2\pi^2}{A} \cdot \frac{\ln 2}{T_{1/2}}, \tag{10}$$

where A is the nuclear factor characterized by Q_β and Z, and $T_{1/2}$ is the half-life of the parent nucleus.

Considering the running time and target mass of a particular experiment, the distributions of the numbers of capture signal and β-decay background events are expressed, respectively, as

$$\begin{aligned} \frac{\mathrm{d}N_{\mathrm{S}}}{\mathrm{d}T_e} &= \frac{1}{\lambda_\beta} \cdot \frac{\mathrm{d}\lambda_\nu}{\mathrm{d}T_e} \cdot \frac{\ln 2}{T_{1/2}} \bar{N}_{\mathrm{T}}\, t, \\ \frac{\mathrm{d}N_{\mathrm{B}}}{\mathrm{d}T_e} &= \frac{1}{\lambda_\beta} \cdot \frac{\mathrm{d}\lambda_\beta}{\mathrm{d}T_e} \cdot \frac{\ln 2}{T_{1/2}} \bar{N}_{\mathrm{T}}\, t, \end{aligned} \tag{11}$$

where $\bar{N}_{\mathrm{T}}$ is the averaged number of target atoms for a given exposure time t. Therefore, $\bar{N}_{\mathrm{T}}\, t$ gives the total target factor in the experiment:

$$\bar{N}_{\mathrm{T}}\, t = N(0) \cdot \frac{T_{1/2}}{\ln 2} \cdot \left(1 - e^{-t \cdot \frac{\ln 2}{T_{1/2}}}\right), \tag{12}$$

with $N(0)$ being the initial target number at $t = 0$.

To get a better signal-to-background ratio, one can investigate different kinds of candidate nuclei by considering factors of the cross-section, half-life, β-decay rate, and the detector energy resolution. The target nuclei should have the half-life $T_{1/2}$ longer than duration of the exposure time, have the maximal possible cross-section times neutrino velocity, and have the minimal possible background rate. An exhaustive survey was done several years ago,[6] and a summary of several candidates of the β^--decaying nuclei is shown in Table 1, from which one can find ^{3}H, ^{106}Ru, and ^{187}Re can be possible promising nuclei.

Table 1. Several candidates of the β^--decaying nuclei.[6]

Isotope	Q_β (keV)	Decay type	Half-life (sec)	$\sigma_{\nu_i} \cdot v_{\nu_i}$ (10^{-41} cm^2)
^{3}H	18.591	β^-	3.8878×10^8	7.84×10^{-4}
^{63}Ni	66.945	β^-	3.1588×10^9	1.38×10^{-6}
^{93}Zr	60.63	β^-	4.952×10^{13}	2.39×10^{-10}
^{106}Ru	39.4	β^-	3.2278×10^7	5.88×10^{-4}
^{107}Pd	33	β^-	2.0512×10^{14}	2.58×10^{-10}
^{187}Re	2.64	β^-	1.3727×10^{18}	4.32×10^{-11}

The β-decay experiments of current generation includes the spectrometer of KATRIN[26] and the calorimeter of MARE.[27] KATRIN uses 50 μg of ^{3}H as the effective target mass, and MARE is planning to deploy 760 grams of ^{187}Re. Therefore, we can estimate the CνB event rates, respectively, as

$$N^\nu(\text{KATRIN}) \simeq 4.2 \times 10^{-6} \times \sum_i |U_{ei}|^2 \zeta_i \quad \text{yr}^{-1}\,, \tag{13}$$

$$N^\nu(\text{MARE}) \simeq 7.6 \times 10^{-8} \times \sum_i |U_{ei}|^2 \zeta_i \quad \text{yr}^{-1}\,. \tag{14}$$

Moreover, a realistic proposal for the CνB detection is the PTOLEMY project,[14] which is designed to employ 100 grams of ^{3}H as the capture target using a combination of a large-area surface-deposition tritium target, the MAC-E filter, the RF tracking, the time-of-flight systems, and the cryogenic calorimetry. Finally, the event rate of PTOLEMY are calculated to reach the observable level:

$$N^\nu(\text{PTOLEMY}) \simeq 8.0 \times \sum_i |U_{ei}|^2 \zeta_i \quad \text{yr}^{-1}\,. \tag{15}$$

3. Flavor Effects

Besides the total capture rates, the CνB detection exhibits interesting properties of flavor effects due to the neutrino mixing. In this section, we shall discuss the effects of the neutrino mass hierarchy, presence of light sterile neutrinos,[28,29] and keV sterile neutrinos as a candidate of warm dark matter.[30]

In our calculation, we adopt best-fit values of the relevant three-neutrino oscillation parameters (i.e. θ_{12}, θ_{13}, Δm^2_{21} and $|\Delta m^2_{31}|$) from Review of Particle Physics,[2] and all the other parameters will be explicitly mentioned when needed. Our default assumption of the target mass is 100 grams of ^{3}H, but will be 10 kg of ^{3}H or 1 ton of ^{106}Ru when we discuss the keV sterile neutrinos.

Figure 2 shows the capture rate of the CνB as a function of the kinetic energy T_e of electrons in the standard three-neutrino scheme with $\Delta m^2_{31} > 0$ (left panel) and $\Delta m^2_{31} < 0$ (right panel). The finite energy resolution Δ (i.e. $\Delta = 2\sqrt{2\ln 2}\,\sigma$)

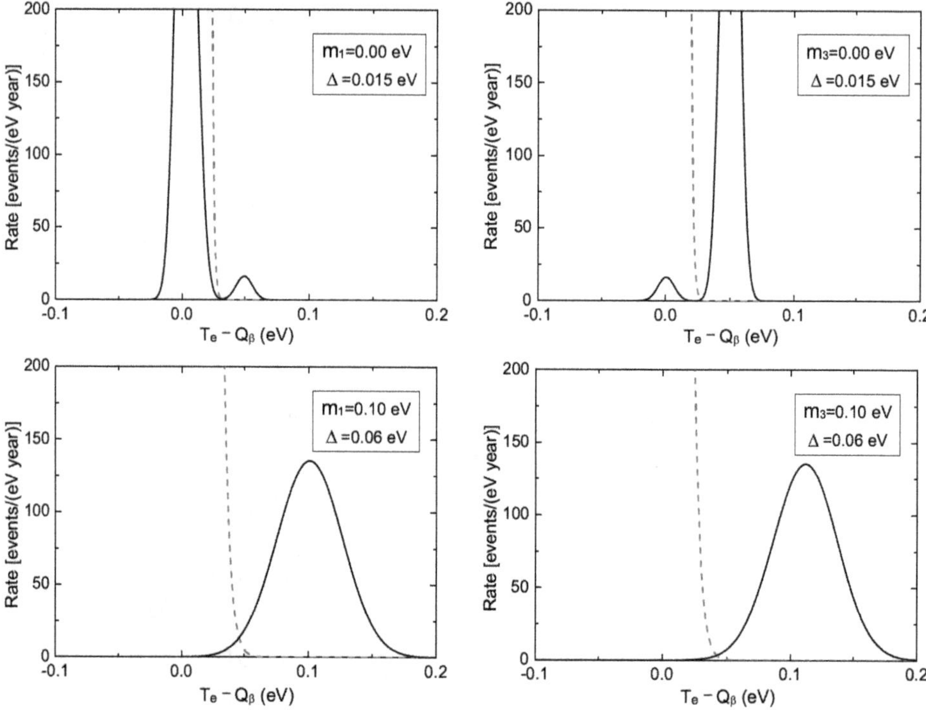

Fig. 2. The relic neutrino capture rate as a function of the kinetic energy of electrons in the standard scheme with $\Delta m^2_{31} > 0$ (left panel) or $\Delta m^2_{31} < 0$ (right panel). The gravitational clustering of three active neutrinos has been neglected for simplicity.

is taken in such a way that only one single peak can be observed beyond the β-decay background. The gravitational clustering of three active neutrinos has been neglected for simplicity. As the lightest neutrino mass (m_1 on the left panel or m_3 on the right panel) increases from 0 to 0.1 eV, the neutrino capture signal moves towards the larger T_e region. Hence the distance between the signal peak and the β-decay background becomes larger for a larger value of the lightest neutrino mass, and therefore the required energy resolution is less stringent. Comparing between the left panel and right panel, one can observe that it is easier to detect the CνB in the $\Delta m^2_{31} < 0$ case, where the capture signal is separated more apparently from the β-decay background. The reason is that the dominant mass eigenstates ν_1 and ν_2 in ν_e have greater eigenvalues than in the $\Delta m^2_{31} > 0$ case.

Next we are going to study the $(3+2)$ mixing scheme with two light sterile neutrinos. Considering the hints of short baseline oscillations,[28,29] we assume $m_4 = 0.2$ eV and $m_5 = 0.4$ eV together with $|U_{e1}| \approx 0.792$, $|U_{e2}| \approx 0.534$, $|U_{e3}| \approx 0.168$, $|U_{e4}| \approx 0.171$ and $|U_{e5}| \approx 0.174$ in the numerical calculations. We also take $m_1 = 0$ or $m_3 = 0$ for simplicity. We illustrate the capture rate of the CνB as a function of the electron's kinetic energy T_e against the corresponding β-decay background for both $\Delta m^2_{31} > 0$ and $\Delta m^2_{31} < 0$ schemes in Fig. 3. To take account

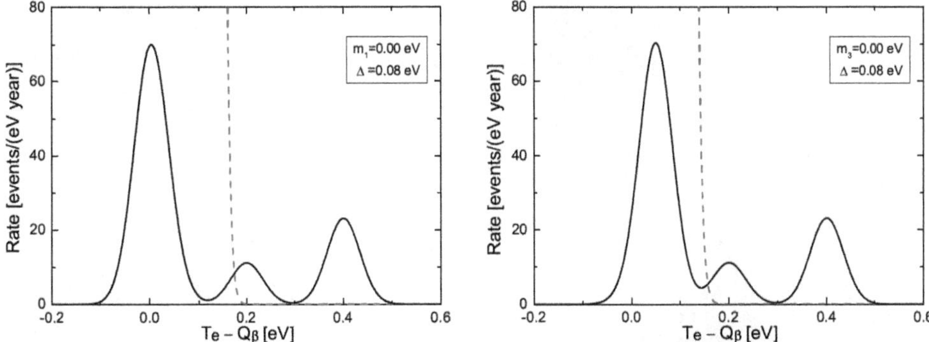

Fig. 3. The capture rate of the CνB as a function of the electron's kinetic energy in the $(3+2)$ mixing scheme with $\Delta m^2_{31} > 0$ (left panel) and $\Delta m^2_{31} < 0$ (right panel).[10] The gravitational clustering of relic sterile neutrinos around the Earth has been illustrated by taking $\zeta_1 = \zeta_2 = \zeta_3 = 1$ and $\zeta_5 = 2\zeta_4 = 10$ for example.

of possible gravitational clustering effects, we assume $\zeta_1 = \zeta_2 = \zeta_3 = 1$ (without clustering effects for three active neutrinos) and $\zeta_5 = 2\zeta_4 = 10$ (with mild clustering effects for two sterile neutrinos). As one can see from Fig. 3, the signals of sterile neutrinos are obviously enhanced because of $\zeta_4 > 1$ and $\zeta_5 > 1$. If the overdensity of relic neutrinos is very significant around the Earth, it will be helpful for the CνB detection through the neutrino capture process.

Finally we want to talk about the detection of keV sterile neutrinos as a candidate of warm dark matter. We shall work in the $(3+1)$ mixing scheme, where the mixing element and mass of the sterile neutrino are severely constrained by cosmological observational data.[30] Here $m_4 = 2$ keV and $|U_{e4}|^2 \simeq 5 \times 10^{-7}$ are assumed for illustration. We also take $\Delta m^2_{31} > 0$ and $m_1 = 0$ for simplicity. From the mass density of dark matter around the Earth,[30] (i.e. $\rho^{\rm local}_{\rm DM} \simeq 0.3$ GeV cm^{-3}), we can calculate the number density of ν_4 as

$$n_{\nu_4} \simeq 10^5 \times \frac{3\ \text{keV}}{m_4}\ \text{cm}^{-3}\,. \tag{16}$$

Our numerical calculations are presented in Fig. 3, where two isotope sources (i.e. 10 kg ^{3}H and 1 ton ^{106}Ru) are illustrated for comparison. One can notice that the required energy resolution is not a problem because of the larger sterile neutrino mass. However, the extremely small active-sterile mixing makes the observability of keV sterile neutrinos rather dim and remote. We also show the half-life effects of both isotopes in Fig. 4. The finite lifetime is negligible for the ^{3}H nuclei, but important for the ^{106}Ru nuclei. It can reduce around 30% of the capture rate on ^{106}Ru. Hence the half-life effect must be considered if duration of the exposure time is comparable with the source half-life.

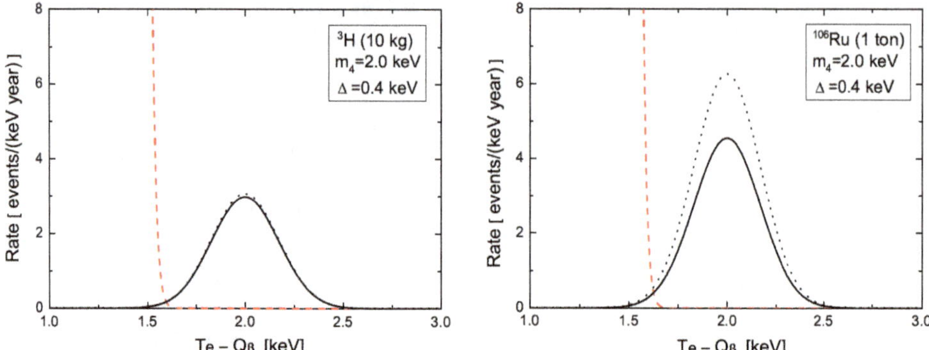

Fig. 4. The keV sterile neutrino capture rate as a function of the kinetic energy of electrons with ^{3}H (left panel) and ^{106}Ru (right panel) as our target sources.[11] The solid (or dotted) curves denote the signals with (or without) the half-life effect.

4. Other Possibilities

The capture on the β-decaying nuclei is one of the most promising methods to detect directly the CνB. However, from Eq. (7) only neutrinos of the electron flavor are relevant for this detection. Therefore, we should consider other possibilities for relic neutrinos of μ and τ flavors and antiparticles of the CνB.

Similar to the process of captures on β-decaying nuclei, the EC-decaying nuclei can be the target of relic antineutrino captures. Here we take the isotope ^{163}Ho as the working example.[15–18] It should be stressed that the structure near the endpoint of the ^{163}Ho EC-decay spectrum is applicable to study the absolute neutrino masses in a similar way as the β-decay. We could distinguish between the inverted and normal mass hierarchies by comparing the right and left panels of Fig. 5. The properties of the relic antineutrino capture against the EC-decaying background are similar to those discussed in Sec. 2. As the order of magnitude estimate, one

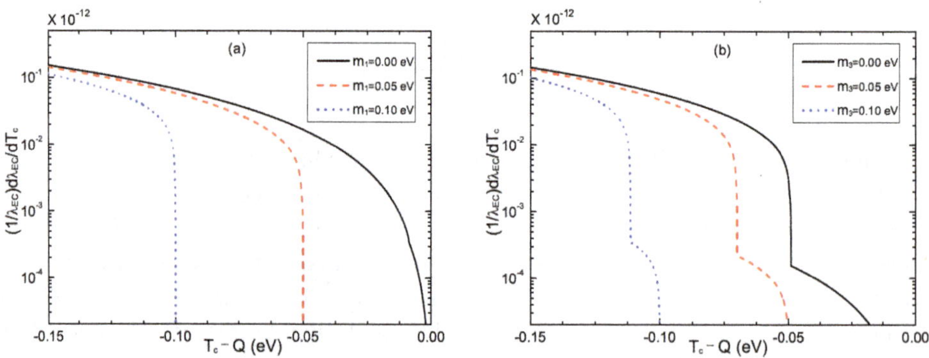

Fig. 5. The fine structure spectrum near the endpoint of the ^{163}Ho EC-decay in the $m^2_{31} > 0$ (left panel) or $m^2_{31} < 0$ (right panel) case.[17]

needs 30 kg ^{163}Ho to obtain one event per year for the relic antineutrino detection, and needs as much as 600 ton ^{163}Ho to get one event per year for the keV sterile antineutrino detection.

Another appealing possibility is the annihilation of EHECνs with the CνB in the vicinity of Z-resonance[19–22] (i.e. $\nu\bar{\nu} \to Z$). The resonance energy of EHECνs associated with each neutrino mass eigenstate can be calculated as

$$E_{0,i}^{\rm res} = \frac{m_Z^2}{2m_i} \simeq 4.2 \times 10^{12} \left(\frac{1~{\rm eV}}{m_{\nu_i}}\right) {\rm GeV}\,, \tag{17}$$

where m_Z denotes the Z boson mass. Since the annihilation cross-section at the resonance is enhanced by several orders of magnitude compared to the non-resonant scattering. Therefore, the absorption dips at the resonance energies are expected for the EHECνs arriving at the Earth,[19–22] which could provide the direct evidence for the existence of the CνB.

Recently there is another interesting method of the CνB detection using the atomic de-excitation process.[23] The de-excitation process of metastable atoms into the emission mode of a single photon and a neutrino pair, is defined as the radiative emission of neutrino pair (RENP),

$$|e\rangle \to |g\rangle + \gamma + \nu_i + \nu_j\,, \tag{18}$$

where $|e\rangle$ and $|g\rangle$ are the respective initial and final states of the atoms, ν_i and ν_j are the neutrino mass eigenstates. The existence of the CνB may distort the photon energy spectrum because of the Pauli exclusion principle, which provides a possible promising way to detect the CνB.[23]

5. Conclusion

The standard Big Bang cosmology predicts the existence of a cosmic neutrino background formed at a temperature of about one MeV and an age of one second after the Big Bang. A direct measurement of the relic neutrinos would open a new window to the early Universe. In this review we have discussed the future prospects for the direct detection of the CνB, with the emphasis on the method of captures on β-decaying nuclei and the PTOLEMY project. We calculated the neutrino capture rate as a function of the kinetic energy of electrons against the corresponding β-decay background, and discussed the possible flavor effects including the neutrino mass hierarchy, light sterile neutrinos, and keV sterile neutrinos as a candidate of warm dark matter. Other possibilities using the EC-decaying nuclei, the annihilation of EHECνs at the Z-resonance, and the atomic de-excitation method are also presented in this review. We stress that such direct measurements of the CνB in the laboratory experiments might not be hopeless in the long term.

Acknowledgments

The author would like to thank the organizers for the kind invitation and warm hospitality in Singapore, where this wonderful conference was held. This work was supported by the National Natural Science Foundation of China under grant Nos. 11135009 and 11305193.

References

1. Z. Z. Xing and S. Zhou, *Neutrinos in Particle Physics, Astronomy and Cosmology* (Zhejiang University Press and Springer-Verlag, 2011).
2. Particle Data Group (K. A. Olive *et al.*), *Chin. Phys. C* **38**, 090001 (2014).
3. A. Ringwald, *Nucl. Phys. A* **827**, 501c (2009).
4. S. Weinberg, *Phys. Rev.* **128**, 1457 (1962).
5. J. M. Irvine and R. Humphreys, *J. Phys. G* **9**, 847 (1983).
6. A. Cocco, G. Mangano and M. Messina, *J. Cosmol. Astropart. Phys.* **0706**, 015 (2007).
7. R. Lazauskas, P. Vogel and C. Volpe, *J. Phys. G* **35**, 025001 (2008).
8. M. Blennow, *Phys. Rev. D* **77**, 113014 (2008).
9. A. Kaboth, J. A. Formaggio and B. Monreal, *Phys. Rev. D* **82**, 062001 (2010).
10. Y. F. Li, Z. Z. Xing and S. Luo, *Phys. Lett. B* **692**, 261 (2010).
11. Y. F. Li and Z. Z. Xing, *Phys. Lett. B* **695**, 205 (2011).
12. W. Liao, *Phys. Rev. D* **82**, 073001 (2010).
13. A. J. Long, C. Lunardini and E. Sabancilar, *J. Cosmol. Astropart. Phys.* **1408**, 038 (2014).
14. PTOLEMY (S. Betts *et al.*), arXiv:1307.4738 [astro-ph].
15. A. G. Cocco, G. Mangano and M. Messina, *Phys. Rev. D* **79**, 053009 (2009).
16. M. Lusignoli and M. Vignati, *Phys. Lett. B* **697**, 11 (2011).
17. Y. F. Li and Z. Z. Xing, *Phys. Lett. B* **698**, 430 (2011).
18. Y. F. Li and Z. Z. Xing, *J. Cosmol. Astropart. Phys.* **1108**, 006 (2011).
19. T. J. Weiler, *Phys. Rev. Lett.* **49**, 234 (1982).
20. T. J. Weiler, *Astrophys. J.* **285**, 495 (1984).
21. B. Eberle *et al.*, *Phys. Rev. D* **70**, 023007 (2004).
22. G. Barenboim, O. Mena Requejo and C. Quigg, *Phys. Rev. D* **71**, 083002 (2005).
23. M. Yoshimura, N. Sasao and M. Tanaka, *Phys. Rev. D* **91**, 063516 (2015).
24. E. W. Otten and C. Weinheimer, *Rept. Prog. Phys.* **71**, 086201 (2008).
25. A. Ringwald and Y. Y. Y. Wong, *J. Cosmol. Astropart. Phys.* **0412**, 005 (2004).
26. KATRIN Collab. (A. Osipowicz *et al.*), arXiv:hep-ex/0109033.
27. MARE Collab. (A. Nucciotti), *Nucl. Phys. B (Proc. Suppl.)* **229**, 155 (2012).
28. C. Giunti *et al.*, *Phys. Rev. D* **88**, 073008 (2013).
29. J. Kopp *et al.*, *JHEP* **1305**, 050 (2013).
30. A. Kusenko, *Phys. Rept.* **481**, 1 (2009).

Supernova Bounds on keV-mass Sterile Neutrinos

Shun Zhou

Institute of High Energy Physics, Chinese Academy of Sciences, Beijing 100049, China

zhoush@ihep.ac.cn

Sterile neutrinos of keV masses are one of the most promising candidates for the warm dark matter, which could solve the small-scale problems encountered in the scenario of cold dark matter. We present a detailed study of the production of such sterile neutrinos in a supernova core, and derive stringent bounds on the active-sterile neutrino mixing angles and sterile neutrino masses based on the standard energy-loss argument.

Keywords: Sterile neutrinos; warm dark matter; supernova bounds.

1. Motivation

The observations of rotational curves of galaxies, the bullet cluster, gravitational lensing effects and cosmic microwave background have provided us with robust evidence that the matter content of our Universe is dominated by the nonbaryonic dark matter. So far, a lot of attention has been focused on the scenario of cold dark matter (CDM), which has a negligible velocity dispersion at the radiation–matter equality and damps structures below the Earth-mass scales.[1] The favorite candidates for CDM stem from the well-motivated theories, which have been proposed to solve fundamental problems of the standard model of elementary particle physics,[2] such as the lightest supersymmetric particle and the axion. However, the CDM scenario suffers from a few serious problems in the galaxy and small-scale structure formation, e.g. the overprediction of the observed satellites in the galaxy-scale halos[3] and the high concentration of dark matter in galaxies.[4] In the scenario of warm dark matter (WDM), a light-mass particle with a large velocity dispersion can suppress the structure formation up to the galaxy scales and thus solve the potential small-scale structure problems.[5]

Sterile neutrinos of keV masses are a promising candidate for the WDM.[6] As a simple but instructive example, Dodelson and Widrow have proposed that right-handed neutrinos with masses $m_s \sim$ keV can be produced via neutrino oscillations in the early Universe and account for all the dark matter,[7] if they mix with the ordinary neutrinos via a tiny mixing angle $\theta \sim 10^{-(4\cdots6)}$ in vacuum. Due to such a small mixing angle, sterile neutrinos can never be in thermal equilibrium. In the presence of a primordial lepton asymmetry, Shi and Fuller have observed that the production rate of sterile neutrinos could be enhanced by the Mikheyev–Smirnov–Wolfenstein (MSW) effect[8,9] and the correct relic abundance can be obtained even for much smaller mixing angles.[10] One possible way to detect sterile-neutrino WDM is to search for the X-rays from their radiative decays in the DM-dominated galaxies.[11] Conversely, the nonobservation of an X-ray line from the local group dwarf galaxies has placed restrictive limits on the mass and mixing angle of sterile neutrinos. Recently, the observation of an X-ray line around 3.5 keV in the Andromeda galaxy and the Perseus galaxy cluster has been claimed by two independent groups.[12,13] However, this claim is criticized by other authors, who attribute this X-ray line to the Potassium and Chlorine emission lines from the plasma in the galactic regions and the intergalactic gas.[14] At the present time, it is better to leave this problem open, and the future X-ray data could hopefully offer us a clue. Other limits can be obtained from the observations of Lyman-alpha forest and Supernova (SN) 1987A.[6] Put all together, the window for the Dodelson–Widrow mechanism of nonresonant production is closed, while the Shi–Fuller mechanism of resonant production is still viable.[15] Roughly speaking, sterile neutrinos with $m_s = 1 \sim 10$ keV and $\theta = 10^{-(4\cdots6)}$ could be WDM. However, it should be noticed that the observational constraints depend crucially on the production mechanisms and evolution of sterile neutrinos in the early Universe, so the constraints can be evaded as in several new-physics models.[16–18]

Moreover, the WDM sterile neutrinos could play an important role in generating the supernova asymmetries for pulsar kicks,[19] and in supporting supernova explosions.[20,21] Hence it is interesting to reexamine the SN bounds on the keV-mass sterile neutrinos by studying in detail the production and propagation of sterile neutrinos in the SN core. On the other hand, as we show later, the phenomenon of neutrino oscillations and interactions in a dense medium is intriguing by itself.

2. General Formalism

The matter density in a SN core $\rho = 3.0 \times 10^{14}$ g cm^{-3} is so high that even the weakly-interacting neutrinos could not escape freely.[22] Hence the production of sterile neutrinos in the dense matter can be very efficient through both neutrino oscillations and repeated scattering of ordinary neutrinos off background particles. In consideration of both coherent flavor oscillations and the decoherence caused by frequent scattering, it is convenient to describe the whole neutrino system in terms of density matrices.[23,24] We assume that neutrino interactions do not affect substantially the medium, whose configuration is mainly determined by the conditions of thermal equilibrium. The time duration of collisions between neutrinos and matter particles is much shorter than the evolution time scale, on which neutrino density matrices experience a significant change. On the other hand, the evolution of density matrices is sufficiently rapid compared to the macroscopic and hydrodynamic time scale.

Under the above assumptions, we follow the formalism mainly developed by Raffelt and Sigl,[23,24] and define the ensemble average of the density matrices of n-flavor neutrinos as $\langle a_j^\dagger(\mathbf{p}) a_i(\mathbf{q})\rangle = (2\pi)^3\delta^3(\mathbf{p}-\mathbf{q})(\rho_{\mathbf{p}})_{ij}$ and $\langle b_i^\dagger(\mathbf{p}) b_j(\mathbf{q})\rangle = (2\pi)^3\delta^3(\mathbf{p}-\mathbf{q})(\bar{\rho}_{\mathbf{p}})_{ij}$ for $i, j = 1, 2, \ldots, n$, where $a_i(\mathbf{p})$ and $b_i(\mathbf{p})$ stand for the annihilation operators for neutrinos and antineutrinos, respectively. The diagonal elements of $\rho_{\mathbf{p}}$ and $\bar{\rho}_{\mathbf{p}}$ are just the occupation numbers, while the off-diagonal elements encode the phase information. The evolution of the matrix of densities $\rho_{\mathbf{p}}$ is governed by[23,24]

$$\begin{aligned}\dot{\rho}_{\mathbf{p}} = {} & -\mathrm{i}\left[\Omega_{\mathbf{p}}, \rho_{\mathbf{p}}\right] + \frac{1}{2}\sum_{i=1}^{n}\left[\{I_i, 1-\rho_{\mathbf{p}}\}\mathcal{P}_{\mathbf{p}}^i - \{I_i, \rho_{\mathbf{p}}\}\mathcal{A}_{\mathbf{p}}^i\right] \\ & + \frac{1}{2}\sum_a \int \frac{\mathrm{d}^3\mathbf{p}'}{(2\pi)^3}\left\{\left[G^a \rho_{\mathbf{p}'} G^a (1-\rho_{\mathbf{p}}) + \mathrm{h.c.}\right]\mathcal{W}^a_{\mathbf{p}'\mathbf{p}}\right. \\ & \left. - \left[\rho_{\mathbf{p}} G^a (1-\rho_{\mathbf{p}'}) G^a + \mathrm{h.c.}\right]\mathcal{W}^a_{\mathbf{p}\mathbf{p}'}\right\}. \end{aligned} \tag{1}$$

On the right-hand side, the first term contains neutrino kinetic energy and matter potential from the coherent forward scattering. The second term arises from the charged-current production and absorption of neutrinos, where I_i is the projection matrix onto the neutrino flavor i, $\mathcal{P}^i_{\mathbf{p}}$ the production rate, and $\mathcal{A}^i_{\mathbf{p}}$ the absorption rate. The third term represents the contribution from scattering processes $\nu_{\mathbf{p}} + a \to \nu_{\mathbf{p}'} + a'$ with a and a' being any kind of background particles, where G^a is a diagonal $n \times n$ matrix of couplings g_i^a for the interaction type a and $(g_i^a)^2\mathcal{W}^a_{\mathbf{p}'\mathbf{p}}$ denotes the transition probability. In the case of antineutrinos, the evolution equation for $\bar{\rho}_{\mathbf{p}}$ can be obtained in a similar way.

Applying this general formalism to an ensemble of a keV-mass sterile neutrino mixing with only one flavor of ordinary neutrinos, we can simplify significantly the Boltzmann-type equations. To further simplify the picture, we consider a homogeneous and isotropic SN core, in which neutrinos are essentially trapped and well in thermal equilibrium with the medium. Due to the isotropy of the medium and neutrino ensemble, the relevant interaction rates depend only on neutrino energies. The next important step is to average $(\rho_E)_{ij}$ and $(\bar{\rho}_E)_{ij}$ over many cycles of flavor oscillations, since the oscillation length is much shorter than neutrino mean free path. More explicitly, we can estimate the oscillation length as

$$\lambda_{\rm osc} \lesssim 0.7~{\rm cm}\left(\frac{E}{30~{\rm MeV}}\right)\left(\frac{10^{-4}}{\sin 2\theta}\right)\left(\frac{10~{\rm keV}}{m_s}\right)^2; \tag{2}$$

and the mean-free-path as

$$\lambda_{\rm mfp} = \frac{1}{N_{\rm B}\sigma_{\nu N}} \approx 1.1\times 10^3~{\rm cm}\left(\frac{30~{\rm MeV}}{E}\right)^2 \rho_{14}^{-1}, \tag{3}$$

where $N_{\rm B}$ denotes the number of baryons, $\sigma_{\nu N} \sim G_{\rm F}^2 E^2/\pi$ the cross-section of neutrino–nucleon interaction and ρ_{14} the matter density in units of 10^{14} g cm^{-3}. For a typical mixing angle $\sin^2 2\theta = 10^{-8}$ and sterile neutrino mass $m_s = 10$ keV, we always have $\lambda_{\rm mfp} \gg \lambda_{\rm osc}$, corresponding to the weak-damping limit. This comparison indicates that neutrinos oscillate a large number of times before a subsequent scattering with nucleons. The averaged matrix of densities $\tilde{\rho}_E$ can now be parametrized in terms of the distribution functions of active and sterile neutrinos, denoted as f_E^α and f_E^s, respectively. If the active neutrino stays in thermal equilibrium and if the mixing angle is so small that sterile neutrinos escape from the SN core immediately after production, the evolution equation of f_E^s is given by[24]

$$\dot{f}_E^s = \frac{1}{4} s_{2\theta_\nu}^2 \left[\mathcal{P}_E + \int \frac{E'^2 {\rm d}E'}{2\pi^2} g_\alpha^2 \mathcal{W}_{E'E} f_{E'}^\alpha\right], \tag{4}$$

where $s_{2\theta_\nu} \equiv \sin 2\theta_\nu$ with θ_ν being the neutrino mixing angle in matter, g_α is the coupling constant for active neutrinos. Therefore, the emission rate of lepton number can be obtained by integrating the above equation over sterile neutrino energy. Similarly, the energy-loss rate can be calculated by convolving Eq. (4) with neutrino energy and integrating them over the whole energy range.

3. Sterile Neutrinos in SN Cores

Sterile neutrinos with masses in the keV range can be copiously produced in the SN core. For $m_s \gtrsim 100$ keV, the vacuum mixing angle of sterile neutrinos is stringently constrained $\sin^2 2\theta \lesssim 10^{-9}$ in order to avoid excessive energy loss.[25–27] For smaller masses, however, the MSW effect on active-sterile neutrino mixing becomes very important and the SN bound on vacuum mixing angle is not that obvious. Note that the bounds on mixing angles depend on which neutrino species the sterile neutrino mixes with. In the following, we shall discuss ν_τ–ν_s and ν_e–ν_s mixing cases, and emphasize the main difference between them.

3.1. ν_τ–ν_s *mixing*

First, we concentrate on the SN bound in the simplest case of ν_τ–ν_s mixing, because ν_τ and $\overline{\nu}_\tau$ only have neutral-current interactions and essentially stay in thermal equilibrium with the ambient matter. In the weak-damping limit, which is always valid for supernova neutrinos mixing with keV-mass sterile neutrinos, the evolution of ν_τ number density is determined by[28]

$$\dot{N}_{\nu_\tau} = -\frac{1}{4} \sum_a \int \frac{E^2 {\rm d}E}{2\pi^2} s_{2\theta_\nu}^2 \int \frac{E'^2 {\rm d}E'}{2\pi^2} \mathcal{W}_{E'E}^a f_{E'}^\tau, \tag{5}$$

where f_E^τ is the occupation number of ν_τ, and $\mathcal{W}_{E'E}^a$ the transition probability for $\nu(E') + a \to \nu(E) + a$ with a being background particles in the SN core. Note that Eq. (5) is actually an immediate consequence of integrating Eq. (4) for ν_τ–ν_s mixing over the sterile neutrino energy. For ν_τ, the coupling constant is just $g_\tau = 1$, so we write the transition probability as $W^a_{E'E} = \mathcal{W}^a_{E'E}$. The interaction types include neutrino–nucleon and neutrino–electron scattering, whereas the former dominates over the latter. In a similar way, we can derive the evolution equation of the $\bar{\nu}_\tau$ number density, involving the mixing angle $\theta_{\bar{\nu}}$, the occupation number $f^{\bar{\tau}}_E$ and the transition probability $\bar{W}^a_{E'E}$.

Due to the MSW effect, the mixing angle of neutrinos in matter is different from that of antineutrinos, i.e.,

$$\sin^2 2\theta_{\nu,\bar{\nu}} = \frac{\sin^2 2\theta}{\sin^2 2\theta + (\cos 2\theta \mp E/E_{\rm r})^2}, \tag{6}$$

where θ denotes the vacuum mixing angle, and the sign "$\mp$" refers to ν and $\bar{\nu}$. However, the actual sign depends on the lepton asymmetries in the matter, i.e. the matter potential V_{ν_τ}. The resonant energy $E_{\rm r} \equiv \Delta m^2/(2|V_{\nu_\tau}|)$ can be written as

$$E_{\rm r} = 3.25\ {\rm MeV}\left(\frac{m_s}{10\ {\rm keV}}\right)^2 \rho_{14}^{-1}|Y_0 - Y_{\nu_\tau}|^{-1}, \tag{7}$$

where $Y_0 \equiv (1 - Y_e - 2Y_{\nu_e})/4$, and $Y_x \equiv (N_x - N_{\bar{x}})/N_{\rm B}$ with $N_{\rm B}$ being the baryon number density, N_x and $N_{\bar{x}}$ being the number densities of particle x and its antiparticle $\bar{x}$. As for tau neutrinos, the matter potential $V_{\nu_\tau} = -(G_{\rm F}/\sqrt{2})N_{\rm B}\,(1 - Y_e - 2Y_{\nu_e} - 4Y_{\nu_\tau})$ is negative if the typical values of $Y_e = 0.3$, $Y_{\nu_e} = 0.07$ and $Y_{\nu_\tau} = 0$ for a SN core are taken. Therefore, the mixing angle for $\bar{\nu}_\tau$ is enhanced by matter effects, and the emission rate for $\bar{\nu}_\tau$ exceeds that for ν_τ, indicating that a ν_τ–$\bar{\nu}_\tau$ asymmetry (i.e. $Y_{\nu_\tau} \neq 0$) will be established. An interesting feedback effect emerges: (i) The chemical potential for tau neutrinos develops and thus changes the occupation numbers of ν_τ and $\bar{\nu}_\tau$; (ii) The ν_τ–$\bar{\nu}_\tau$ asymmetry shifts the resonant energy $E_{\rm r}$, and thus modifies the mixing angles θ_ν and $\theta_{\bar{\nu}}$; (iii) Both effects in (i) and (ii) will feed back on the emission rates. Hence a stationary state of this active-sterile neutrino system could be achieved if the emission rates for neutrinos and antineutrinos become equal to each other.[28] However, whether such a stationary state can be really reached depends on the mixing angle and sterile neutrino mass. In any case, it is straightforward to calculate the energy-loss rate by following the time evolution of the neutrino system.

3.2. ν_e–ν_s *mixing*

Since there is a great similarity between muon and tauon neutrino interactions in a SN core, we expect that ν_μ–ν_s mixing case shares the same features as the previous case. Then, we turn to the ν_e–ν_s mixing case, where the physical processes for the production and evolution of sterile neutrinos are quite different.

First, the charged-current interaction $e^- + p \rightleftharpoons \nu_e + n$ leads to a direct production of ν_e, and an absorption of ν_e by the medium as well. In addition to the charged-current interaction, neutrino scatterings off degenerate electrons and nondegenerate nucleons are also important. Therefore, both terms on the right-hand side of Eq. (4) exist for the case of ν_e–ν_s mixing.

Second, the emission of ν_s reduces the electron lepton number in the SN core, which is in thermal and beta equilibrium. From the latter condition, we can get a relationship $\mu_e = \mu_{\nu_e} + \hat{\mu}$ among the electron chemical potential μ_e, the electron–neutrino chemical potential μ_{ν_e} and the difference $\hat{\mu} \equiv \mu_n - \mu_p$ between the neutron μ_n and proton μ_p chemical potentials. The parameter $\hat{\mu} = 50 \sim 100$ MeV is determinable from the equation of state of the nuclear matter. In our discussions, we assume the beta equilibrium is always guaranteed and the equation of state is not significantly modified by the emission of sterile neutrinos. Under these assumptions, $\hat{\mu} = 55$ MeV is taken to be valid throughout. The emission of sterile neutrinos affects the number density of ν_e, and thus μ_{ν_e} and μ_e, which in turn modify the matter potential for neutrinos and the mixing angle in matter. The modification of neutrino mixing angle obviously feeds back on the emission rate.

Third, since ν_τ and $\bar{\nu}_\tau$ are always generated in pairs, there is no asymmetry between them, namely $Y_{\nu_\tau} = 0$, and likewise for ν_μ and $\bar{\nu}_\mu$. As a result, the resonant energy $E_{\rm r} = \Delta m^2/(2|V_{\nu_e}|)$ for neutrino oscillations is found to be

$$E_{\rm r} = 13 \text{ MeV} \left(\frac{m_s}{10 \text{ keV}}\right)^2 \rho_{14}^{-1} |1 - 3Y_e - 4Y_{\nu_e}|^{-1}, \tag{8}$$

where $Y_{\nu_\mu} = Y_{\nu_\tau} = 0$ is implemented. For the initial conditions $\mu_e = 218$ MeV and $\mu_{\nu_e} = 163$ MeV, one obtains $Y_e = 0.3$ and $Y_{\nu_e} = 0.07$ for the temperature $T = 30$ MeV and matter density $\rho = 3.0 \times 10^{14}$ g cm^{-3}. Unlike in the ν_τ–ν_s mixing case, the matter potential V_{ν_e} is positive, implying the resonance is initially in the neutrino sector. Therefore, the emission rate of ν_s is at the beginning much larger than that of $\bar{\nu}_s$, and the lepton number decreases rapidly. But even small changes of Y_e and Y_{ν_e} are able to drive $E_{\rm r}$ to infinity, where $1 - 3Y_e - 4Y_{\nu_e} = 0$ is reached.

It is worthwhile to mention that a large chemical potential μ_{ν_e} indicates the population of $\bar{\nu}_e$ is rare. The initial emission of ν_s and the corresponding energy loss in this period become crucial for us to draw any SN limits based on the energy-loss argument. When $E_{\rm r}$ is approaching infinity, the mixing angles in matter for neutrinos and antineutrinos are equal, and both of them are close to the vacuum mixing angle. During the evolution, the matter potential flips its sign and the resonance moves to the antineutrino sector. However, the $\bar{\nu}_e$ number density is largely suppressed due to the large chemical potential, even the enhancement of antineutrino mixing angle cannot enlarge the total emission rate remarkably.

4. SN Bound on Sterile Neutrinos

Given the sterile neutrino mass m_s and vacuum mixing angle θ, the energy loss rate $\mathcal{E}(t)$ due to sterile neutrino emission can be calculated by following the evolution of ν_τ–$\bar{\nu}_\tau$ asymmetry $Y_{\nu_\tau}(t)$. It has been found that the stationary state can be reached within one second and the feedback effect is very important for

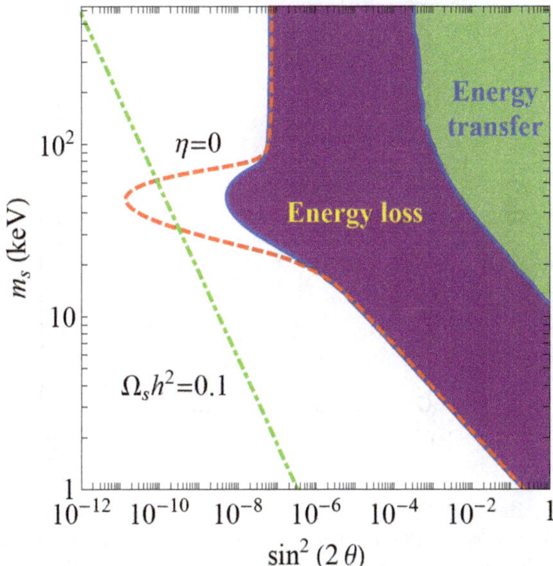

Fig. 1. (Color online) Supernova bound on sterile neutrino masses m_s and active-sterile mixing angles θ in the ν_τ–ν_s mixing case, where the shaded purple region is excluded by the energy-loss argument while the shaded green one by the energy-transfer argument.[28] The excluded region will be extended to the dashed (red) line if the build-up of degeneracy parameter is ignored, i.e. $\eta(t) = 0$. The dot-dashed (green) line represents the keV-mass sterile neutrinos as dark matter with the correct relic abundance $\Omega_s h^2 = 0.1$.

20 keV $\lesssim m_s \lesssim$ 80 keV and $10^{-9} \lesssim \sin^2 2\theta \lesssim 10^{-4}$. To avoid excessive energy losses, we require that the average energy-loss rate $\langle \mathcal{E} \rangle \equiv \int_0^{\tau_\text{d}} \mathcal{E}(t)\text{d}t$ with $\tau_\text{d} = 1$ s should be $\langle \mathcal{E} \rangle \lesssim 3.0 \times 10^{33}$ erg cm^{-3} s^{-1}. Otherwise, the duration of neutrino burst from SN 1987A would have been significantly reduced.

In Fig. 1, we show the contours of energy-loss rates in the $(\sin^2 2\theta, m_s)$-plane, where we have assumed a homogeneous and isotropic core with matter density $\rho = 3.0 \times 10^{14}$ g cm^{-3} and temperature $T = 30$ MeV. Based on the energy-loss argument, the shaded purple region has been excluded. The most stringent bound $\sin^2 2\theta \lesssim 10^{-8}$ arises for $m_s = 50$ keV. For the large-mixing angle region, the energy-loss rate is actually small, because sterile neutrinos have been trapped in the core and cannot carry energies away. However, the mean free path of sterile neutrinos is comparable to or even larger than that of ordinary neutrinos, indicating that they may transfer energies in a more efficient way. As a consequence, the duration of neutrino burst will be shortened by emitting neutrinos more rapidly. In this sense, the excessive energy transfer should be as dangerous as the excessive energy loss. Therefore, the large-mixing angle region is excluded when the energy-transfer argument is applied. The green dot-dashed line in Fig. 1 indicates the relic abundance of dark matter $\Omega_s h^2 = 0.1$, where keV-mass sterile neutrinos are warm dark matter and the nonresonant production mechanism is assumed. If we

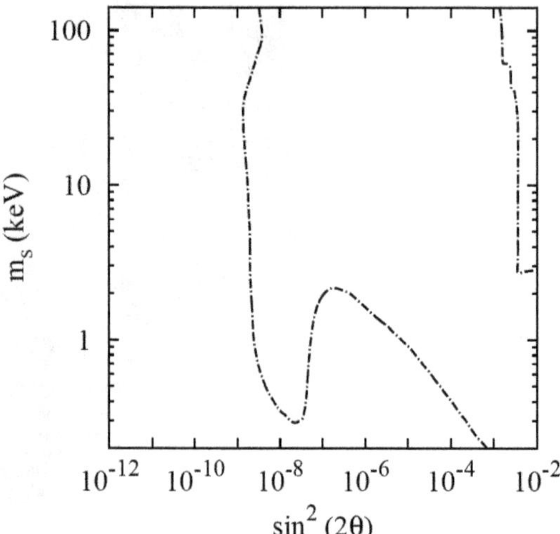

Fig. 2. Supernova bound on sterile neutrino masses m_s and active-sterile mixing angles θ in the ν_e–ν_s mixing case. Based on the energy-loss argument, the parameter space between two dot-dashed curves is excluded. The region of large mixing angles at the upper-right corner is also disfavored if the energy-transfer argument is implemented.

ignore the feedback effect (i.e. a vanishing chemical potential for tau neutrinos $\eta = \mu_{\nu_\tau}/T = 0$), the excluded region will extend to the red dashed line, which overlaps the relic-abundance line. However, the mixing angles are essentially unconstrained in the favored warm-dark-matter mass range 1 keV $\lesssim m_s \lesssim$ 10 keV.

As for the ν_e–ν_s-mixing case, we have included all the scattering processes of neutrinos, and solved the Boltzmann-like equation of the occupation numbers. In assumption of thermal equilibrium of ν_e, one can derive the evolution equation of the degeneracy parameter $\eta = \mu_{\nu_e}/T$. It is straightforward to verify that the infinite $E_{\rm r}$ is reached quickly, and the crossing from a negative to a positive matter potential takes place. In Fig. 2, using the same method as before, we have presented the contour plot of the energy-loss rate in the parameter plane of sterile neutrino masses and active-sterile neutrino mixing angles.[29] On the upper-right corner, the energy-transfer argument can be applied to exclude this large-mixing angle region. For sterile neutrinos of masses above 1 keV, the leftmost exclusion line is nearly vertical around $\sin^2 2\theta = 10^{-9}$. For much smaller masses $m_s \ll 1$ keV, the emission rate is highly suppressed by the matter effect, so the bound is rather loose.

5. Summary and Outlook

Since sterile neutrinos of keV masses as a promising candidate for warm dark matter are interesting in astrophysics and cosmology, we have considered their production in a SN core, and derived restrictive bounds on their masses and mixing angles by requiring the excessive energy loss caused by sterile neutrinos to be under control.

The main idea is to follow the time evolution of occupation numbers of neutrinos, and compute the emission rates of lepton number and sterile neutrino energy. Our final results about SN bounds are presented in Figs. 1 and 2. We have also emphasized the feedback effects in both ν_τ–ν_s and ν_e–ν_s mixing cases: The emission of sterile neutrinos leads to changes in the lepton number asymmetries and thus the neutrino matter potentials, which in turn affect the emission rates via the MSW effects on neutrino mixing angles. In the case of ν_τ–ν_s mixing, the MSW resonance occurs in the antineutrino sector, and thus an asymmetry between ν_τ and $\bar{\nu}_\tau$ number densities will be developed. This asymmetry contributes to the matter potential, intending to modify the mixing angles in matter and equating the emission rates of ν_s and $\bar{\nu}_s$. In the case of ν_e–ν_s mixing, the matter potential V_{ν_e} is rapidly driven to zero, implying that the neutrino and antineutrino mixing angles are close to the one in vacuum, which is in agreement with the result found in the literature.[27] However, for much smaller masses, the emission rates will be suppressed by the matter effects, and the SN bound becomes very weak.

It is worth pointing out that keV-mass sterile neutrinos could also be produced in the collapsing phase of massive stars.[20,21] Such sterile neutrinos after production propagate from the inner core to the outer one, and oscillate into electron neutrinos, which could deposit their energies in the out-layer matter and support SN explosions. A dedicated study of keV-mass sterile neutrinos in both core-collapse and cooling phases is certainly necessary and intriguing, and left for future works.

Acknowledgments

The author is grateful to Harald Fritzsch for kind invitation to this interesting conference, and to Georg Raffelt and Bo Cao for enjoyable collaboration. This work was in part supported by the Innovation Program of the Institute of High Energy Physics under Grant No. Y4515570U1 and by the CAS Center for Excellence in Particle Physics (CCEPP).

References

1. S. Hofmann, D. J. Schwarz and H. Stoecker, *Phys. Rev. D* **64**, 083507 (2001).
2. F. D. Steffen, *Eur. Phys. J. C* **59**, 557 (2009).
3. G. Kauffmann, S. D. M. White and B. Guiderdoni, *Mon. Not. R. Astron. Soc.* **264**, 201 (1993).
4. P. J. E. Peebles, arXiv:astro-ph/0101127.
5. S. Colombi, S. Dodelson and L. M. Widrow, *Astrophys. J.* **458**, 1 (1996), arXiv:astro-ph/9505029.
6. A. Kusenko, *Phys. Rep.* **481**, 1 (2009).
7. S. Dodelson and L. M. Widrow, *Phys. Rev. Lett.* **72**, 17 (1994).
8. L. Wolfenstein, *Phys. Rev. D* **17**, 2369 (1978).
9. S. P. Mikheyev and A. Yu. Smirnov, *Sov. J. Nucl. Phys.* **42**, 913 (1985).
10. X. D. Shi and G. M. Fuller, *Phys. Rev. Lett.* **82**, 2832 (1999).
11. K. N. Abazajian, arXiv:0903.2040 [astro-ph.CO].

12. E. Bulbul, M. Markevitch, A. Foster, R. K. Smith, M. Loewenstein and S. W. Randall, *Astrophys. J.* **789**, 13 (2014), arXiv:1402.2301 [astro-ph.CO].
13. A. Boyarsky, O. Ruchayskiy, D. Iakubovskyi and J. Franse, *Phys. Rev. Lett.* **113**, 251301 (2014), arXiv:1402.4119 [astro-ph.CO].
14. T. E. Jeltema and S. Profumo, arXiv:1408.1699 [astro-ph.HE].
15. A. Boyarsky *et al.*, *Phys. Rev. Lett.* **102**, 201304 (2009).
16. A. Kusenko, *Phys. Rev. Lett.* **97**, 241301 (2006).
17. F. Bezrukov, H. Hettmansperger and M. Lindner, *Phys. Rev. D* **81**, 085032 (2010).
18. A. Merle, *Int. J. Mod. Phys. D* **22**, 1330020 (2013), arXiv:1302.2625 [hep-ph].
19. A. Kusenko and G. Segre, *Phys. Lett. B* **396**, 197 (1997).
20. J. Hidaka and G. M. Fuller, *Phys. Rev. D* **74**, 125015 (2006).
21. J. Hidaka and G. M. Fuller, *Phys. Rev. D* **76**, 083516 (2007).
22. H. A. Bethe, *Rev. Mod. Phys.* **62**, 801 (1990).
23. G. Sigl and G. Raffelt, *Nucl. Phys. B* **406**, 423 (1993).
24. G. Raffelt and G. Sigl, *Astropart. Phys.* **1**, 165 (1993).
25. K. Kainulainen, J. Maalampi and J. T. Peltoniemi, *Nucl. Phys. B* **358**, 435 (1991).
26. X. Shi and G. Sigl, *Phys. Lett. B* **323**, 360 (1994).
27. K. Abazajian, G. M. Fuller and M. Patel, *Phys. Rev. D* **64**, 023501 (2001).
28. G. G. Raffelt and S. Zhou, *Phys. Rev. D* **83**, 093014 (2011).
29. B. Cao and S. Zhou, in preparation.

Precision Calculations for Supersymmetric Higgs Bosons

W. Hollik

Max Planck Institute for Physics,
Föhringer Ring 6, 80805 Munich, Germany
hollik@mpp.mpg.de www.mpp.mpg.de

Recent progress is presented on higher-order calculations for the mass spectrum of Higgs particles in the CP-conserving and CP-violating MSSM, covering diagrammatic two-loop calculations for neutral and charged Higgs bosons as well as all-order resummation of large logarithms arising from the strong and Yukawa coupling sectors.

Keywords: Higgs bosons; MSSM; higher-order calculations; mass spectrum.

1. Introduction

The neutral Higgs-like particle with a mass at 125 GeV, discovered by the ATLAS and CMS experiments,[1,2] behaves within the presently still sizeable experimental uncertainties like the Higgs boson of the Standard Model (see Ref. 3 for the latest results), but on the other hand leaves ample room for interpretations within extended models with a richer spectrum. A scenario of particular interest thereby is the Minimal Supersymmetric Standard Model (MSSM) with two scalar doublets accommodating five physical Higgs bosons, at lowest order given by the light and heavy CP-even h and H, the CP-odd A, and the charged $H^{\pm}$ Higgs bosons. At lowest order, their masses can be parametrized in terms of the A-boson mass M_A and the ratio of the two vacuum expectation values, $\tan\beta \equiv v_2/v_1$. Higher-order contributions, however, give in general substantial corrections to the tree-level relations. A review of the calculations and an illustration of the recent results are presented in this talk.

2. The Higgs-Boson Masses of the MSSM

The parameters entering the MSSM Higgs-boson spectrum at lowest order can be chosen as the gauge-boson masses $M_{W,Z}$, the mass M_A, and $\tan\beta$. The other neutral and charged Higgs-boson masses are predicted in terms of these quantities:

$$m_{H^\pm}^2 = M_A^2 + M_W^2\,, \tag{1}$$

$$m_{h,H}^2 = \frac{1}{2}\Big(M_A^2 + M_Z^2 \mp \sqrt{\left(M_A^2 + M_Z^2\right)^2 - 4M_A^2 M_Z^2\cos^2(2\beta)}\,\Big)\,. \tag{2}$$

Higher-order contributions, however, give in general substantial corrections to the tree-level relations, with the dominant one-loop terms arise from the Yukawa sector. However, all the other sectors of the MSSM also contribute in a nonnegligible way in particular to the mass of the lightest Higgs particle, making thus m_h a valuable observable that is sensitive to the still unknown SUSY particles. Precision tests of the MSSM will therefore require precise calculations of the mass spectrum including higher-order terms.

3. Higher-Order Calculations for the Higgs-Boson Masses

The status of higher-order corrections to the masses and mixing angles in the neutral Higgs sector is quite advanced. A remarkable amount of work has been done for higher-order calculations of the mass spectrum, for real SUSY parameters[4–36] as well as for complex parameters.[37–46] They are based on full one-loop calculations improved by higher-order contributions to the leading terms from the Yukawa sector involving the top and bottom Yukawa couplings α_t, α_b and the strong coupling constant α_s. Recently, also the $\mathcal{O}(\alpha_t^2)$ terms for the complex version of the MSSM were computed.[45,46] All the available higher-order terms have been implemented into the public program `FeynHiggs`.[47–49]

Also the mass of the charged Higgs boson is affected by higher-order corrections when expressed in terms of M_A. The status is, however, somewhat less advanced as compared to the neutral Higgs bosons. Approximate one-loop corrections were already derived in Refs. 50–55. The first complete one-loop calculation in the Feynman-diagrammatic approach was done in Ref. 56, and more recently the corrections were re-evaluated in Refs. 44, 57 and 58. At the two-loop level, important ingredients for the leading corrections are the $\mathcal{O}(\alpha_t\alpha_s)$ and $\mathcal{O}(\alpha_t^2)$ contributions to the charged $H^\pm$ self-energy. The $\mathcal{O}(\alpha_t\alpha_s)$ part was obtained in Ref. 43 for the complex MSSM, where it is required for predicting the neutral Higgs-boson spectrum in the presence of CP-violating mixing of all three neutral CP eigenstates and the charged Higgs-boson mass used as an independent (on-shell) input parameter instead of M_A. In the CP-conserving case, on the other hand, with M_A conventionally chosen as an independent input quantity, the corresponding self-energy contribution has been exploited for obtaining corrections of $\mathcal{O}(\alpha_t\alpha_s)$ to the mass of the charged Higgs boson.[58] In an analogous way, the recently calculated $\mathcal{O}(\alpha_t^2)$ part of the $H^\pm$ self-energy in the complex MSSM,[45,46] has been utilized for the real, CP-conserving, case to derive the $\mathcal{O}(\alpha_t^2)$ corrections to the charged Higgs-boson mass as well.[59]

4. The Spectrum of the Real MSSM

A higher-orders, replacing (2), the masses M_h and M_H of the neutral Higgs-bosons are obtained from the poles of the propagator matrix $\Delta(q^2)$ dressed by the diagonal and nondiagonal self-energies,

$$\left(\Delta(q^2)\right)^{-1} = \begin{pmatrix} q^2 - m_H^2 + \hat{\Sigma}_H(q^2) & \hat{\Sigma}_{hH}(q^2) \\ \hat{\Sigma}_{Hh}(q^2) & q^2 - m_h^2 + \hat{\Sigma}_h(q^2) \end{pmatrix}, \tag{3}$$

where the self-energies $\hat{\Sigma}$ have been renormalized by adding the appropriate counter-terms. As far as the Higgs sector is concerned, the parameters M_A and $\tan\beta$ acquire renormalization, providing the counterterms δM_A^2 and $\delta\tan\beta$. In the hybrid on-shell/$\overline{\rm DR}$ scheme, they are obtained from the (unrenormalized) A-boson self-energy and the $\overline{\rm DR}$ field-renormalization constants of the two Higgs doublets,

$$\delta M_A^2 = \Sigma_A(M_A^2)\,, \quad \delta\tan\beta = \tan\beta(\delta Z_{H_2} - \delta Z_{H_1})\,, \tag{4}$$

preserving the mass M_A as the on-shell mass of the neutral A boson. Moreover, the loop contributions to the one-point functions, the h, H tadpoles, have to be canceled by tadpole counterterms δT_h, δT_H ensuring the vacua v_1, v_2 to remain the true vacua also at higher orders:

$$T_h + \delta T_h = 0\,, \quad T_H + \delta T_H = 0\,. \tag{5}$$

4.1. *Two-loop momentum-dependent self-energies*

In most of the higher-order calculations mentioned above, the self-energies beyond one loop are evaluated for vanishing external momenta, either in the Feynman-diagrammic approach or applying the effective-potential method. Going beyond the zero-momentum approximation, the momentum-dependence of the self-energies and the impact on the masses of h and H were obtained[12] at $\mathcal{O}(\alpha_t\alpha_s)$ for the real MSSM in the on-shell/$\overline{\rm DR}$ renormalization scheme as described above extended to the two-loop order. As a new element, one-loop subrenormalization (Fig. 1) is required for the colored sector of $\mathcal{O}(\alpha_s)$: for the top-quark and top-squark masses as well as the trilinear coupling A_t, with counterterms determined by on-shell conditions as formulated, e.g. in Ref. 43. In addition, a calculation with the colored sector renormalized in the $\overline{\rm DR}$ scheme was performed in Ref. 13. For illustration of the effects we display in Fig. 2 the mass shift ΔM_h of the h boson resulting exclusively from the q^2-dependence of the two-loop $\mathcal{O}(\alpha_t\alpha_s)$ part of the self-energies in (3), versus M_A and the gluino mass $M_{\tilde{g}}$, as obtained in Ref. 12 for the $M_h^{\rm max}$ scenario.[60]

4.2. *Resummation of large logarithms*

For high SUSY-breaking mass scales M_S in the scalar–quark sector, the logarithmically enhanced loop contributions $\sim \alpha_t L$, $\alpha_t^2 L^2$, $\alpha_t\alpha_s L^2$, $\alpha_t^2 L$, $\alpha_t\alpha_s L, \ldots$, with $L = \log\left(M_S^2/m_t^2\right)$, constitute the dominant part of the higher-order corrections

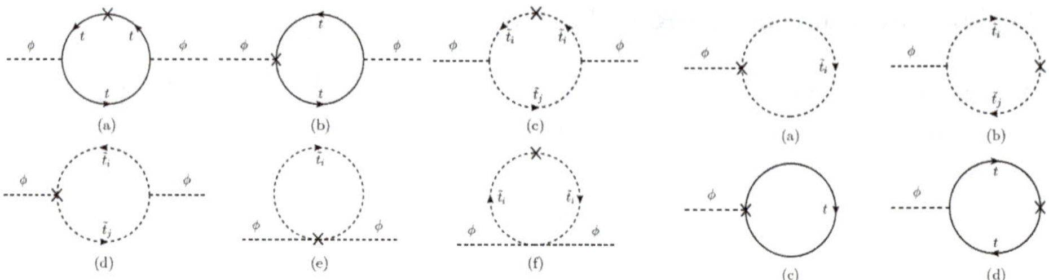

Fig. 1. Two-loop diagrams with one-loop counterterm insertions ($\phi = h, H, A$).

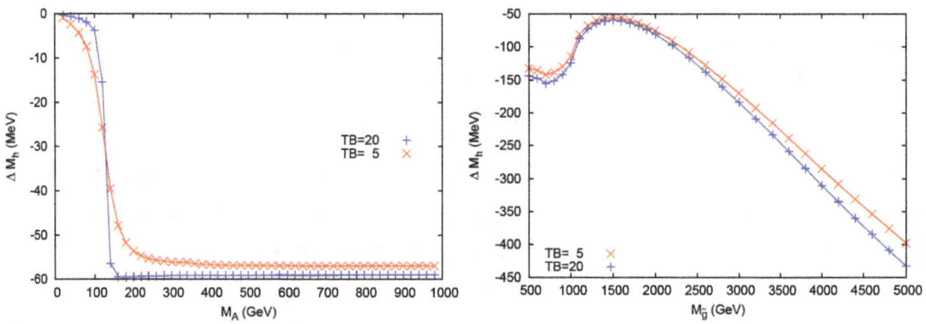

Fig. 2. Mass shift of the h boson arising from the two-loop $\mathcal{O}(\alpha_t\alpha_s)$ momentum dependence, for $\tan\beta = 5$ and 20 in the $M_h^{\max}$ scenario. $M_{\tilde{g}} = 1500$ GeV (left), $M_A = 250$ GeV (right).

ΔM_h^2 to the light Higgs-boson mass and have to be taken into account beyond the two-loop level for a reliable prediction. The fixed-order result obtained by the diagrammatic calculation, as available via `FeynHiggs`, has been combined[61] with an all-order resummation of the leading and subleading logarithmic contributions from the top/stop sector, derived via renormalization group equations for α_t, α_s, and the quartic Higgs coupling in the Standard Model as the effective theory below the large scale $M_S \gg M_Z$ (see also Refs. 62 and 63 for similar RGE studies).

In the combination of the logarithmic contributions obtained from solving the RGEs with the fixed-order result in `FeynHiggs`, double counting has to be avoided for the logarithmic contributions up to the two-loop level and the different schemes (on-shell versus $\overline{\text{MS}}$) employed in the diagrammatic and RGE approach have to be taken into account by converting the parameters properly. Figure 3 illustrates the strong impact of the resummation for large SUSY scales.

4.3. *The $M_{H^\pm} - M_A$ correlation at two-loop order*

In the real MSSM, the mass M_A of the CP-odd A boson is conventionally chosen as a free input parameter, fixing also the mass of the charged Higgs boson uniquely

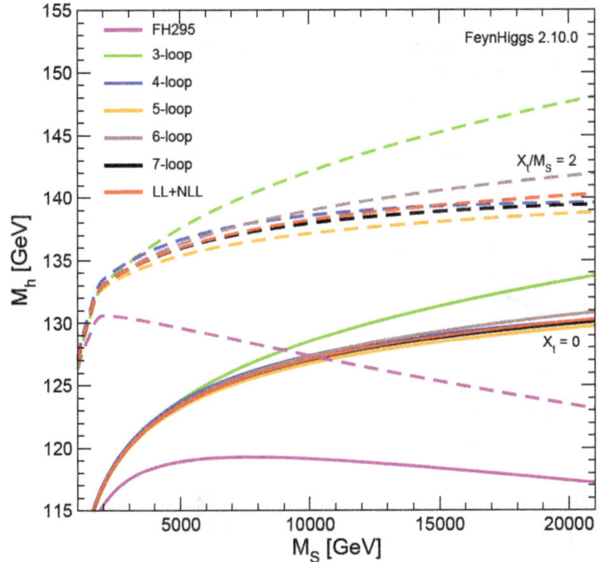

Fig. 3. M_h as a function of M_S for unmixed top-squarks ($X_t = 0$) and for L–R mixed squarks with $X_t/M_s = 2$ ($X_t = A_t - \mu \cot\beta$): Full result (LL+NLL) compared with the fixed-order result of `FeynHiggs` and with 3-loop, ... higher-order results containing the logarithmic contributions. $\tan\beta = 10$, $M_A = M_2 = \mu = 1000$ GeV, $M_{\tilde{g}} = 1500$ GeV.

at lowest order by Eq. (1). The correlation between the two masses is a decisive feature of the MSSM and thus phenomenologically important for testing the model; it is, however, subject to loop corrections as well, yielding the charged on-shell mass via $M^2_{H^\pm} = m^2_{H^\pm} + \Delta M^2_{H^\pm}$ with the correction $\Delta M^2_{H^\pm}$ derived from the charged Higgs-boson self-energy. At present, the renormalized self-energy,

$$\hat{\Sigma}_{H^\pm}(q^2) = \hat{\Sigma}^{(\text{1-loop})}_{H^\pm}(q^2) + \hat{\Sigma}^{(\alpha_t\alpha_s)}_{H^\pm}(0) + \hat{\Sigma}^{(\alpha_t^2)}_{H^\pm}(0)\,, \tag{6}$$

contains the full one-loop result and the leading two-loop terms from the Yukawa and strong interactions in the zero-momentum approximation[58,59] renormalized in the on-shell scheme. Thereby, the subloop renormalization for the colored sector is performed as in Subsec. 4.1 at $\mathcal{O}(\alpha_s)$, and augmented to $\mathcal{O}(\alpha_t)$ as described in Ref. 46. The numerical impact of the various higher-order terms are illustrated in Fig. 4. More details can be found in Ref. 59.

5. The Spectrum of the Complex MSSM

At lowest order, the MSSM Higgs sector is CP conserving; CP-violation is induced via loop contributions involving complex parameters from other SUSY sectors leading to mixing between h, H and A in the mass eigenstates.[64,65] The propagator matrix for the neutral Higgs fields becomes a 3×3 matrix, augmenting (3) by the diagonal and nondiagonal self-energies $\hat{\Sigma}_{\{A,Ah,AH\}}$, from which the masses are

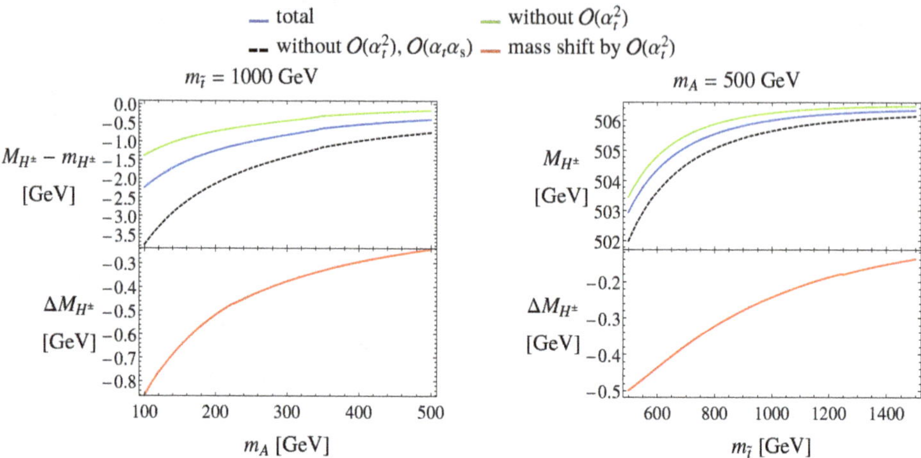

Fig. 4. Charged Higgs-boson mass $M_{H\pm}$ with all available terms and mass shift $\Delta M_{H\pm}$ exclusively from $\mathcal{O}(\alpha_t^2)$. $\tan\beta = 10$, $M_{\tilde{g}} = 1500$ GeV, $\mu = 2000$ GeV, $A_t = 1.5 m_{\tilde{t}}$.

obtained via the real parts of the complex poles. Differently from the real MSSM, M_A cannot be used as an input parameter; instead, the charged Higgs-boson mass $M_{H\pm}$ is used, which can be renormalized independently by an on-shell condition replacing the condition for M_A in (4). Moreover, a third tadpole condition for the A-field with an independent counterterm δT_A has to be added to the set (5). The self-energies in the complex MSSM are completely known at the one-loop level[44] and presently include the leading $\mathcal{O}(\alpha_t\alpha_s)$ and $\mathcal{O}(\alpha_t^2)$ contributions in the approximation of zero external momentum.[43,45,46]

The mass eigenvalues show a significant dependence on the phases of the SUSY parameters, especially of A_t and μ. The variation of the spectrum with the phase ϕ_{A_t} (and other parameters real) is illustrated in Fig. 5, showing mass shifts of typically 5 GeV. For $\phi_{A_t} \neq 0$, the lightest Higgs particle remains almost completely CP-even, whereas the two heavier ones develop large CP mixing.[46] This is of particular interest for $m_{h_1} \simeq 125$ GeV, where h_1 is Standard-Model like, but the features of the heavier Higgs bosons deviate substantially from those of the real MSSM.

Also the coefficient of the bilinear term of the superpotential, the Higgsino mass parameter μ, is in general a complex quantity. The phase of μ is severely constrained by the experimental limits on the electric dipole moments of electron and neutron. These bounds can, however, be circumvented in principle by a specific fine-tuning of the phases of μ and of the nonuniversal SUSY parameters,[66–73] leaving room also for a nonvanishing phase ϕ_μ. Potential effects of $\phi_\mu \neq 0$ are illustrated in terms of an example in Fig. 6, showing the influence of the phases ϕ_μ and ϕ_{X_t} (with $X_t = A_t^* - \mu/\tan\beta$) on the mass of the lightest Higgs boson.[46] The quantity Δm_{h_1} in the right panel is the difference $m_{h_1}(\phi_{X_t}, \phi_\mu) - m_{h_1}(0,0)$ resulting exclusively from the $\mathcal{O}(\alpha_t^2)$ terms. In the depicted scenario the variation of m_{h_1} with ϕ_μ is of

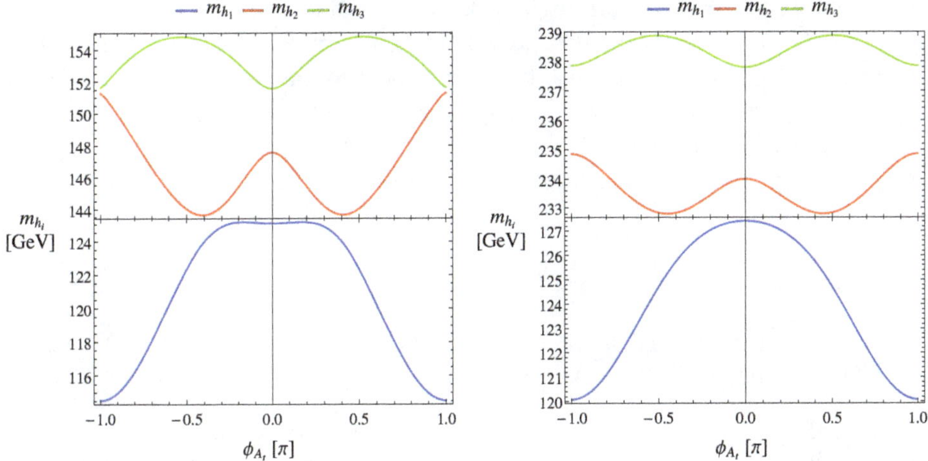

Fig. 5. The neutral Higgs-boson mass eigenvalues m_{h_1}, m_{h_2}, m_{h_3}: dependence on the phase of the complex A_t parameter. Left: $M_{H^\pm} = 170$ GeV, right: $M_{H^\pm} = 250$ GeV. The other parameters are: $\tan\beta = 10$, $m_{\tilde{t}} = 1500$ GeV, $|A_t| = 2m_{\tilde{t}}$, $\mu = 2500$ GeV, $M_{\tilde{g}} = 1500$ GeV.

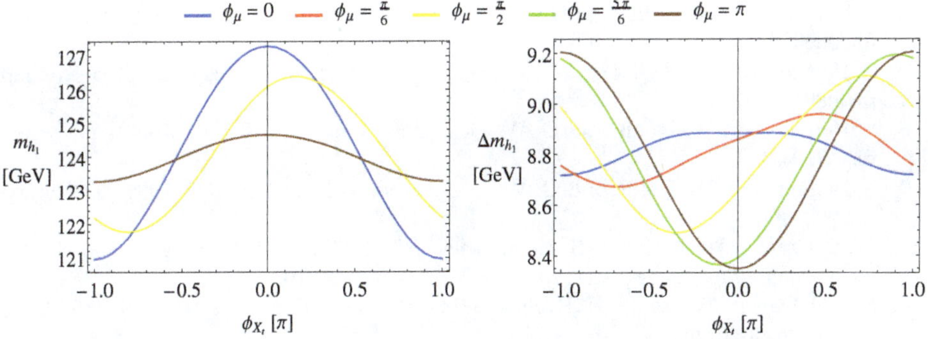

Fig. 6. Dependence of the lightest neutral Higgs-boson mass on the phases ϕ_{X_t} and ϕ_μ. Left: prediction for m_{h_1} including all available contributions, with the phase dependence arising from one-loop, $\mathcal{O}(\alpha_t\alpha_s)$ and $\mathcal{O}(\alpha_t^2)$ terms. Right: the contribution Δm_{h_1} to m_{h_1} owing exclusively to the $\mathcal{O}(\alpha_t^2)$ terms, for different phases. The input parameters are $M_{H^\pm} = 200$ GeV, $\tan\beta = 10$, $|\mu| = 2500$ GeV, $m_{\tilde{t}} = 1500$ GeV, $|X_t| = 2m_{\tilde{t}}$, $M_{\tilde{g}} = 2000$ GeV.

the order of 0.5 GeV. Reversing the sign of ϕ_μ mirrors the graphs at the central axis $\phi_{X_t} = 0$.

6. Conclusions

With the observed Higgs-boson like particle with a mass of 125 GeV at the LHC, a wide range of interpretations in various models beyond the Standard Model is still the current reality. Since its mass is meanwhile a precisely measured quantity, in models like the MSSM where the mass follows as a prediction depending basically on the entire SUSY spectrum, it may serve as an additional precision observable for dedicated tests of the model, providing thus a sensitive probe of other SUSY sectors. A large amount of theoretical work has been done to achieve predictions with the required accuracy; recent progress and sample results have been reported in this talk. The precise predictions will become of special interest if one or more further Higgs particles would be discovered in the upcoming run of the LHC.

References

1. ATLAS Collab. (G. Aad *et al.*), *Phys. Lett. B* **716**, 1 (2012), arXiv:1207.7214 [hep-ex].
2. CMS Collab. (S. Chatrchyan *et al.*), *Phys. Lett. B* **716**, 30 (2012), arXiv:1207.7235 [hep-ex].
3. ATLAS and CMS Collabs. (M. Duehrssen), Talk given at the *50th Rencontre de Moriond on Electroweak Interactions and Unified Theories*, La Thuile, France, March 2015.
4. J. A. Casas, J. R. Espinosa, M. Quiros and A. Riotto, *Nucl. Phys. B* **436**, 3 (1995) [Erratum: *ibid.* **439**, 466 (1995)], arXiv:hep-ph/9407389.
5. M. S. Carena, J. R. Espinosa, M. Quiros and C. E. M. Wagner, *Phys. Lett. B* **355**, 209 (1995), arXiv:hep-ph/9504316.
6. S. Heinemeyer, W. Hollik and G. Weiglein, *Phys. Rev. D* **58**, 091701 (1998), arXiv:hep-ph/9803277.
7. S. Heinemeyer, W. Hollik and G. Weiglein, *Phys. Lett. B* **440**, 296 (1998), arXiv:hep-ph/9807423.
8. S. Heinemeyer, W. Hollik and G. Weiglein, *Eur. Phys. J. C* **9**, 343 (1999), arXiv:hep-ph/9812472.
9. S. Heinemeyer, W. Hollik and G. Weiglein, *Phys. Lett. B* **455**, 179 (1999), arXiv:hep-ph/9903404.
10. M. S. Carena, H. E. Haber, S. Heinemeyer, W. Hollik, C. E. M. Wagner and G. Weiglein, *Nucl. Phys. B* **580**, 29 (2000), arXiv:hep-ph/0001002.
11. S. Heinemeyer, W. Hollik, H. Rzehak and G. Weiglein, *Eur. Phys. J. C* **39**, 465 (2005), arXiv:hep-ph/0411114.
12. S. Borowka, T. Hahn, S. Heinemeyer, G. Heinrich and W. Hollik, *Eur. Phys. J. C* **74**, 2994 (2014), arXiv:1404.7074 [hep-ph].
13. G. Degrassi, S. Di Vita and P. Slavich, arXiv:1410.3432 [hep-ph].
14. R. Harlander, P. Kant, L. Mihaila and M. Steinhauser, *Phys. Rev. Lett.* **100**, 191602 (2008) [Erratum: *ibid.* **101**, 039901 (2008)], arXiv:0803.0672 [hep-ph].
15. R. Harlander, P. Kant, L. Mihaila and M. Steinhauser, *J. High Energy Phys.* **1008**, 104 (2010), arXiv:1005.5709 [hep-ph].
16. R.-J. Zhang, *Phys. Lett. B* **447**, 89 (1999), arXiv:hep-ph/9808299.
17. J. R. Espinosa and R.-J. Zhang, *Nucl. Phys. B* **586**, 3 (2000), arXiv:hep-ph/0003246.
18. J. R. Espinosa and R.-J. Zhang, *J. High Energy Phys.* **0003**, 026 (2000), arXiv:hep-ph/9912236.
19. J. R. Espinosa and I. Navarro, *Nucl. Phys. B* **615**, 82 (2001), arXiv:hep-ph/0104047.
20. G. Degrassi, P. Slavich and F. Zwirner, *Nucl. Phys. B* **611**, 403 (2001), arXiv:hep-ph/0105096.
21. R. Hempfling and A. H. Hoang, *Phys. Lett. B* **331**, 99 (1994), arXiv:hep-ph/9401219.
22. A. Brignole, G. Degrassi, P. Slavich and F. Zwirner, *Nucl. Phys. B* **643**, 79 (2002), arXiv:hep-ph/0206101.

23. A. Dedes, G. Degrassi and P. Slavich, *Nucl. Phys. B* **672**, 144 (2003), arXiv:hep-ph/0305127.
24. J. R. Espinosa and R.-J. Zhang, *Nucl. Phys. B* **586**, 3 (2000), arXiv:hep-ph/0003246.
25. A. Brignole, G. Degrassi, P. Slavich and F. Zwirner, *Nucl. Phys. B* **631**, 195 (2002), arXiv:hep-ph/0112177.
26. G. Degrassi, S. Heinemeyer, W. Hollik, P. Slavich and G. Weiglein, *Eur. Phys. J. C* **28**, 133 (2003), arXiv:hep-ph/0212020.
27. S. Heinemeyer, W. Hollik and G. Weiglein, *Phys. Rep.* **425**, 265 (2006), arXiv:hep-ph/0412214.
28. B. C. Allanach, A. Djouadi, J. L. Kneur, W. Porod and P. Slavich, *J. High Energy Phys.* **0409**, 044 (2004), arXiv:hep-ph/0406166.
29. S. P. Martin, *Phys. Rev. D* **65**, 116003 (2002), arXiv:hep-ph/0111209.
30. S. P. Martin, *Phys. Rev. D* **66**, 096001 (2002), arXiv:hep-ph/0206136.
31. S. P. Martin, *Phys. Rev. D* **67**, 095012 (2003), arXiv:hep-ph/0211366.
32. S. P. Martin, *Phys. Rev. D* **68**, 075002 (2003), arXiv:hep-ph/0307101.
33. S. P. Martin, *Phys. Rev. D* **70**, 016005 (2004), arXiv:hep-ph/0312092.
34. S. P. Martin, *Phys. Rev. D* **71**, 016012 (2005), arXiv:hep-ph/0405022.
35. S. P. Martin, *Phys. Rev. D* **71**, 116004 (2005), arXiv:hep-ph/0502168.
36. S. P. Martin and D. G. Robertson, *Comput. Phys. Commun.* **174**, 133 (2006), arXiv:hep-ph/0501132.
37. D. A. Demir, *Phys. Rev. D* **60**, 055006 (1999), arXiv:hep-ph/9901389.
38. S. Y. Choi, M. Drees and J. S. Lee, *Phys. Lett. B* **481**, 57 (2000), arXiv:hep-ph/0002287.
39. T. Ibrahim and P. Nath, *Phys. Rev. D* **63**, 035009 (2001), arXiv:hep-ph/0008237.
40. T. Ibrahim and P. Nath, *Phys. Rev. D* **66**, 015005 (2002), arXiv:hep-ph/0204092.
41. A. Pilaftsis and C. E. M. Wagner, *Nucl. Phys. B* **553**, 3 (1999), arXiv:hep-ph/9902371.
42. M. S. Carena, J. R. Ellis, A. Pilaftsis and C. E. M. Wagner, *Nucl. Phys. B* **586**, 92 (2000), arXiv:hep-ph/0003180.
43. S. Heinemeyer, W. Hollik, H. Rzehak and G. Weiglein, *Phys. Lett. B* **652**, 300 (2007), arXiv:0705.0746 [hep-ph].
44. M. Frank, T. Hahn, S. Heinemeyer, W. Hollik, H. Rzehak and G. Weiglein, *J. High Energy Phys.* **0702**, 047 (2007), arXiv:hep-ph/0611326.
45. W. Hollik and S. Paßehr, *Phys. Lett. B* **733**, 144 (2014), arXiv:1401.8275 [hep-ph].
46. W. Hollik and S. Paßehr, *J. High Energy Phys.* **1410**, 171 (2014), arXiv:1409.1687 [hep-ph].
47. S. Heinemeyer, W. Hollik and G. Weiglein, *Comput. Phys. Commun.* **124**, 76 (2000), arXiv:hep-ph/9812320.
48. T. Hahn, S. Heinemeyer, W. Hollik, H. Rzehak and G. Weiglein, *Nucl. Phys. B* (*Proc. Suppl.*) **205-206**, 152 (2010), arXiv:1007.0956 [hep-ph].
49. T. Hahn, S. Heinemeyer, W. Hollik, H. Rzehak and G. Weiglein, *Comput. Phys. Commun.* **180**, 1426 (2009).
50. J. Gunion and A. Turski, *Phys. Rev. D* **39**, 2701 (1989).
51. J. Gunion and A. Turski, *Phys. Rev. D* **40**, 2333 (1989).
52. A. Brignole, J. Ellis, G. Ridolfi and F. Zwirner, *Phys. Lett. B* **271**, 123 (1991).
53. A. Brignole, *Phys. Lett. B* **277**, 313 (1992).
54. M. Diaz and H. Haber, *Phys. Rev. D* **45**, 4246 (1992).
55. M. Diaz, Ph.D. thesis: Radiative corrections to Higgs masses in the MSSM, University of California, Santa Cruz, 1992, SCIPP-92/13.
56. P. Chankowski, S. Pokorski and J. Rosiek, *Phys. Lett.* **274**, 191 (1992).

57. M. Frank, Ph.D. thesis: Radiative corrections in the Higgs sector of the MSSM with *CP* violation, University of Karlsruhe, 2002, ISBN 3-937231-01-3.
58. M. Frank, L. Galeta, T. Hahn, S. Heinemeyer, W. Hollik, H. Rzehak and G. Weiglein, *Phys. Rev. D* **88**, 055013 (2013), arXiv:1306.1156 [hep-ph].
59. W. Hollik and S. Paßehr, arXiv:1502.02394 [hep-ph].
60. M. Carena, S. Heinemeyer, O. Stål, C. E. M. Wagner and G. Weiglein, *Eur. Phys. J. C* **73**, 2552 (2013), arXiv:1302.7033 [hep-ph].
61. T. Hahn, S. Heinemeyer, W. Hollik, H. Rzehak and G. Weiglein, *Phys. Rev. Lett.* **112**, 141801 (2014), arXiv:1312.4937 [hep-ph].
62. P. Draper, G. Lee and C. E. M. Wagner, *Phys. Rev. D* **89**, 055023 (2014), arXiv:1312.5743 [hep-ph].
63. E. Bagnaschi, G. F. Giudice, P. Slavich and A. Strumia, *J. High Energy Phys.* **1409**, 092 (2014), arXiv:1407.4081 [hep-ph].
64. A. Pilaftsis, *Phys. Rev. D* **58**, 096010 (1998), arXiv:hep-ph/9803297.
65. A. Pilaftsis, *Phys. Lett. B* **435**, 88 (1998), arXiv:hep-ph/9805373.
66. A. Masiero and L. Silvestrini, *Perspectives on Supersymmetry*, ed. G. L. Kane (World Scientific, 1997), pp. 423–441, arXiv:hep-ph/9709242.
67. M. Brhlik, G. J. Good and G. L. Kane, *Phys. Rev. D* **59**, 115004 (1999), arXiv:hep-ph/9810457.
68. M. Brhlik, L. L. Everett, G. L. Kane and J. D. Lykken, *Phys. Rev. Lett.* **83**, 2124 (1999), arXiv:hep-ph/9905215.
69. T. Ibrahim and P. Nath, *Phys. Lett. B* **418**, 98 (1998), arXiv:hep-ph/9707409.
70. T. Ibrahim and P. Nath, *Phys. Rev. D* **58**, 111301 (1998) [Erratum: *ibid.* **60**, 099902 (1999)], arXiv:hep-ph/9807501.
71. T. Ibrahim and P. Nath, *Phys. Rev. D* **61**, 093004 (2000), arXiv:hep-ph/9910553.
72. T. Ibrahim and P. Nath, *Phys. Rev. D* **57**, 478 (1998) [Erratum: *ibid.* **58**, 019901 (1998); *ibid.* **60**, 079903 (1999); *ibid.* **60**, 119901 (1999)], arXiv:hep-ph/9708456.
73. A. Bartl, T. Gajdosik, W. Porod, P. Stockinger and H. Stremnitzer, *Phys. Rev. D* **60**, 073003 (1999), arXiv:hep-ph/9903402.

Minimal Supersymmetric Standard Model with Gauged Baryon and Lepton Numbers

Bartosz Fornal

Department of Physics and Astronomy, University of California, Irvine, CA 92697, USA

bfornal@uci.edu

A simple extension of the minimal supersymmetric standard model in which baryon and lepton numbers are local gauge symmetries spontaneously broken at the supersymmetry scale is reported. This theory provides a natural explanation for proton stability. Despite violating R-parity, it contains a dark matter candidate carrying baryon number that can be searched for in direct detection experiments. The model accommodates a light active neutrino spectrum and predicts one heavy and two light sterile neutrinos. It also allows for lepton number violating processes testable at the Large Hadron Collider.

Keywords: Baryon number violation; dark matter; neutrinos.

1. Introduction

There have been several attempts in the 70s, 80s and 90s at promoting the accidental baryon and lepton number global symmetries of the standard model (SM) to local gauge symmetries (see Refs. 1–5). The first complete nonsupersymmetric model with gauged baryon (B) and lepton number (L) was constructed by P. Fileviez Pérez and M. B. Wise in Ref. 6. More realistic nonsupersymmetric models avoiding all current experimental constraints were introduced by M. Duerr, P. Fileviez Pérez and M. B. Wise in Ref. 7 and by P. Fileviez Pérez, S. Ohmer and H. H. Patel in Ref. 8. The former model was further generalized and analyzed by M. Duerr and P. Fileviez Pérez in Ref. 9. Its full supersymmetric version was constructed in Ref. 10 within the collaboration between P. Fileviez Pérez, S. Spinner, J. M. Arnold and the author, and is the subject of this paper.

The minimal supersymmetric standard model (MSSM) extended by right-handed (RH) neutrino superfields is one of the most promising candidates for physics beyond the standard model. It has a number of attractive features: it does not suffer from a large hierarchy problem, contains a dark matter (DM) candidate, and explains gauge coupling unification. However, despite all of those virtues, the MSSM also has its weaknesses. For instance, the discrete symmetry, R-parity,

$$R = (-1)^{3(B-L)+2s}\,, \tag{1}$$

needed to forbid proton decay at tree level and assure the stability of DM, is simply imposed by hand. Another uncomfortable aspect of the MSSM is connected to proton decay through nonrenormalizable operators. In particular, even after enforcing R-parity, there exist dangerous dimension-5 operators triggering proton decay:

$$\frac{c_1}{\Lambda}\,\hat{Q}\hat{Q}\hat{Q}\hat{L}\,, \quad \frac{c_2}{\Lambda}\,\hat{u}^c\hat{u}^c\hat{d}^c\hat{e}^c\,. \tag{2}$$

The constraint on the coefficient c_1 is model-dependent. However, assuming that all the couplings participating in the interaction are of order one sets the following upper bound on this coefficient:

$$|c_1| \lesssim 10^{-25}\,\frac{m_{\rm soft}M_{\rm GUT}}{(1\ {\rm TeV})^2} \approx 10^{-9}\,. \tag{3}$$

This number is small enough to be viewed as a problem. It turns out that gauging baryon and lepton number solves both of those issues.

2. The Model

The theory is based on the SM gauge group extended by the additional baryon and lepton number gauge symmetries:

$$SU(3)_c \times SU(2)_L \times U(1)_Y \times U(1)_B \times U(1)_L\,. \tag{4}$$

Without new particles, such a theory with just the MSSM particle spectrum is inconsistent, since it contains gauge anomalies. There are many ways to add new fields to make the theory anomaly-free. However, driven by the requirement of minimality, we choose to cancel the anomalies with the set of new fields given in Table 1, which introduces the minimum possible number of new degrees of freedom.

Apart from the RH neutrino superfields, we add six new leptobaryonic superfields, each containing a fermionic and scalar component. They all carry baryon and lepton quantum numbers satisfying the conditions:

$$B_1 + B_2 = -3 \quad \text{and} \quad L_1 + L_2 = -3\,, \tag{5}$$

which are required by anomaly cancellation. The new superfields are vector-like with respect to the SM. This is why they can have vector-like masses and avoid the stringent experimental constraints on chiral matter. Now, since the $\hat{X}$ superfield is a SM singlet, its fermionic and scalar components are possible candidates for DM. The other fields should be unstable since they carry nonzero electric charge.

In order for those fields to actually obtain vector-like masses, we have to introduce new Higgs superfields into the theory. Because of the anomaly cancellation conditions, those new Higgs superfields, apart from being singlets under the SM gauge group, must carry baryon and lepton number 3 or -3:

$$\hat{S}_1 = (1,1,0,3,3) \quad \text{and} \quad \hat{S}_2 = (1,1,0,-3,-3)\,. \tag{6}$$

Table 1. New particle content of the model.

Field	$SU(3)$	$SU(2)_L$	$U(1)_Y$	$U(1)_B$	$U(1)_L$
$\hat{\nu}_{iR}$	1	1	0	0	1
$\hat{\Psi}$	1	2	$-1/2$	B_1	L_1
$\hat{\Psi}^c$	1	2	$1/2$	B_2	L_2
$\hat{\eta}$	1	1	-1	$-B_2$	$-L_2$
$\hat{\eta}^c$	1	1	1	$-B_1$	$-L_1$
$\hat{X}$	1	1	0	$-B_2$	$-L_2$
$\hat{X}^c$	1	1	0	$-B_1$	$-L_1$

The scalar components of those superfields get vacuum expectation values (VEVs) below the B and L breaking scale. Because of their charges, baryon and lepton number can only be broken by a multiple of three units, so the theory predicts that:

Proton is stable.

More generally, there are no interactions violating baryon or lepton number at the renormalizable level. The least suppressed nonrenormalizable operators generating B and L violation are dimension-14:

$$W_{14} = \frac{1}{\Lambda^{10}}\left[c_1\hat{S}_1(\hat{u}^c\hat{u}^c\hat{d}^c\hat{e}^c)^3 + c_2\hat{S}_1(\hat{u}^c\hat{d}^c\hat{d}^c\hat{\nu}^c)^3 + c_3\hat{S}_2(\hat{Q}\hat{Q}\hat{Q}\hat{L})^3\right]. \qquad (7)$$

The superpotential generating masses for the particles in our model is the following:

$$\begin{aligned}\mathcal{W} &= Y_u\hat{Q}\hat{H}_u\hat{u}^c + Y_d\hat{Q}\hat{H}_d\hat{d}^c + Y_e\hat{L}\hat{H}_d\hat{e}^c + Y_\nu\hat{L}\hat{H}_u\hat{\nu}^c + \mu\hat{H}_u\hat{H}_d\\ &+ Y_1\hat{\Psi}\hat{H}_d\hat{\eta}^c + Y_2\hat{\Psi}\hat{H}_u\hat{X}^c + Y_3\hat{\Psi}^c\hat{H}_u\hat{\eta} + Y_4\hat{\Psi}^c\hat{H}_d\hat{X}\\ &+ \lambda_1\hat{\Psi}\hat{\Psi}^c\hat{S}_1 + \lambda_2\hat{\eta}\hat{\eta}^c\hat{S}_2 + \lambda_3\hat{X}\hat{X}^c\hat{S}_2 + \mu_{BL}\hat{S}_1\hat{S}_2\,. \end{aligned} \qquad (8)$$

The first line consists of the usual Yukawa and Higgs terms responsible for the MSSM masses. The second line contains new Yukawa interactions between the leptobaryons. The third line provides vector-like masses for the leptobaryons and the new Higgses. There are no terms mixing the new sector with the MSSM. The new fields couple to the MSSM only through the gauge sector. Since there are no operators violating baryon or lepton number, R-parity does not have to be imposed to avoid proton decay.

It turns out that both before and after B and L breaking, the Lagrangian has a discrete Z_2 symmetry — it is invariant with respect to the following transformation:

$$\hat{\Psi} \to -\hat{\Psi}\,,\quad \hat{\Psi}^c \to -\hat{\Psi}^c\,,\quad \hat{\eta} \to -\hat{\eta}\,,\quad \hat{\eta}^c \to -\hat{\eta}^c\,,\quad \hat{X} \to -\hat{X}\,,\quad \hat{X}^c \to -\hat{X}^c\,, \qquad (9)$$

under which all leptobaryons are odd and all MSSM fields are even. The consequence of this accidental Z_2 symmetry is that the lightest leptobaryon is stable, since it cannot decay to MSSM particles. If the lightest leptobaryon is also electrically neutral, like the $\hat{X}$ superfield, its fermionic and scalar components are possible DM candidates. In our analysis, we assumed that the fermionic component of $\hat{X}$, the $\tilde{X}$, is the DM. It interacts with the MSSM only through the gauge bosons and its mass, given by the VEV of the scalar component of $\hat{S}_2$, is set at the baryon and lepton number breaking scale.

3. Symmetry Breaking

The scalar potential relevant for B and L breaking is

$$\begin{aligned} V &= \left(M_1^2 + |\mu_{BL}|^2\right)|S_1|^2 + \left(M_2^2 + |\mu_{BL}|^2\right)|S_2|^2 \\ &\quad + \frac{9}{2}g_B^2\left(|S_1|^2 - |S_2|^2\right)^2 + \frac{1}{2}g_L^2\left(3|S_1|^2 - 3|S_2|^2\right)^2 - \left(b_{BL}S_1S_2 + \text{h.c.}\right). \end{aligned} \tag{10}$$

After the scalar components of the new Higgs superfields $\hat{S}_1$ and $\hat{S}_2$ get their VEVs, $\langle S_1\rangle \equiv v_1$ and $\langle S_2\rangle \equiv v_2$, the resulting mass matrix for the new Z' gauge bosons has a zero determinant. This signalizes the existence of a new massless gauge boson and is ruled out by experiment. It is simple to understand why this happens — after B and L breaking there is still a residual $U(1)_{B-L}$ gauge symmetry,

$$U(1)_B \times U(1)_L \to U(1)_{B-L}\,, \tag{11}$$

with a Z' gauge boson that does not get a mass through the Higgs superfields. A natural solution to this problem is to introduce a new superfield whose scalar component does get a VEV which breaks $B-L$. This is clearly an option, however, in order to keep the model as minimal as possible, one can use a field already existing in the theory. A perfect candidate is the RH sneutrino, since it is a SM singlet and carries nonzero lepton charge, so it obviously breaks $B-L$. A nonzero VEV for the RH sneutrino,

$$\langle\tilde{\nu}_R\rangle \equiv v_R \neq 0\,, \tag{12}$$

produces additional terms in the scalar potential,

$$\begin{aligned} V &= \left(M_1^2 + |\mu_{BL}|^2\right)|S_1|^2 + \left(M_2^2 + |\mu_{BL}|^2\right)|S_2|^2 + \frac{9}{2}g_B^2\left(|S_1|^2 - |S_2|^2\right)^2 \\ &\quad + M_{\tilde{\nu}^c}^2|\tilde{\nu}^c|^2 + \frac{1}{2}g_L^2\left(3|S_1|^2 - 3|S_2|^2 - |\tilde{\nu}^c|^2\right)^2 - \left(b_{BL}S_1S_2 + \text{h.c.}\right), \end{aligned} \tag{13}$$

which now gives a mass matrix for the Z' gauge bosons of the form:

$$\mathcal{M}_{Z'}^2 = 9\begin{pmatrix} g_B^2(v_1^2+v_2^2) & g_Bg_L(v_1^2+v_2^2) \\ g_Bg_L(v_1^2+v_2^2) & g_L^2(v_1^2+v_2^2) + \frac{1}{9}g_L^2v_R^2 \end{pmatrix}, \tag{14}$$

with both eigenvalues positive.

The breaking of $B-L$ implies that R-parity, which is a discrete subgroup of $U(1)_{B-L}$, is also spontaneously broken. However, proton is still stable since the VEV of the RH sneutrino breaks only lepton number. Analyzing the symmetry breaking conditions for the scalar potential reveals that the VEVs: v_1, v_2 and v_R

have to be of the same order as the supersymmetry breaking soft terms, simply because they enter the potential on equal footing. Another way to phrase this is that $U(1)_B$, $U(1)_L$ and R-parity have to be broken at the SUSY scale, which we set around a TeV. The violation of R-parity at the TeV scale opens the door to many interesting and exotic signatures at the LHC, including decays of the LSPs. On top of that, since our model has a DM candidate not affected by R-parity violation, one expects an additional presence of missing energy signatures.

4. Neutrino Sector

After B and L breaking the model reduces to an effective $B-L$ gauged model. It turns out that the breaking of the remnant $U(1)_{B-L}$ gauge symmetry by the VEV of the RH sneutrino has very interesting consequences for neutrino physics. Those were already studied in the case of a pure $B-L$ gauged extension of the MSSM in Refs. 11–15. The predictions for the neutrino sector are very similar here and we discuss them following the analysis done by V. Barger, P. Fileviez Pérez and S. Spinner in Ref. 12.

First of all, the existence of RH neutrinos in our model allows for standard Dirac mass terms through Yukawa couplings. In addition, R-parity violation triggers mixing between the neutrinos and neutralinos, since after $B-L$ breaking they have common mass terms. This generates Majorana masses for the left-handed (LH) neutrinos.

One can show that only one generation of RH sneutrinos can attain a significant VEV. Since this VEV is at the SUSY breaking scale, we expect it to be around a TeV. The other RH sneutrino VEVs have to be very small and we ignore them:

$$\langle \tilde{\nu}_{R1} \rangle \equiv v_R \sim \text{TeV}\,, \quad \langle \tilde{\nu}_{R2} \rangle = \langle \tilde{\nu}_{R3} \rangle = 0\,. \tag{15}$$

The mixing between neutrinos and neutralinos comes from the gauge sector term $g_{BL} v_R (\nu_R^c \tilde{B}')$ and the superpotential term $Y_\nu v_R (l^T i\sigma_2 \tilde{H}_u)$. In principle, the situation is a little more complicated, since the LH sneutrinos can also develop VEVs. These VEVs, however, have to be very small, since they produce mixing between the LH neutrinos and neutralinos, which directly affects the active neutrino masses. This mixing arises from the terms: $g_{BL} v_L (\nu_L \tilde{B}')$, $g_1 v_L (\nu_L \tilde{B})$, $g_2 v_L (\nu_L \tilde{W}^0)$ and $Y_\nu v_L (\nu_R^c \tilde{H}_u)$. Consistency with experiment requires:

$$\langle \tilde{\nu}_{Li} \rangle \equiv v_{Li} \lesssim \text{MeV}\,. \tag{16}$$

The contributions to the neutrino masses can be described by the 11×11 neutrino–neutralino mass matrix. In the basis $(\nu_{iL}, \nu_{jR}^c, \tilde{B}', \tilde{B}^0, \tilde{W}^0, \tilde{H}_d^0, \tilde{H}_u^0)$, i.e. the three active LH neutrinos, three sterile RH neutrinos and the five heavy gauginos and Higgsinos, this matrix takes the form:

$$\begin{pmatrix} 0_{3\times3} & \frac{v_u}{\sqrt{2}}Y_\nu^{j1} & \frac{v_u}{\sqrt{2}}Y_\nu^{j2} & \frac{v_u}{\sqrt{2}}Y_\nu^{j3} & -\frac{1}{2}g_{BL}v_{Li} & -\frac{1}{2}g_1v_{Li} & \frac{1}{2}g_2v_{Li} & 0 & \frac{v_R}{\sqrt{2}}Y_\nu^{i3} \\ \frac{v_u}{\sqrt{2}}Y_\nu^{1j} & 0 & 0 & 0 & 0 & 0 & 0 & 0 & \frac{v_{Lj}}{\sqrt{2}}Y_\nu^{1j} \\ \frac{v_u}{\sqrt{2}}Y_\nu^{2j} & 0 & 0 & 0 & 0 & 0 & 0 & 0 & \frac{v_{Lj}}{\sqrt{2}}Y_\nu^{2j} \\ \frac{v_u}{\sqrt{2}}Y_\nu^{3j} & 0 & 0 & 0 & \frac{1}{2}g_{BL}v_R & 0 & 0 & 0 & \frac{v_{Lj}}{\sqrt{2}}Y_\nu^{3j} \\ -\frac{1}{2}g_{BL}v_{Li} & 0 & 0 & \frac{1}{2}g_{BL}v_R & M_{BL} & 0 & 0 & 0 & 0 \\ -\frac{1}{2}g_1v_{Li} & 0 & 0 & 0 & 0 & M_1 & 0 & -\frac{1}{2}g_1v_d & \frac{1}{2}g_1v_u \\ \frac{1}{2}g_2v_{Li} & 0 & 0 & 0 & 0 & 0 & M_2 & \frac{1}{2}g_2v_d & -\frac{1}{2}g_2v_u \\ 0 & 0 & 0 & 0 & 0 & -\frac{1}{2}g_1v_d & \frac{1}{2}g_2v_d & 0 & -\mu \\ \frac{v_R}{\sqrt{2}}Y_\nu^{3i} & \frac{v_{Lj}}{\sqrt{2}}Y_\nu^{j1} & \frac{v_{Lj}}{\sqrt{2}}Y_\nu^{j2} & \frac{v_{Lj}}{\sqrt{2}}Y_\nu^{j3} & 0 & \frac{1}{2}g_1v_u & -\frac{1}{2}g_2v_u & -\mu & 0 \end{pmatrix}. \tag{17}$$

The 6×6 subblock involving the LH and RH neutrinos contains just Dirac mass terms. Since only one generation of RH sneutrinos develops a TeV-scale VEV, this implies that only one of the RH neutrinos obtains a TeV mass through the seesaw mechanism. Thus, effectively, there is a 5×5 subblock for the light states (three active and two sterile neutrinos) on the order of 0.1 eV, and a 6×6 TeV-scale subblock involving the neutralinos and the heavy RH neutrino.

The off-diagonal blocks take part in generating Majorana masses for the active neutrinos through an effective type I seesaw mechanism. Experimental limits on the active neutrino spectrum constrain those two subblocks to be at the MeV scale at most. All those constraints can be translated into upper bounds on the Yukawa couplings and the aforementioned VEV of the LH sneutrino:

$$Y_\nu^{ij} \lesssim 10^{-12}, \quad Y_\nu^{i3} \lesssim 10^{-6}, \quad v_{Li} \lesssim 1 \text{ MeV}. \tag{18}$$

The hierarchy of scales in the full 11×11 mass matrix:

$$\mathcal{M} = \begin{pmatrix} (m)_{5\times5} & (m_D)_{5\times6} \\ (m_D)^T_{6\times5} & (M_{\text{heavy}})_{6\times6} \end{pmatrix} \sim \begin{pmatrix} \lesssim 0.1 \text{ eV} & \lesssim \text{MeV} \\ \lesssim \text{MeV} & \sim \text{TeV} \end{pmatrix} \tag{19}$$

enables us to diagonalize the 5×5 light neutrino mass matrix perturbatively, using the standard seesaw formula:

$$\mathcal{M}_{\text{light}} = m - m_D(M_{\text{heavy}})^{-1}m_D^T. \tag{20}$$

Keeping only terms with a maximum product of two small flavorful parameters:

$$\frac{v_{Li}}{m_{\text{soft}}}, \frac{Y_\nu^{i3}v_u}{m_{\text{soft}}} \lesssim 10^{-3}, \tag{21}$$

yields the following mass matrix for the five light neutrino states:

$$\mathcal{M}_{\text{light}} = \begin{pmatrix} Av_{Li}v_{Lj} + Bv_u(Y_\nu^{i3}v_{Lj} + Y_\nu^{j3}v_{Li}) + Cv_u^2Y_\nu^{i3}Y_\nu^{j3} & \frac{1}{\sqrt{2}}Y_\nu^{i\beta}v_u \\ \frac{1}{\sqrt{2}}Y_\nu^{\alpha i}v_u & 0_{2\times2} \end{pmatrix}. \tag{22}$$

The sterile neutrinos have no Majorana masses in this approximation. In the active neutrino 3×3 subblock, the parameters A, B, C are on the order of $1/m_{\rm soft}$ which is essentially 1/TeV. Surprisingly, the determinant of this 3×3 matrix is zero due to its flavor structure. This indicates that one of the active neutrinos has a vanishing Majorana mass. Therefore, after partial diagonalization, we can write the 5×5 light neutrino mass matrix as

$$\mathcal{M}^D_{\rm light} = \left(\begin{array}{ccc|c} m_1 & 0 & 0 & \\ 0 & m_2 & 0 & \frac{v_u}{\sqrt{2}} U^{ji} Y_\nu^{i\beta} \\ 0 & 0 & 0 & \\ \hline \multicolumn{3}{c|}{\frac{v_u}{\sqrt{2}} U^{ij} Y_\nu^{\alpha i}} & \begin{matrix} 0 & 0 \\ 0 & 0 \end{matrix} \end{array}\right). \tag{23}$$

The U_{ij} are the elements of the 3×3 unitary matrix diagonalizing the active neutrino subblock. Therefore, without having to assume unnatural cancellations:

The model implies the existence of two light sterile neutrinos.

Depending on the magnitude of m_1, m_2 and the Yukawa couplings there are three possibilities:

(a) If the Majorana contributions are negligible, neutrinos obtain their masses only through Dirac terms. This case is generic for many models. However, for this particular model it is in tension with oscillation experiments since the active–sterile neutrino mixing is maximal and not all mass differences between the active and sterile neutrinos can be zero due to the different number of light LH and RH fields.

(b) The second possibility involves comparable Dirac and Majorana masses. Most of those cases are ruled out by oscillation experiments since they lead to large active–sterile mixing with significant left–right mass differences. There are some special cases, for example:

$$\mathcal{M}^D_{\rm light} = \begin{pmatrix} m_1 & 0 & 0 & 0 & 0 \\ 0 & m_2 & 0 & 0 & 0 \\ 0 & 0 & 0 & \frac{v_u}{\sqrt{2}} U^{3i} Y_\nu^{i1} & 0 \\ 0 & 0 & \frac{v_u}{\sqrt{2}} U^{i3} Y_\nu^{1i} & 0 & 0 \\ 0 & 0 & 0 & 0 & 0 \end{pmatrix}, \tag{24}$$

which avoids a large mixing, but it requires an unnatural hierarchy between the Yukawas.

(c) The final possibility is when there exist only Majorana mass terms, i.e. the Yukawas are negligible. In this case our model has the most interesting consequences. One does not have to worry about constraints from oscillation experiments, since there is no active–sterile mixing. In addition, this situation is interesting from a collider point of view — the Majorana masses are generated as a result of R-parity violation, which leads to lepton number violating signatures. Although this possibility requires all Yukawas to be quite small, there might exist an underlying, not yet discovered symmetry responsible for this. We therefore concentrate on this particular case below.

Since our model predicts one massless active neutrino, one can use the experimental values of solar and atmospheric neutrino mass splittings to write down the exact active neutrino spectrum, given the choice of hierarchy:

- Normal hierarchy:

$$m_1 = 0\,, \quad m_2 = \sqrt{\Delta m^2_{\rm sol}} \approx 9\ {\rm meV}\,, \quad m_3 = \sqrt{\Delta m^2_{\rm atm}} \approx 50\ {\rm meV}\,.$$

- Inverted hierarchy:

$$m_3 = 0\,, \quad m_1 = \sqrt{\Delta m^2_{\rm atm}} \approx 50\ {\rm meV}\,, \quad m_2 = \sqrt{\Delta m^2_{\rm atm} + \Delta m^2_{\rm sol}} \approx 50\ {\rm meV}\,.$$

Regarding the two light sterile neutrinos, without sizable Dirac terms the model predicts that they have extremely small masses. As explained above, the spectrum contains also a TeV-scale sterile neutrino:

$$m_4, m_5 \ll \max\{m_1, m_2, m_3\}\,, \quad m_6 \sim {\rm TeV}\,. \tag{25}$$

Finally, although the model does not make concrete predictions for the active neutrino mixing, one can express the values of the mixing angles θ_{ij} in terms of the LH sneutrino VEVs and the Yukawa couplings:

$$\theta_{ij} \leftrightarrow v_{Li},\ Y_\nu^{i3}\,. \tag{26}$$

There exists a concrete set of values for those six parameters which reproduces the experimental results for the mixing angles.

5. Leptobaryonic Dark Matter

As emphasized earlier, the DM candidate in the model is the fermionic component of the $\hat{X}$ superfield. Because of its quantum numbers, it couples to the MSSM only through the gauge bosons (new Z's). In our analysis we assumed that it interacts predominantly with the leptophobic gauge boson Z_B. There are two main constraints which have to be fulfilled.

If one requires thermal production of DM, the $\tilde{X}$ particle must have an appropriate annihilation cross section to satisfy the relic density bounds. The process responsible for this is shown in Fig. 1. The other constraint comes from direct detection experiments, which look for DM scattering off nuclei. This process is described by a similar diagram (see Fig. 2).

Having calculated the two cross sections, we performed a scan over a range of parameter values of the model to see which points fulfill the experimental constraints. A plot of such a scan showing the spin-independent direct detection cross section versus the DM mass is shown in Fig. 3. For this particular parameter scan,

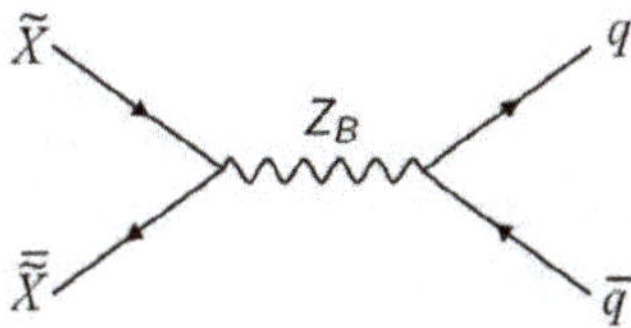

Fig. 1. Dark matter annihilation diagram $\bar{\tilde{X}}\tilde{X} \to Z_B \to \bar{q}q$.

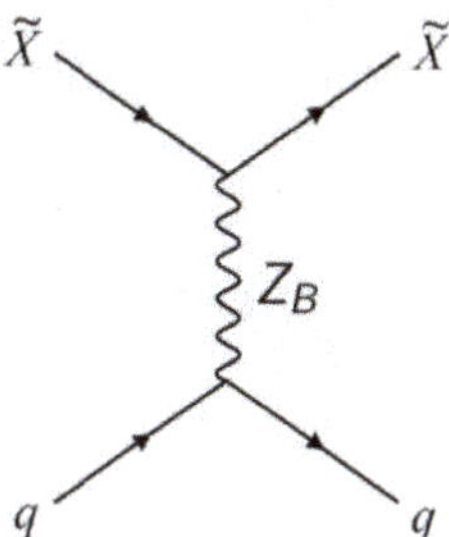

Fig. 2. Direct detection diagram $\tilde{X}N \to Z_B \to \tilde{X}N$.

we chose a concrete combination of couplings and quantum numbers equal to 1 (its most natural value), we varied the baryon number gauge coupling $0.1 \leq g_B \leq 0.3$, the Z_B gauge boson mass 2.5 TeV $\leq M_{Z_B} \leq$ 5 TeV, and the relic density in the experimentally favored range $0.11 < \Omega_{\tilde{X}}h^2 < 0.13$. The purple line is the latest DM bound coming from the LUX experiment and does not exclude any of the points. However, the XENON1T sensitivity projected for 2017 and the reach of XENON10T predicted for 2023 (red lines) will probe the majority of the scan points on the plot. Therefore, if there is no direct detection signal discovery within the next decade, the model will require a higher SUSY breaking scale in order to explain DM.

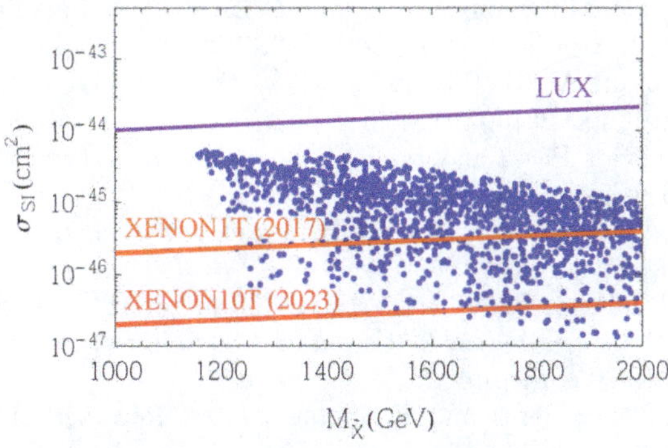

Fig. 3. Direct detection cross section versus DM mass for a scan over the parameters of the model (details are provided in the text) with overplotted current and future experimental bounds.

6. Summary

We have discussed an extension of the MSSM with gauged baryon and lepton numbers. This model has a number of attractive features. Most importantly, due to the extended gauge structure, it entirely forbids proton decay, even at the nonrenormalizable level. Baryon and lepton number symmetries are broken at the supersymmetry breaking scale. An experimentally viable scenario requires also violation of R-parity. This leads to many interesting lepton number violating signatures at colliders. Although the lightest supersymmetric particle is no longer a good dark matter candidate, the model contains a new type of dark matter particle at the TeV scale carrying both baryon and lepton numbers. Its existence can be tested in dark matter experiments, as well as through missing energy signals at the LHC. We showed that the model accommodates a light active neutrino spectrum consistent with observation and predicts the existence two light sterile neutrinos. The measured values of mixing angles can be accommodated as well, however, the predictive power of the model still has to be strengthened in this regard.

Acknowledgments

I am grateful to my collaborators: Pavel Fileviez Pérez, Sogee Spinner and Jonathan Arnold, without whom the work I discussed would not have been completed. I would also like to thank the organizers of the International Conference on Massive Neutrinos in Singapore, especially the chairman, Harald Fritzsch, for the invitation, warm hospitality, and a fantastic scientific and social atmosphere during the conference.

References

1. A. Pais, *Phys. Rev. D* **8**, 1844 (1973).
2. S. Rajpoot, *Int. J. Theor. Phys.* **27**, 689 (1988).
3. R. Foot, G. C. Joshi and H. Lew, *Phys. Rev. D* **40**, 2487 (1989).
4. C. D. Carone and H. Murayama, *Phys. Rev. D* **52**, 484 (1995), arXiv:hep-ph/9501220.
5. H. Georgi and S. L. Glashow, *Phys. Lett. B* **387**, 341 (1996), arXiv:hep-ph/9607202.
6. P. Fileviez Pérez and M. B. Wise, *Phys. Rev. D* **82**, 011901 (2010) [Erratum: *ibid.* **82**, 079901 (2010)], arXiv:1002.1754 [hep-ph].
7. M. Duerr, P. Fileviez Pérez and M. B. Wise, *Phys. Rev. Lett.* **110**, 231801 (2013), arXiv:1304.0576 [hep-ph].
8. P. Fileviez Pérez, S. Ohmer and H. H. Patel, *Phys. Lett. B* **735**, 283 (2014), arXiv:1403.8029 [hep-ph].
9. M. Duerr and P. Fileviez Pérez, arXiv:1409.8165 [hep-ph].
10. J. M. Arnold, P. Fileviez Pérez, B. Fornal and S. Spinner, *Phys. Rev. D* **88**, 115009 (2013), arXiv:1310.7052 [hep-ph].
11. V. Barger, P. Fileviez Pérez and S. Spinner, *Phys. Rev. Lett.* **102**, 181802 (2009), arXiv:0812.3661 [hep-ph].
12. V. Barger, P. Fileviez Pérez and S. Spinner, *Phys. Lett. B* **696**, 509 (2011), arXiv:1010.4023 [hep-ph].

13. D. K. Ghosh, G. Senjanović and Y. Zhang, *Phys. Lett. B* **698**, 420 (2011), arXiv:1010.3968 [hep-ph].
14. P. Fileviez Pérez and S. Spinner, *J. High Energy Phys.* **1204**, 118 (2012), arXiv:1201.5923 [hep-ph].
15. Z. Marshall, B. A. Ovrut, A. Purves and S. Spinner, *Phys. Rev. D* **90**, 015034 (2014), arXiv:1402.5434 [hep-ph].

From the Fourth Color to Spin-Charge Separation — Neutrinos and Spinons

Chi Xiong

Institute of Advanced Studies, Nanyang Technological University,
60 Nanyang View, Singapore 639673
xiongchi@ntu.edu.sg

We introduce the spin-charge separation mechanism to the quark–lepton unification models which consider the lepton number as the fourth color. In certain finite-density systems, quarks and leptons are decomposed into spinons and chargons, which carry the spin and charge degrees of freedom respectively. Neutrinos can be related to the spinons with respect to the electric-charge and spin separation in the early universe or other circumstances. Some effective, probably universal couplings between the spinon sector and the chargon sector are derived and a phenomenological description for the chargon condensate is proposed. It is then demonstrated that the spinon current can induce vorticity in the chargon condensate, and spinon zero modes are trapped in the vortices, forming spinon-vortex bound states. In cosmology this configuration may lead to the emission of extremely high energy neutrinos when vortices split and reconnect.

Keywords: Massive neutrinos; spin-charge separation; quark–lepton unification.

1. The Fourth Color and the Haplon Model

Neutrinos are one of the most important probes to the quark–lepton unification. It has been proposed forty years ago that the lepton number can be considered as the fourth "color"[1] in the frameworks based on the gauge group $SU(2)_L \times SU(2)_R \times SU(4)^c$, and the gauge group $SO(10)$,[2] respectively. Right-handed neutrinos are introduced in both models. The descents from the $SO(10)$ are through the $SU(5)$ group or the $SU(2)_L \times SU(2)_R \times SU(4)^c$ group. In the latter case the usual color group $SU(3)^c$ is extended to the group $SU(4)^c$ and the leptons are interpreted as the fourth color column. Using the original notations in Ref. 1, the first two generations of fermions are written as

$$\Psi_{L,R} = \begin{pmatrix} \mathcal{P}_a & \mathcal{P}_b & \mathcal{P}_c & \mathcal{P}_d = \nu_e \\ \mathcal{N}_a & \mathcal{N}_b & \mathcal{N}_c & \mathcal{N}_d = e^- \\ \lambda_a & \lambda_b & \lambda_c & \lambda_d = \mu^- \\ \chi_a & \chi_b & \chi_c & \chi_d = \nu_\mu \end{pmatrix}_{L,R} \tag{1}$$

where $(\mathcal{P}, \mathcal{N}, \lambda, \chi)$ indicates valency and (a, b, c, d) indicates color degrees of freedom. In Ref. 1 the matrices (1) are also written in a composite way, with the more familiar notations

$$\Psi_{L,R} = \begin{pmatrix} u_1 & u_2 & u_3 & u_4 = \nu_e \\ d_1 & d_2 & d_3 & d_4 = e^- \\ c_1 & c_2 & c_3 & c_4 = \nu_\mu \\ s_1 & s_2 & s_3 & s_4 = \mu^- \end{pmatrix}_{L,R} = \begin{pmatrix} \mathcal{F}_1 \\ \mathcal{F}_2 \\ \mathcal{F}_3 \\ \mathcal{F}_4 \end{pmatrix}_{L,R} \otimes (\mathcal{B}_1, \mathcal{B}_2, \mathcal{B}_3, \mathcal{B}_4) \tag{2}$$

where $\mathcal{F}$ are spinors carrying the spin of Ψ while $\mathcal{B}$ are scalars carrying the (color) charge of Ψ. As mentioned in Ref. 1, it is attractive to consider the components $(\mathcal{F}, \mathcal{B})$ in the decomposition

$$\Psi = \mathcal{F}\mathcal{B}, \quad \mathcal{F} = (\mathcal{F}_1, \mathcal{F}_2, \mathcal{F}_3, \mathcal{F}_4)^T, \quad \mathcal{B} = (\mathcal{B}_1, \mathcal{B}_2, \mathcal{B}_3, \mathcal{B}_4), \tag{3}$$

as *fundamental fields* and Ψ as *composite* ones. Similar ideas has been used in some preon models, for example, the "haplon" model[3] which consider quarks and leptons as bound states of some more fundamental particles (preons) called haplons. In the haplon model the preons are a weak SU(2) doublet of colorless fermions (α, β), and a quartet of scalars $(x^i, y), i = R, G, B$ and y carries the fourth color (lepton number), thus leading to an $SU(4)$ symmetry. The first generation of fermions

$$\nu = (\alpha y), \; e^- = (\beta y), \; u = (\alpha x), \; d = (\beta x) \tag{4}$$

which has the same spin-charge separation pattern $\Psi = \mathcal{F}\mathcal{B}$ as in Eq. (3). The vector bosons are also composite ones

$$\begin{aligned} W^+ &= (\alpha\bar{\beta}), \; W^3 = \frac{1}{\sqrt{2}}(\alpha\bar{\alpha} + \beta\bar{\beta}), \\ W^- &= (\bar{\alpha}\beta), \; Y^0 = \frac{1}{\sqrt{2}}(\alpha\bar{\alpha} - \beta\bar{\beta}), \end{aligned} \tag{5}$$

However, we will not address here the decomposition of gauge field. The spin-charge separation of non-abelian gauge fields is a more complicated issue (see for example Refs. 4 and 5). For simplicity we may consider them as fundamental fields at least in this paper.

2. Spin-Charge Separation from Condensed Matter Physics

In the previous section we have seen how the spin-charge separation may happen in particle physics models. In condensed matter physics, spin-charge separation describes electrons in some materials as "bound states" of spinon and chargon (or holons), which carry the spin and charge of electrons respectively. Under certain conditions, such as the high-temperature cuprate superconductivity, the "composite" electrons can have a deconfinement phase and the spinon and chargon becomes independent particles. Many elaborations of this idea followed in the studies of high T_c superconductors (see e.g. Ref. 6 for a review). There are experimental observations and computer simulations supporting this idea — the first direct observations of spinons and holons was reported in Ref. 7; Simulations on spin-charge separation via quantum computing has been performed in Ref. 8. To demonstrate the basic idea we take the slave-boson formalism in the t-J model as an example.[9] It is well-known that the low-energy physics of the high-temperature cuprates can be described by the t-J model

$$H = \sum_{ij} J(\mathbf{S}_i \cdot \mathbf{S}_j - \frac{1}{4}n_i n_j) - \sum_{ij,\sigma} t_{ij}(c^\dagger_{i\sigma}c_{j\sigma} + \text{h.c.}), \tag{6}$$

where $c^\dagger_{i\sigma}, c_{i\sigma}$ are the projected electron operators with constraint $\sum_\sigma c^\dagger_{i\sigma}c_{i\sigma} \leqslant 1$, which can be treated with the slave-boson approach by writing

$$c^\dagger_{i\sigma} = f^\dagger_{i\sigma} b_i \tag{7}$$

where the operators $f^\dagger_{i\sigma}$ creates a chargeless spin 1/2 fermion state — "spinon" and b_i creates a charged spin 0 boson state — "holon", respectively. Again, Eq. (7) shows the same decomposition pattern as in Eq. (3) and Eq. (4) although it is at the operator level. Note that for the (anti)commutator relations to work out correctly a constraint:

$$f^\dagger_{i\uparrow}f_{i\uparrow} + f^\dagger_{i\downarrow}f_{i\downarrow} + b^\dagger_i b_i = 1 \tag{8}$$

must be satisfied. What is more, a U(1) gauge symmetry

$$b_i \to e^{i\theta}b_i, \quad f_{i\sigma} \to e^{i\theta}f_{i\sigma} \tag{9}$$

emerges. The d-wave high-T_c superconducting phase appears when holons condense $\left\langle b^\dagger_i b_i \right\rangle \neq 0$. We emphasize on that, as Eq. (8) suggests, to apply spin-charge separation we need a finite-density system. An isolated electron cannot be decomposed into the spin and charge components. Besides the high T_c cuprate superconductors in the condensed matter physics, the early universe and the inner cores of the compact stars may provided such a finite-density environment in cosmology and astrophysics.

3. Spinon and Chargon: General Couplings

The examples from the 4th-color, the haplon model and the high-T_c superconductors tell us that a general spin-charge separation can be written as

$$\Psi = \mathcal{F}\mathcal{B}. \tag{10}$$

Extra constraint(s) like Eq. (8) might be needed to form the correct (anti) commutation relations for the operators, and to match the degrees of freedom before and after the spin-charge separation. For simplicity and for the purpose of studying neutrinos, we will only consider the spin-electric-charge separation. In contrast to most studies of electrons in the condensed matter physics, in cosmology and particle physics we have a very special type of particles — the neutrinos. Being electrically neutral, neutrinos seem to be the right candidate for the spinon from the spin-charge separation point of view. This will be discussed in the next section.

From the bound state or the confined phase to the deconfined phase in which the spin and charge degrees of freedom are decoupled, it merits some study to see how the effective interaction between spinon and chargon changes. In the confined phase, the component fields $\mathcal{F}$ and $\mathcal{B}$ are coupled, even for a free fermion field Ψ. Plugging Eq.(10) into the kinetic term of a Dirac spinor,

$$i\bar{\Psi}\gamma^\mu \overleftrightarrow{\partial}_\mu \Psi, \tag{11}$$

we obtain an effective coupling between the spinor current and the chargon current

$$\sim g_J\, J_{\text{spinon}} \cdot J_{\text{chargon}}, \tag{12}$$

where g_J is an effective coupling constant. The spinor current and the chargon current are defined as

$$\begin{aligned} J^\mu_{\text{spinon}} &= \bar{\mathcal{F}}\gamma^\mu \mathcal{F}, \\ J^\mu_{\text{chargon}} &= \frac{1}{2i}\left(\mathcal{B}^*\partial^\mu\mathcal{B} - \mathcal{B}\partial^\mu\mathcal{B}^*\right). \end{aligned} \tag{13}$$

We can do similar spin-charge separation in the Weyl representation[a]

$$\Psi = |\mathcal{B}_L|e^{i\alpha_L/2}\mathcal{F}_L + |\mathcal{B}_R|e^{i\alpha_R/2}\mathcal{F}_R. \tag{14}$$

With constraints like (8) and $\alpha_L = -\alpha_R$, a Dirac mass term corresponds to a Yukawa-type coupling

$$\bar{\mathcal{F}}(\text{Re}\mathcal{B} + i\gamma_5\text{Im}\mathcal{B})\mathcal{F}. \tag{15}$$

The Majorana representation and the Majorana mass term will be discussed in details in a separate publication.[10]

To phenomenologically describe the chargon condensate, we introduce a non-linear Klein–Gordon equation (for simplicity we only consider the U(1) case)

$$\partial^\mu\partial_\mu\mathcal{B} - \lambda\left(|\mathcal{B}|^2 - \mathcal{B}_0^2\right)\mathcal{B} - ig_J J^\mu_{\text{spinon}}\partial_\mu\mathcal{B} = 0, \tag{16}$$

derived from a Higgs potential $V(\mathcal{B}^*, \mathcal{B})$ and the current–current interaction (12). As the chargon current is proportional to the phase of the $\mathcal{B}$ field ($\mathcal{B} \equiv |\mathcal{B}|e^{i\alpha}$), it is natural to consider the possibility of generating vorticity in the chargon condensate by the spinon current. Here is a simple and interesting example simulating that a rotating spinon current induces vorticity in the chargon condensate[b]

$$J^\mu_{\text{spinon}} = \left(\rho, \vec{J}\right), \quad \vec{J} = \rho\vec{\Omega} \times \vec{r} \tag{17}$$

[a] Reality condition and chiral projection do not combine in four-dimensional Minkowski spacetime. Therefore one should have two Majorana spinors plus other scalars for the spin-charge separation. Extra degrees of freedoms should be removed by constraints like (8).

[b] Equation (16) has been numerically solved in Refs. 16 and 17 and vortex patterns have been observed. Nevertheless, the complex scalar field in Refs. 16 and 17, being the order parameter for describing a superfluid vacuum, has a different physical meaning from the chargon condensate in the present paper.

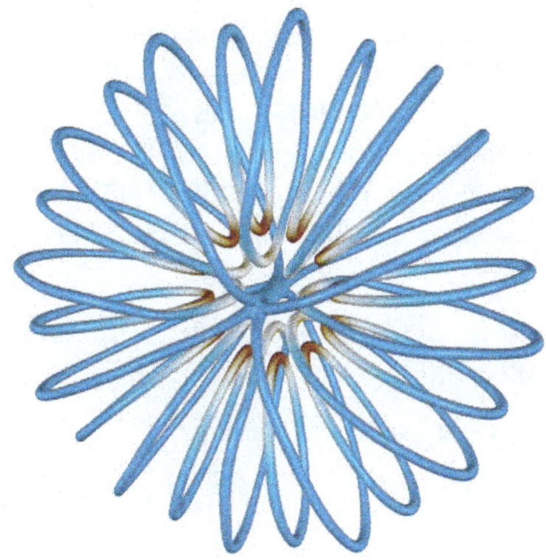

Fig. 1. A numerical solution to the chargon equation (16). It is a three-dimensional vortex ring lattice, induced by the spinon-chargon coupling (12) (see Refs. 16 and 17 for two-dimensional vortex lattice solutions and more computations on vortex dynamics). Such topological configurations can trap light chiral fermions and release them quickly when the vortex lines split and reconnect. This mechanism may be used to explain the emission of extremely high energy neutrino from cosmic strings.

where ρ is a density distribution and Ω is an angular velocity. The current–current interaction with (17) mimics the local rotation effects of the Coriolis force. It can also simulate the case in which the vorticity is induced by an external magnetic background with an axial symmetry. Fig. 1 shows a three-dimensional vortex-ring lattice as a numerical solution to Eq. (16). As it can be seen in the next section, such a topological configuration in the chargon condensate can trap spinon zero-mode due to the Yukawa type coupling (15).

4. Neutrinos and Spinons

Can neutrinos be considered as spinons with respect to the electric-charge-spin separation? It will depend on the material environment. A more meaningful question might be: Is it useful to think neutrinos as spinons from the spin-charge separation? The answer seems to be yes from the point of view of neutrino condensation[11,13] and neutrino superfluidity,[12] as well as their applications in dark matter and other cosmological issues. In Ref. 12 massive neutrinos are shown to display BCS superfluidity by forming Cooper pairs through the exchange of attractive scalar Higgs boson between left- and right-handed neutrinos.

Now we study a system where the spinon $\mathcal{F}$ interacts with the chargon $\mathcal{B}$ and another gauge field A_μ.[c] The vortex-ring lattice in Fig. 1 is quite complicated for further studies, so we assume that the holon condensate provides a single, straight vortex line background $\mathcal{B} = |\mathcal{B}|(\rho)e^{im\theta}$($m$ is the winding number and ρ, θ are the polar coordinates in the transverse plane), and A_μ provides a gauge background $A_\mu = (A_0, A_1, 0, 0)$. As the current–current interaction (12) decouples in the chargon condensation phase and the Yukawa coupling (15) dominates, the system is then reduced to the case studied in Refs. 18 and 19.[d] The equations of motion of the spinons read (for $m = 1$)[18,19]

$$\begin{aligned} i\gamma^i(\partial_i - igA_i)\mathcal{F}_L + i(\gamma^2\cos\theta + \gamma^3\sin\theta)\partial_\rho\mathcal{F}_L &= -|\mathcal{B}|(\rho)e^{-i\theta}\mathcal{F}_R, \\ i\gamma^i(\partial_i - igA_i)\mathcal{F}_R + i(\gamma^2\cos\theta + \gamma^3\sin\theta)\partial_\rho\mathcal{F}_R &= -|\mathcal{B}|(\rho)e^{+i\theta}\mathcal{F}_L. \end{aligned} \tag{18}$$

The spinon zero-mode has an exponential profile[18,19]

$$\mathcal{F}_L = \chi_L \exp\Big[-\int_0^\rho |\mathcal{B}|(\rho')d\rho'\Big] \tag{19}$$

and $\mathcal{F}_R = i\gamma^2\mathcal{F}_L$. χ_L is a two-dimensional spinor satisfying

$$i\gamma^i(\partial_i - igA_i)\chi_L = 0. \tag{20}$$

The exponential profile of $\mathcal{F}_L$ in Eq. (19) shows that chiral zero-modes of spinons are localized on a vortex, which can be considered as a fermion-vortex bound-state. If we take the spinons to be the neutrinos, and think about more complicated configurations than the one shown in Fig. 1, such as a vortex tangle or quantum turbulence in the chargon condensate, then the dynamical vortex lines or rings will split and reconnect, the trapped neutrinos will be emitted quickly during the process of reconnection. The resulting phenomena might be similar to the cases in which cosmic strings emit extremely high energy neutrinos.[14] Note that such mechanism has been used in the scalar field cosmology[15–17] where a superfluid cosmological vacuum plays the role of the chargon condensate here.

5. Conclusions and Discussions

Inspired by the spin-charge separation scenario in difference physics models, such as the fourth-color model, the haplon model, and the t-J model in particle physics or condensed matter physics, we give a general picture about how the composite models split into spin and charge sectors under certain environment condition (e.g. finite density), and derive some simple but probably universal couplings. This provides a

[c]The gauge field A_μ is not necessary for the formation of the spinon-string bound state. It is included to the system because in our case the spinon is only neutral to the separated electric charge, it may still carry other quantum numbers like the color charge or weak isospin. When it is minimally coupled to other gauge fields, the current–current coupling (12) can be absorbed by a gauge transformation, then what is left is the Yukawa type coupling.

[d]It is reasonable to expect that the current–current interaction (12) decouples in the deconfined phase, since it comes from the kinetic term of the composite particle which is only meaningful in the bound state or confined phase. The Yukawa type coupling, however, might survive in the deconfined phase due to the potential nature of the mass term.

new understanding about the lepton–quark unification and suggests the similarity between the spinons and the neutrinos (with respect to the electric-charge-spin separation). We also give a simple phenomenological description for the chargon condensate, and show that it supports vorticity, which leads to the localization of the spinons and hence the formation of spinon-string bound states. This may provide an alternative explanation for the emission of extremely high energy neutrinos from cosmic strings.

Acknowledgments

I thank Peter Minkowski, Harald Fritzsch and Kerson Huang for valuable discussions, and Xiaopei Liu and Yulong Guo for their support on the simulation and the visualization of superfluid vorticity. This work is supported by the research funds from the Institute of Advanced Studies, Nanyang Technological University, Singapore.

References

1. J. C. Pati and A. Salam, "Lepton Number as the Fourth Color," *Phys. Rev. D* **10**, 275 (1974) [Erratum-*ibid.* **11**, 703 (1975)].
2. H. Fritzsch and P. Minkowski, "Unified Interactions of Leptons and Hadrons," *Annals Phys.* **93**, 193 (1975).
3. H. Fritzsch and G. Mandelbaum, "Weak Interactions as Manifestations of the Substructure of Leptons and Quarks," *Phys. Lett. B* **102**, 319 (1981).
4. A. J. Niemi and N. R. Walet, "Splitting the gluon?," *Phys. Rev. D* **72**, 054007 (2005) [hep-ph/0504034].
5. L. D. Faddeev and A. J. Niemi, "Spin-Charge Separation, Conformal Covariance and the SU(2) Yang-Mills Theory," *Nucl. Phys. B* **776**, 38 (2007) [hep-th/0608111].
6. P. A. Lee, N. Nagaosa and X. G. Wen, "Doping a Mott insulator: Physics of high-temperature superconductivity ", *Rev. Mod. Phys.* **78**, 17 (2006).
7. B. J. Kim et al, "Distinct spinon and holon dispersions in photo emission spectral functions from one-dimensional $SrCuO_2$", *Nature Physics* **2**, 397–401 (2006).
8. D. G. Angelakis, M-X. Huo, E. Kyoseva and L. C. Kwek, "Luttinger Liquid of Photons and Spin-Charge Separation in Hollow-Core Fibers", *Phys. Rev. Lett.* **106**, 153601, (2011).
9. S. E. Barnes, "New method for the Anderson model", *J. Phys. F: Metal Phys.* **6**, 1375 (1976).
10. C. Xiong, manuscript in preparation.
11. D. G. Caldi and A. Chodos, "Cosmological neutrino condensates," hep-ph/9903416.
12. J. I. Kapusta, "Neutrino superfluidity," *Phys. Rev. Lett.* **93**, 251801 (2004) [hep-th/0407164].
13. R. Horvat, P. Minkowski and J. Trampetic, "Dark consequences from light neutrino condensations," *Phys. Lett. B* **671**, 51 (2009) [arXiv:0809.0582 [hep-ph]].
14. C. Lunardini and E. Sabancilar, "Cosmic strings as emitters of extremely high energy neutrinos", *Phys. Rev. D* **86**, 085008 (2012).
15. K. Huang, H. B. Low and R. S. Tung, "Scalar Field Cosmology II: Superfluidity and

Quantum Turbulence," *Int. J. Mod. Phys. A* **27**, 1250154 (2012). [arXiv:1106.5283 [gr-qc]].

16. K. Huang, C. Xiong and X. Zhao, "Scalar-field theory of dark matter," *Int. J. Mod. Phys. A* **29**, 1450074 (2014) [arXiv:1304.1595 [gr-qc]].
17. C. Xiong, M. Good, Y. Guo, X. Liu and K. Huang, "Relativistic superfluidity and vorticity from the nonlinear Klein-Gordon equation," *Phys. Rev. D* **90**, 125019 (2014). [arXiv:1408.0779 [hep-th]].
18. C. Callan and J. Harvey, "Anomalies and Fermion Zero Modes on Strings and Domain Walls", *Nucl. Phys. B* **250**, 427 (1985).
19. C. Xiong, "QCD Flux Tubes and Anomaly Inflow," *Phys. Rev. D* **88**, 025042 (2013), [arXiv:1302.7312 [hep-th]].

Measurement of the Underlying Event Activity Using Charged-Particle Jets in Proton–Proton Collisions at $\sqrt{s} = 2.76$ TeV

W. Y. Wang*, A. H. Chan and C. H. Oh
On behalf of the CMS Collaboration

Department of Physics, National University of Singapore, 117551, Singapore
www.nus.edu.sg
**w.y.wang@u.nus.edu*

A measurement of the underlying event (UE) activity in proton–proton collisions is performed using events with a charged-particle jet produced at central pseudorapidity ($|\eta^{\text{jet}}| < 2$) and with transverse momentum $1 \leq p_{\mathrm{T}}^{\text{jet}} < 100$ GeV. The analysis uses a data sample collected by the CMS experiment at the LHC at a centre-of-mass energy of 2.76 TeV, corresponding to an integrated luminosity of 0.3 nb^{-1}. The UE activity is measured as a function of $p_{\mathrm{T}}^{\text{jet}}$ in terms of the average multiplicity and scalar-p_{T} sum of charged particles, with $|\eta| < 2$ and $p_{\mathrm{T}} > 0.5$ GeV, in the azimuthal region transverse to the highest-p_{T} jet direction. By further dividing the transverse region into two halves of smaller and larger activity respectively, valuable information on different UE contributions is obtained. The measurements are compared to previous results at 0.9 and 7 TeV, and to predictions of several Monte Carlo event generators, providing constraints on the theoretical modelling of the UE dynamics.

1. Introduction

Hadron production in high-energy proton–proton (pp) collisions originates from multiple scatterings of the partonic constituents of the protons at central rapidities, and from "spectator" (non-colliding) partons emitted in the very forward direction. The produced partons reduce their virtuality through gluon radiation and quark–antiquark splittings, and finally fragment into hadrons at scales approaching $\Lambda_{\text{QCD}} \approx 0.2$ GeV. Usually, one separates the produced hadrons into two classes: those directly coming from the fragmentation of partons with the largest momentum transfer ("hard scattering") in the event, and the rest ("underlying event", or UE). The UE, thus, consists of hadrons coming from (i) initial- and final-state radiation (ISR, FSR) from the hard scattering, (ii) softer partonic scatters in the same collision (multiple parton interactions, or MPI), and (iii) beam remnants concentrated along the beam direction.

An accurate understanding of the UE is required for precise measurements of standard model processes at high energies and searches of new physics. Indeed, the UE affects measurements of isolated high-p_T leptons or photons, and it dominates most of the hadronic activity from the overlapping "minimum bias" pp collisions taking place in a given bunch crossing at high LHC luminosities. The semihard and low-momentum processes, which dominate the UE, cannot be theoretically described with perturbative QCD (pQCD) methods alone, and require a phenomenological description containing parameters that must be tuned to the data.

The topological structure of pp interactions with a hard scattering can be used to define experimental observables which are sensitive to the UE. One example is the study of particle properties in regions away from the direction of the products of the hard scattering. At the Tevatron, the CDF experiment measured UE observables using inclusive jet and Drell–Yan (DY) events in $p\bar{p}$ collisions at centre-of-mass energies $\sqrt{s}$ = 0.63, 1.8, and 1.96 TeV.[1–3] In pp collisions at the LHC, the CMS, ATLAS, and ALICE experiments have carried out UE measurements at $\sqrt{s} = 0.9$ and 7 TeV using events containing a leading (highest-p_T) charged-particle jet[4,5] or a leading charged particle,[6,7] and in DY events.[8] In this paper, we study the UE activity by measuring the average multiplicity and scalar transverse momentum sum (Σp_T) densities of charged particles in the azimuthal transverse region with respect to the leading charged particle jet in pp collisions at $\sqrt{s} = 2.76$ TeV.

At a given centre-of-mass energy, the UE activity is expected to increase with the hard scale of the interaction. On average, increasingly harder parton interactions result from pp collisions with decreasing impact parameters which, in turn, enhance the overall hadronic activity originating from MPI up to a saturation reached for central collisions with maximum overlap.[9,10] At the same time, the activity related to the ISR and FSR components continues to increase with the hard scale. For events with the same hard scale, e.g. given by the p_T of jets or DY pairs, the MPI activity rises with $\sqrt{s}$, as denser parton distributions in the protons are probed at increasingly smaller parton fractional momenta $x \sim 2p_T/\sqrt{s}$.[9,10] Hence, studying the UE as a function of the hard scale at several centre-of-mass energies provides an insight into the UE dynamics and its evolution with the collision energy, further constraining the model parameters.

The paper is organised as follows. Section 2 presents the main features of the Monte Carlo (MC) event generators used in this study, that are relevant for the description of the UE properties. Section 3 briefly describes the experimental methods, observables, event and track selection, as well as the corrections and systematic uncertainties of the measurements. The results are presented in Section 4, and summarised in Section 5.

2. Monte Carlo Event Generators

In this analysis, the PYTHIA6,[11] PYTHIA8,[12] and HERWIG++[13] MC event generators are used, with various tunes that are described below. In PYTHIA, the $2 \to 2$ non-diffractive processes, including multiple partonic interactions, are described by lowest-order perturbative QCD with the divergence of the cross-section as $p_T \to 0$ regulated with a phenomenological model. There are various tunable parameters that control, among other things, the behaviour of this regularisation, the transverse matter profile distribution of partons within the hadrons, and final state colour reconnection effects among the produced partons. When QCD radiation is modelled via a p_T-ordered evolution, the MPI and parton shower are interleaved in one common sequence of decreasing p_T values.[14] For the latest version of PYTHIA6, the interleaving is between the initial-state shower and MPI only, while for PYTHIA8 it also includes final-state showers. The final non-perturbative transition of partons to hadrons is described by the Lund string fragmentation model.[15]

HERWIG++ is another general purpose generator like PYTHIA, but it uses angular-ordered parton showers and the cluster model[13] for hadronisation. It has an MPI model similar to the one used by PYTHIA, with tunable parameters for regularising the partonic cross sections at low momentum transfer, but does not include the interleaved evolution with ISR and FSR. While HERWIG++ has no explicit model for diffractive interactions, diffractive-like events are simulated by applying the MPI model to events in which only beam remnants are produced, with no other activity between them.

Both MC models incorporate multiple parton collisions "perturbatively" — i.e. based on a "regularisation" of underlying pQCD subprocess cross sections — but require a non-perturbative ansatz for the impact-parameter profile of the colliding protons. MPIs are then generated by assuming a Poissonian distribution of elementary partonic interactions over the overlapping pp volume, with an average number depending on the impact parameter of the hadronic collision.[9,10] The MPI cross section is dominated by scatterings with semi-hard momentum transfers, O(1–2 GeV) in the low-x regime of the parton densities, and thus show a stronger dependence on the incoming parton fluxes than the single hard-scattering interactions,[9,10] and on the evolution of the low-p_T infrared cutoff, p_{T_0}. In PYTHIA6, PYTHIA8 and HERWIG++, the energy dependence of MPIs is mostly controlled by the energy evolution of the p_{T_0} parameter, which follows a (tunable) power-law dependence on the centre-of-mass energy.[11–13] The dependence of the UE activity on the energy scale is well described by Monte Carlo event generators,[4–8] illustrating the universality of MPIs in different event topologies and hard-scattering production processes. Such a universality is confirmed by the similarity between the UE activity measured in DY[8] and jet-dominated events,[4–7] despite their different underlying parton radiation patterns.

In this analysis, several event generators tunes are used for comparison with the data. These are the PYTHIA6 version 6.426[11] tune Z2*[16] and tune CUETP6S1,[17] PYTHIA8 version 8.175[12] tune 4C[18] and CUETP8S1,[17] and HERWIG++ 2.7 tune UE-EE-5C.[13,19] These event generators and tunes differ in the treatment of initial- and final-state radiation, hadronisation, colour reconnections, and cutoff values for the MPI mechanism. These tunes were mostly obtained from comparisons between predictions and data aimed at providing a reasonable description of existing UE data, especially those measured from LHC pp collisions. In addition, minimum bias observables measured at lower (Tevatron) energies were also used.

3. Experimental Methods

The central feature of the CMS apparatus is a superconducting solenoid of 6 m internal diameter, providing a magnetic field of 3.8 T. Within the superconducting solenoid volume there are several complementary detectors: a silicon pixel and strip tracker, a lead tungstate crystal electromagnetic calorimeter, and a brass/scintillator hadron calorimeter, each composed of a barrel and two endcap sections. The silicon tracker measures charged particles within the pseudorapidity range $|\eta| < 2.5$. For non-isolated particles of $1 < p_{\mathrm{T}} < 10$ GeVand $|\eta| < 1.4$, the track resolutions are typically 1.5% in p_{T}, and 25–90 (45–150) μm in the transverse (longitudinal) impact parameter.[20] Two of the CMS subdetectors acting as LHC beam monitors, the Beam Scintillation Counters (BSC) and the Beam Pick-up Timing for the eXperiments (BPTX) devices, were used to trigger the detector readout. The BSC are located along the beam line on each side of the IP at a distance of 10.86 m and cover the pseudorapidity range $3.23 < |\eta| < 4.65$. The two BPTX devices, located inside the beam pipe at distances of 175 m from the IP, are designed to provide precise information on the bunch structure and timing of the incoming beams, with a time resolution better than 0.2 ns. A more detailed description of the CMS detector, together with a definition of the coordinate system used and the relevant kinematic variables, can be found in Ref. 21.

3.1. *Observables*

In this analysis we follow the same methodology as in the previous studies of the UE activity in events with a leading charged-particle jet, carried out at $\sqrt{s} =$ 0.9 and 7 TeV.[4] Charged-particle jets and charged particles produced at central pseudorapidity ($|\eta| < 2$) with transverse momentum larger than 1 GeVand 0.5 GeV, respectively, are used to study the UE properties. The direction of the leading charged-particle jet in the event is used to select charged particles in the azimuthal transverse region defined by $60° < |\Delta\phi| < 120°$, where $\Delta\phi$ is the relative azimuthal distance between a charged particle and the leading jet. The UE is measured in terms of particle and Σp_{T} densities as a function of the leading jet transverse momentum $p_{\mathrm{T}}^{\mathrm{jet}}$, which is used as a proxy for the hard scale of the interaction. The particle density ($\langle N_{\mathrm{ch}}\rangle / [\Delta\eta\Delta(\Delta\phi)]$) and Σp_{T} density ($\langle\sum p_{\mathrm{T}}\rangle / [\Delta\eta\Delta(\Delta\phi)]$) are computed, respectively, as the average number of primary charged particles, and the average of the scalar sum of their transverse momenta, each per unit of η and per unit of $\Delta\phi$.

As suggested in Ref. 22, the transverse region can be studied in more detail by separating — independently for the particle multiplicity and for the p_{T} sum —

the $60° < \Delta\phi < 120°$ and the $-120° < \Delta\phi < -60°$ ranges, and identifying the region with higher and lower activity, as transMAX and transMIN, respectively. The difference between the transMAX and transMIN activities is used to construct the transDIF activity. The resulting particle and Σp_T densities are expected to be sensitive to different components of the UE activity.

For events with large initial- or final-state radiation, the transMAX region contains the third jet, while both transMAX and transMIN regions receive contributions from the MPI and beam remnants. The transMIN activity is therefore sensitive to MPI and beam remnants, and the transDIF activity is sensitive to harder initial- and final-state radiation. Such an approach extends the methodology employed in Ref.4.

3.2. *Event and track selection*

The present analysis is performed with a data sample of proton–proton collisions collected by the CMS detector at $\sqrt{s} = 2.76$ TeVduring a dedicated run in March 2011, corresponding to an integrated luminosity of 0.3 nb^{-1}. In 6.2% of the events, there is an extra (pileup) pp collision, corresponding to an average of 0.12 overlapping pp collisions. Minimum bias events were recorded by requiring activity in both BSC counters in coincidence with signals from both BPTX devices (in contrast to Ref. 4 which requires only one of the BPTX devices). To reduce the statistical uncertainty for large p_T^{jet}, jet triggers based on information from the calorimeters, with p_T thresholds at 20 and 40 GeV, were also used to collect data (at variance with Ref. 4 which uses thresholds of 30 and 50 GeV). Events identified as originating from beam-halo background are removed from the triggered events.[23] Event selection requires exactly one primary vertex with more than four degrees of freedom and no more than ±10 cm from the centre of the luminous region (beamspot) in the z-direction.

For each selected event the reconstructed track collection needs to be cleaned up from undesired tracks, namely secondaries and background from track combinatorics and beam halo associated tracks.. Tracks not corresponding to actual charged particles (fake tracks) are reduced by imposing the *highPurity* selection criteria.[20] Secondary decays are reduced by requiring that the impact parameter significance $d_0/\sigma(d_0)$ and the significance of z separation between the track and primary vertex $d_z/\sigma(d_z)$, to be each less than 3. In order to remove tracks with poor momentum measurement, we require the calculated relative uncertainty of the momentum measurement $\sigma(p_T)/p_T$ to be less than 5%. Average reconstruction efficiency for the selected tracks is about 85%, which drops to 75% for low p_T tracks and large η, while the fake track rate is about 2%, increasing for tracks with small p_T or large $|\eta|$.

The event energy scale and reference direction, for the identification of the UE sensitive region, are defined using leading "track jets".[24] The track jets used in this study are reconstructed from charged-particle tracks with $p_T > 0.5$ GeV and

$|\eta| < 2.5$ using the SISCone[25] algorithm with distance parameter of 0.5. Although anti-k_T[26] is now the most preferred algorithm at the LHC, the SISCone algorithm is chosen in this analysis for compatibility with previous results.[4] A comparison of the UE activity obtained at generator level using SISCone and anti-k_T[26] algorithms was performed, finding only few percent differences at low p_T^{jet}. From all reconstructed track jets with $|\eta| < 2$ and $p_T > 1.0$ GeV, the one with largest p_T^{jet} is selected. Events without such a jet are not considered in the analysis. Jets are reconstructed with a matching efficiency of 80% at the lowest p_T^{jet} and up to 95% for $p_T^{jet} > 20$ GeV. Trigger conditions are matched to keep the efficiency as uniform as possible and close to 100%. For $1 < p_T^{jet} < 25$ GeV, $25 < p_T^{jet} < 50$ GeV, and $p_T^{jet} > 50$ GeV, we use respectively the minimum-bias and the two-jet triggered samples, corresponding to about 11M, 50k and 23k finally selected events.

3.3. *Corrections and systematic uncertainties*

The UE observables ($\langle N_{ch}\rangle / [\Delta\eta\Delta(\Delta\phi)]$and $\langle\sum p_T\rangle / [\Delta\eta\Delta(\Delta\phi)]$) are reconstructed from all selected tracks, with $p_T > 0.5$ GeV and $|\eta| < 2$, in the transverse region to the leading track jet. These measured observables are corrected for detector effects and selections efficiencies to reflect the primary charged-particle activity corresponding to events with the same kinematic properties. An iterative unfolding technique[27] is used based on 4-dimensional response matrices. These matrices are constructed from the UE observables as a function of p_T^{jet} at (MC generator level) particle and reconstructed levels based on simulated events from the PYTHIA6 Z2 tune, accounting for the response of the observables to detector effects and inefficiencies. An unfolding of the PYTHIA8 4C reconstructed events based on response matrices obtained from Z2 is used to estimate the systematic uncertainties related to the correction procedure. These vary between 0.2% and 4% depending on the observables and p_T^{jet}.

Several other sources of systematic uncertainties may affect the results. These include the implementation of the simulation of track and vertex selection criteria, tracker alignment and material content, background contamination, trigger conditions, and pileup contributions. The uncertainty in the simulation of track selection has been evaluated by applying various sets of selection criteria and comparing their effects to the data and to simulated events. The impact-parameter significance ranges are varied by one unit around the nominal window resulting in an effect on the densities of 0.6 to 4%. It has also been checked that replacing the *highPurity* selection by a simple cut of $N_{\text{layers}} \geq 4$ and $N_{\text{pixel layers}} \geq 2$ for silicon strip and pixel detector layers respectively, has an effect of up to 0.8%. Varying the fraction of fake tracks by 50% affects the densities by 0.4 to 0.6%. The description in the simulation of inactive tracker material has been found to be adequate to within 5%, and increasing the material densities by 5% in the simulation induces a change in the observables of 1%. The effects of tracker misalignment, precision in the interaction point position, and dead channels, evaluated by varying the detector conditions in the MC simulation, are all found to change the results by 0.1 to 0.3%. The effect of varying the trigger and vertex efficiencies within their uncertainties, as well as the effect of pileup contributions, have all been found to lead to a negligible effect.

Systematic uncertainties are largely independent of one another, but they are correlated among data points in each observable. Table 1 shows a summary of the systematic uncertainties.

Table 1. Summary of the systematic uncertainties (in percentage) due to various sources.

Source	Systematic (%)
Unfolding procedure	0.2–4
Impact parameter signif.	0.6–4
Fraction of fake tracks	0.4–0.6
Track selection	0.1–0.8
Material density	1.0
Dead channels	0.1
Tracker alignment	0.2–0.3
Interaction point position	0.2
Total	1.9–5.8

4. Results

In Fig. 1, the (a) particle and (b) Σp_T densities, after unfolding, are shown in the transverse region, relative to the leading charged-particle jet, as a function of the transverse momentum p_T^{jet}. A steep rise of the underlying event activity in the transverse region is seen up to $p_T^{\mathrm{jet}} \sim 8$ GeV, followed by a "saturation" (plateau-like) region, with nearly constant multiplicity and small Σp_T density increase. In Fig. 2, the (a)(c) particle and (b)(d) Σp_T densities after unfolding are shown as a function of p_T^{jet} in the transverse region with maximum activity (transMAX) and with minimum activity (transMIN). In the transMIN region, the amount of UE activity is roughly twice smaller compared to transMAX. The shape is also quite different in both categories. At high-p_T the distributions show a slow rise in transMAX, while in transMIN the plateau-like region is more pronounced. The corresponding distributions in the difference between the transMAX and transMIN regions (transDIF) are presented in Fig. 3. The particle and Σp_T densities both show a rise with p_T^{jet}, and the plateau-like region above $p_T^{\mathrm{jet}} \approx 8$ GeV — seen in the previous transMAX and transMIN distributions — is substituted by less pronounced rises as a function of jet transverse momentum.

The rapid increase of the UE activity with p_T^{jet} in the region below ~ 8 GeV is mainly attributed to the increase of MPI activity as the scale of the interaction increases.[10] This fast rise is followed by a saturation region, with nearly constant multiplicity and small Σp_T density increase. This behaviour is expected in the transverse impact parameter picture, as a consequence of a nearly full overlap of the colliding protons in interactions yielding the hardest parton–parton scatterings. Once the most central pp collisions are reached, the amount of MPI activity saturates.[9,10] Such a distinct p_T^{jet}-dependent pattern in the amount of UE activity (sharp rise plus plateau above the $p_T^{jet} \approx 8$ GeVtransition) is clearly seen for all the observables presented in the paper. However the transMAX and transDIF distributions show a continuous rise with p_T^{jet} also in the high-p_T tails. This is expected to be caused by contributions from initial- and final-state radiation in the tranverse region.[22] Assuming such an interpretation, the present results provide valuable constraints on the modelling of the different UE components.

The results are compared to recent tunes of PYTHIA and HERWIG++ event generators. In general, all PYTHIA6 and PYTHIA8 tunes predict the distinctive change of the amount of activity as a function of the leading jet p_T within 5–10%. The HERWIG++ UE-EE-5C tune also provides a fair description of the data. In general, the data-model agreement improves in the transDIF region. The continuous increase observed at high-p_T^{jet} in the transDIF distributions is well reproduced by all MC tunes, corroborating the hypothesis of increased contributions of QCD radiation from the hardest scattered partons. The same trend is observed in Ref. 3. The latest PYTHIA6 (PYTHIA8) tune CUETP6S1 (CUETP8S1) improves the description of the data in comparison to the results obtained with the parameters of the previous Z2* (4C) tunes.

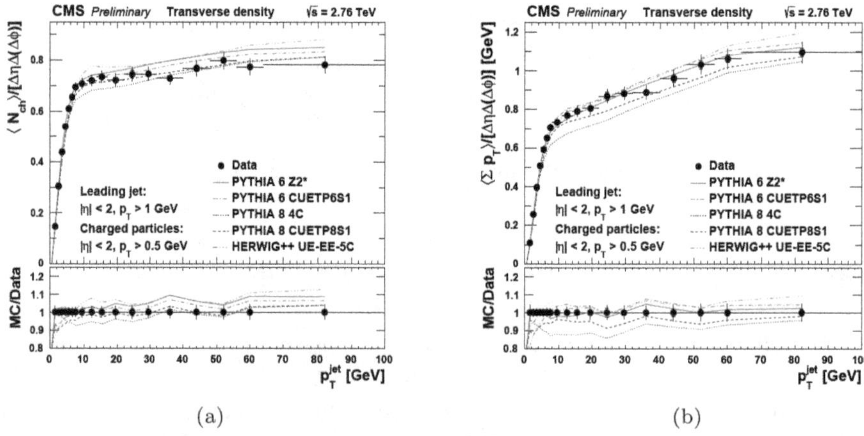

Fig. 1. Measured (a) particle density, and (b) Σp_T density, in the transverse region relative to the leading charged-particle jet in the event ($|\eta| < 2, 60° < |\Delta\varphi| < 120°$), as a function p_T^{jet}. The data (symbols) are compared to various Monte Carlo predictions (curves). The ratios of MC predictions to the measurements are shown in the bottom panels. The inner error bars correspond to the statistical uncertainties, and the outer error bars represent the statistical and systematic uncertainties added in quadrature.

The centre-of-mass energy dependence of the UE activity in the transverse region

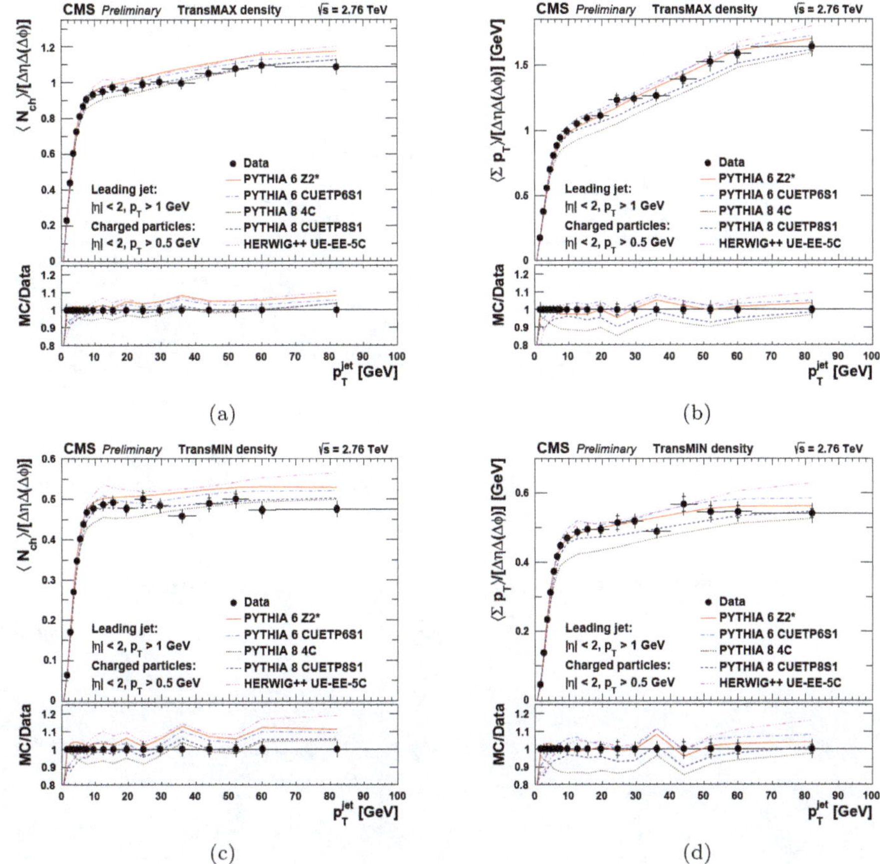

Fig. 2. Measured (a)(c) particle density, and (b)(d) Σp_{T} density, in the transMAX and transMIN region ($60° < |\Delta\varphi| < 120°$, relative to the leading charged-particle jet in the event, with maximum/minimum UE activity), as a function of $p_{\mathrm{T}}^{\mathrm{jet}}$. The definitions of symbols and error bars is the same as Fig. 1.

is presented in Fig. 4 as a function of $p_{\mathrm{T}}^{\mathrm{jet}}$ for $\sqrt{s} = 0.9$, 2.76 and 7 TeV.[4,5] A fast rise with increasing centre-of-mass energy of the activity in the transverse region is observed for the same value of the leading charged-particle jet p_{T}. All tunes predict a centre-of-mass energy dependence of the UE activity, which is very consistent with that found in the data.

5. Summary

The measurement of the underlying event (UE) activity in proton–proton collisions at $\sqrt{s} = 2.76$ TeV has been presented using events with a charged-particle jet produced at central pseudorapidity ($|\eta^{\mathrm{jet}}| < 2$) with transverse momenta $1 < p_{\mathrm{T}}^{\mathrm{jet}} < 100$ GeV. Such analysis complements the results of previous similar measurements

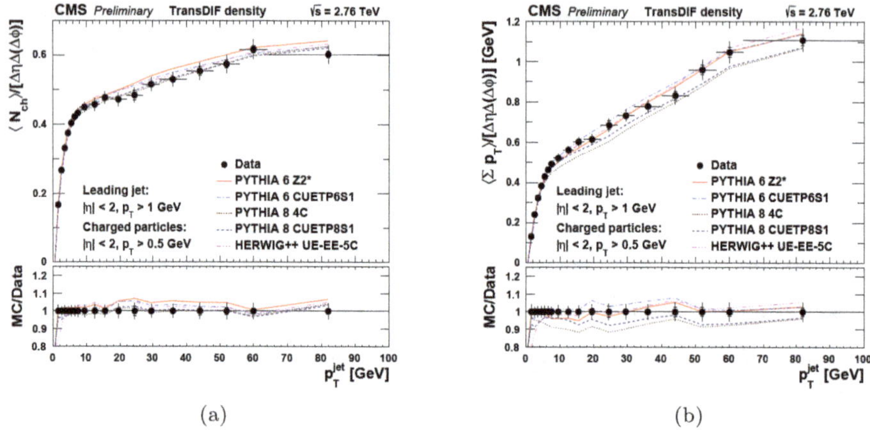

(a) (b)

Fig. 3. Measured transDIF activity (see text for its definition) for (a) particle density, and (b) Σp_{T} density, as a function of $p_{\mathrm{T}}^{\mathrm{jet}}$. The definitions of symbols and error bars is the same as Fig. 1.

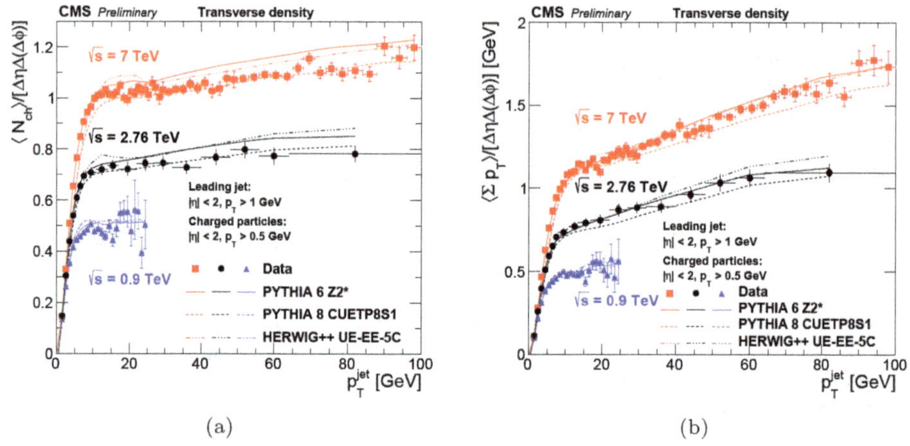

(a) (b)

Fig. 4. Comparison of UE activity at $\sqrt{s} = 0.9$, 2.76, and 7 TeV for (a) particle density, and (b) Σp_{T} density, as a function of $p_{\mathrm{T}}^{\mathrm{jet}}$.[4,5] The data (symbols) are compared to various MC simulations (curves). The definitions of the error bars is the same as Fig. 1.

at $\sqrt{s} = 900$ GeV and 7 TeV.

The leading charged-particle jet in each event is used to set a reference for the energy scale of the hardest partonic scattering, and to define different regions of hadronic activity in azimuth space. The transverse region to the jet's direction is mostly sensitive to UE contributions, and the corresponding charged-hadron activity is studied by measuring its particle and Σp_{T} densities. Furthermore, by dividing the transverse region into minimum and maximum activity areas, the UE contributions from initial- and final-state parton radiation (ISR and FSR) from the hardest scatter can be better factorised out from those originating in multiparton interactions (MPI) and "spectator" partons (beam remnants).

A steep rise of the underlying activity in the transverse region is seen with increasing leading jet p_{T}. This is followed by a plateau above $p_{\mathrm{T}}^{\mathrm{jet}} \sim 8$ GeV, with nearly constant multiplicity and small Σp_{T} density increase. The fast rise followed by a plateau of the UE hadronic activity is clearly seen for all the observables, and confirms the impact-parameter picture of pp collisions featuring an increasing number of MPIs for increasing overlap followed by a saturation of hadron production once the hardest most-central collisions are reached. The transDIF activity shows an increase of the activity with $p_{\mathrm{T}}^{\mathrm{jet}}$ corroborating the hypothesis of a rising contribution from ISR and FSR from the increasingly harder parton-parton scatter.

By comparing data taken at $\sqrt{s}$= 0.9, 2.76, and 7 TeV, a strong growth of the UE activity with higher centre-of-mass energy is observed for the same value of the leading charged-particle $p_{\mathrm{T}}^{\mathrm{jet}}$ as expected from the denser parton densities probed at low-x in the protons, and the larger phase space available for parton radiation.

The results are compared to recent tunes of PYTHIA and HERWIG++ Monte Carlo event generators. PYTHIA6, PYTHIA8 and HERWIG++ tunes describe the data within 5 to 10%. All MC tunes correctly predict the collision-energy dependence of the hadronic activity that is very similar to that observed in the data. The measurements presented here provide valuable constraints for the development and tuning of the underlying event description implemented in hadronic MC models. They will allow one to improve the modelling of key ingredients — such as multiparton interactions, QCD radiation, energy evolution of the transverse proton profile, etc. — which will play an increasing role at higher proton-proton collision energies.

Acknowledgments

We thank our colleagues at CERN, the CMS collaboration and the University of Antwerp for the support given to this work. We are also thankful for the funding provided by Lee Foundation and the NUS Research Grant WBS: R-144-000-178-112.

References

1. T. Affolder *et al.*, Charged jet evolution and the underlying event in proton-antiproton collisions at 1.8 TeV, *Phys. Rev. D* **65**, 092002 (2002).
2. D. Acosta *et al.*, Underlying event in hard interactions at the Fermilab Tevatron $\overline{p}p$ collider, *Phys. Rev. D* **70**, 072002 (2004).
3. T. Aaltonen *et al.*, Studying the underlying event in Drell-Yan and high transverse momentum jet production at the Tevatron, *Phys. Rev. D* **82**, 034001 (2010).
4. S. Chatrchyan *et al.*, Measurement of the underlying event activity at the LHC with $\sqrt{s} = 7$ TeV and comparison with $\sqrt{s} = 0.9$ TeV, *JHEP* **09**, 109 (2011).
5. V. Khachatryan *et al.*, First measurement of the underlying event activity at the LHC with $\sqrt{s} = 0.9$ TeV, *Eur. Phys. J. C* **70**, 555 (2010).

6. B. Abelev *et al.*, Underlying event measurements in *pp* collisions at $\sqrt{s} = 0.9$ and 7 TeV with the ALICE experiment at the LHC, *JHEP* **07**, 116 (2012).
7. G. Aad *et al.*, Measurement of underlying event characteristics using charged particles in *pp* collisions at $\sqrt{s} = 900$ GeV and 7 TeV with the ATLAS detector, *Phys. Rev. D* **83**, 112001 (2011).
8. S. Chatrchyan *et al.*, Measurement of the underlying event in the Drell-Yan process in proton−proton collisions at $\sqrt{s} = 7$ TeV, *Eur. Phys. J. C* **72**, 2080 (2012).
9. T. Sjöstrand and M. V. Zijl, Multiple parton-parton interactions in an impact parameter picture, *Phys. Lett. B* **188**, 149 (1987).
10. L. Frankfurt, M. Strikman and C. Weiss, Transverse nucleon structure and diagnostics of hard parton-parton processes at LHC, *Phys. Rev. D* **83**, 054012 (2011).
11. T. Sjöstrand, S. Mrenna and P. Skands, PYTHIA 6.4 physics and manual, *JHEP* **05**, 026 (2006).
12. T. Sjöstrand, S. Mrenna and P. Z. Skands, A Brief Introduction to PYTHIA 8.1, *Comput.Phys.Commun.* **178**, 852 (2008).
13. M. Bähr, S. Gieseke, M. A. Gigg, D. Grellscheid, K. Hamilton, O. Latunde-Dada, S. Plätzer, P. Richardson, M. H. Seymour, A. Shrestnev and B. R. Webbers, Herwig++ physics and manual, *Eur. Phys. J. C* **58**, 639 (2008).
14. T. Sjostrand and P. Z. Skands, Transverse-momentum-ordered showers and interleaved multiple interactions, *Eur. Phys. J. C* **39**, 129 (2005).
15. B. Andersson, G. Gustafson, G. Ingelman, and T. Sjöstrand, Parton fragmentation and string dynamics, *Phys. Rept.* **97**, 31 (1983).
16. CMS Collaboration, Measurement of energy flow at large pseudorapidities in *pp* collisions at $\sqrt{s} = 0.9$ and 7 TeV, *JHEP* **11**, 148 (2011).
17. CMS Collaboration, *Underlying Event Tunes and Double Parton Scattering*, CMS Physics Analysis Summary CMS-PAS-GEN-14-001 (2014).
18. R. Corke and T. Sjöstrand, Interleaved parton showers and tuning prospects, *JHEP* **03**, 032 (2011).
19. M. H. Seymour and A. Siodmok, Constraining MPI models using σ_{eff} and recent Tevatron and LHC Underlying Event data, *JHEP* **10**, 113 (2013).
20. S. Chatrchyan *et al.*, Description and performance of track and primary-vertex reconstruction with the CMS tracker, *JINST* **9**, P10009 (2014).
21. S. Chatrchyan *et al.*, The CMS experiment at the CERN LHC, *JINST* **03**, S08004 (2008).
22. J. Pumplin, Hard underlying event correction to inclusive jet cross sections, *Phys. Rev. D* **57**, 5787 (1998).
23. V. Khachatryan *et al.*, CMS tracking performance results from early LHC operation, *Eur. Phys. J. C* **70**, 1165 (2010).
24. CMS Collaboration, *Performance of Jet Reconstruction with Charged Tracks only*, CMS Physics Analysis Summary CMS-PAS-JME-08-001 (2008).
25. G. P. Salam and G. Soyez, A practical seedless infrared-safe cone jet algorithm, *JHEP* **05**, 086 (2007).
26. M. Cacciari, G. P. Salam and G. Soyez, The anti-k_t jet clustering algorithm, *JHEP* **04**, 063 (2008).
27. G. D'Agostini, A multidimensional unfolding method based on Bayes' theorem, *Nucl. Instr. Meth. Phys. Res. A* **362**, 487 (1995).

www.ingramcontent.com/pod-product-compliance
Lightning Source LLC
Chambersburg PA
CBHW061926190326
41458CB00009B/2666
9789814704762